Critical Praise for *The Nurture Assumption*

"Important . . . lively anecdotes about real children suffuse this book. . . . Harris writes beautifully, in a tone both persuasive and conversational."

—Carol Tavris, *The New York Times Book Review*

"Harris's book is well written, toughly argued, filled with telling anecdotes and biting wit."

—Howard Gardner, *The New York Review of Books*

"Granted, parents can shape behavior within the home. But in the wider world, Harris argues, the child is a different person, and there lie the roots of the budding adult. Harris's clever and witty book makes this argument with power. . . . Harris's core, convincing message—that many parents wildly overestimate their influence—may usefully calm some nerves in this age of high-anxiety parenting."

—Robert Wright, *Time*

"[Harris] is eloquent and entertaining, she makes people sit up and pay attention, and she opens our eyes to important considerations."

—Sir Michael Rutter, *The London Times Higher Education Supplement*

"A cool compress for feverish parents who fear their every action . . . will mark their child's psyche for life."

—Lynn Smith, *Los Angeles Times*

"A leading tome on child development published in 1934 didn't even include a chapter on parents. . . . With an impish wit and a chatty style, Harris spins a persuasive argument that the 1934 book got it right."

—Sharon Begley, *Newsweek*

"Mixing logic-chopping rigor and wise-cracking humor, Harris turns academic overviews and her own sleuthing into a brisk tour of controversial data collection and interpretation. She deftly leads her readers through the inadequacies of socialization research."

—Ann Hulbert, *The New Republic*

"Occasionally, The Great American Intellectual Hype Machine trumpets a book well worth reading. . . . I'm pleased to welcome Mrs. Harris and her impressive rationality, serious scholarship, sardonic humor, and vivid prose to the ranks of realists."

—Steve Sailer, *The National Review*

"[Harris] presents her arguments in a style that is engaging and fun to read. People who raise children, teach children, and treat children will want to read this book."

—Dr. William Bernet, *Journal of the American Medical Association*

"A sea-changing book."

—Ellen Goodman, *The Boston Globe*

"Her conclusions have rocked the world of child development."

—Susan Reimer, *The Baltimore Sun*

"Ms. Harris takes to bits the assumption which has dominated developmental psychology for almost half a century. . . . Her book is an extraordinary feat. . . . She writes with unusual clarity and irreverent wit."

—*The Economist*

"Judith Rich Harris is a fiery iconoclast who offers relief. If you accept the central thesis of *The Nurture Assumption,* you can at last relax about raising your children. . . . Her book is worth reading if only for the pleasure of watching an acknowledged outsider taking on the conventional wisdom with such chutzpah."

—Jack O'Sullivan, *The Independent*

"The maverick writer and theoretician believes that peers, not parents, determine our personalities, and her unorthodox views have made the very real world of psychology sit up and take notice."

—Annie Murphy Paul, *Psychology Today*

"*The Nurture Assumption* is a hoot. [Harris] is a witty and articulate writer who clearly and systematically explains her refutations of commonly held assumptions in social psychology and behavioral genetics. . . . It's a very readable . . . entertaining book."

—Dr. Marilyn Heins, *The Arizona Daily Star*

"Harris . . . has razor-sharp common sense, perhaps the greatest gift of all."

—Wendy Orent, *The Atlanta Journal and Constitution*

"This is a fascinating, wildly entertaining, and in many ways persuasive book. Harris is a wonderful writer who doesn't stop at drawing research from fields as varied as behavioral psychology, ethnology, evolution, and sociology; she also draws cultural allusion from sources as disparate as *Little House on the Prairie,* Darwin, and Dave Barry."

—Marjorie Williams, *The Washington Monthly*

"As iconoclastic contribution to conventional psychology, *The Nurture Assumption* may also be a window on the future, triggering a shift away from a century of thinking that elevates early parental influence over all else."

—Cate Terwilliger, *The Denver Post*

"What Harris proposes is nothing short of breathtaking . . . her ideas might easily be dismissed, but Harris has done some serious research in psychology, sociology, and anthropology, backing her theory with dozens of articles and studies. . . . She also has the wit to write about them in a breezy and often entertaining manner."

—Peter Jensen, *The Baltimore Sun*

"An extraordinarily ambitious attempt to reexamine, from the ground up, an entire century's worth of findings on the forces that mold the child of today into the adult of tomorrow. . . . Most of what Harris writes is not only illuminating, but thoroughly persuasive."

—Mary Eberstadt, *Commentary*

"Her ideas make fascinating reading, and her work clearly deserves attention from developmental psychologists and other scholars of child development."

—Wendy M. Williams, *The Chronicle of Higher Education*

". . . shockingly persuasive . . . Harris has an impressive breadth of knowledge, and entertainingly leads the reader from social development to genetics, from neuropsychology to criminology, and from social anthropology to linguistics and child-care."

—Simon Baron-Cohen, *Nature*

"*The Nurture Assumption* is a stunning book. . . . Judith Harris shows how in thinking about child development we are trapped in a maze created by our uncritical acceptance of entrenched beliefs and biases. . . . The result is a new perspective that provides a thread we can follow to escape the maze."

—John T. Bruer, president of the James S. McDonnell Foundation; author of *The Myth of the First Three Years*

"Judith Harris's *The Nurture Assumption* is a paradigm shifter, which sounds like heavy work and yet she somehow makes it fun."

—David T. Lykken, professor of psychology, University of Minnesota; author of *The Antisocial Personalities and Happiness: What Studies of Twins Show Us About Nature, Nurture, and the Happiness Set Point*

"*The Nurture Assumption* is a rare book: clear, well informed, occasionally hilarious, and rich with compelling examples."

—David G. Myers, professor of psychology, Hope College; author of *The Pursuit of Happiness: Who Is Happy—and Why*

"The book is based on solid science, analyzed with a piercing style that's not afraid to take on the leading orthodoxy, and communicated in a clear, accessible, terrifically witty way."

—Robert M. Sapolsky, professor of neuroscience and biology, Stanford University; author of *The Trouble with Testosterone* and *Why Zebras Don't Get Ulcers*

"Truly revolutionary ideas turn topsy-turvy our most cherished ways of viewing the world and ourselves. . . . This is essential reading if you want to know how you became who you are—and what your children will grow up to be."

—Dean Keith Simonton, professor of psychology, University of California at Davis; author of *Scientific Genius and Greatness: Who Makes History and Why*

The NURTURE ASSUMPTION

Why Children Turn Out the Way They Do

JUDITH RICH HARRIS

A TOUCHSTONE BOOK

Published by Simon & Schuster

For Charlie, Nomi, and Elaine

TOUCHSTONE
Rockefeller Center
1230 Avenue of the Americas
New York, NY 10020

First Touchstone Edition 1999

TOUCHSTONE and colophon are registered trademarks
of Simon & Schuster Inc.

Designed by Carla Bolte

Manufactured in the United States of America

10 9 8 7 6 5 4 3 2 1

Library of Congress Cataloging-in-Publication Data

Harris, Judith Rich.
The nurture assumption : why children turn out the way they do / Judith Rich Harris.
p. cm.
Includes bibliographical references and index.
1. Child development—United States. 2. Child rearing—United States. 3. Nature and nurture—United States. 4. Environment and children—United States. I. Title.
HQ772.H353 1998
305.231—dc21
98-34824
CIP

ISBN 0-684-84409-5
0-684-85707-3 (Pbk)

Your children are not your children.
They are the sons and daughters of Life's longing for itself.
They come through you but not from you,
And though they are with you yet they belong not to you.
You may give them your love but not your thoughts,
For they have their own thoughts.
You may house their bodies but not their souls,
For their souls dwell in the house of tomorrow, which you cannot
visit, not even in your dreams.
You may strive to be like them, but seek not to make them
like you.
For life goes not backward nor tarries with yesterday.

—*Kahlil Gibran*

CONTENTS

FOREWORD

Three years ago an article in the *Psychological Review* forever changed the way I think about childhood and children. Like most psychologists, I have argued a lot about the relative roles of genetic endowment and parental upbringing. We all take it for granted that what doesn't come from the genes must come from the parents. But here was an article by someone named Judith Rich Harris, with no university affiliation under her name, saying that parents don't matter. What matters, other than genes, is a child's *peer group.* It sounded weird, but Harris soon persuaded me with facts that I knew to be true but had filed away in that mental folder we all keep for undeniable truths that do not fit into our belief systems.

I study language development: how children acquire a grammatical rule system from the parental input, as we say in the business. A strange factoid in our True-but-Inconvenient file is that children always end up with the language and accent of their peers, not of their parents. No one in psycholinguistics has ever called attention to this fact, let alone explained it. But here was a theory that did.

Other facts about language fit Harris's theory, too. Children learn a language even in the many cultures in which adults don't speak to them; they do just fine listening to their slightly older peers. Children who are not exposed to a full-blown grammatical language from adults can *create* one among themselves. And children of immigrants pick up language from the playground so well that they are soon ridiculing their parents' grammatical errors.

Acquiring the particulars of a native language is an example of cultural learning. Children in Japan speak Japanese, children in Italy speak Italian, and these differences have nothing to do with their genes. If these differences also have nothing to do with what they learn from their parents, then maybe, Harris pointed out, we need to rethink cultural learning in general. It had always seemed obvious to me that children are socialized by their parents. But among the items stashed away in my True-but-Ignored file was the fact that many successful people—my own father among them—were children of immigrants who were not handicapped in the least by culturally inept parents who never acquired the language, customs, or know-how of their adopted land.

Harris's article had more than just a neat idea and some everyday truths. She backed up her theory with technical literature from psychology, anthropology, cultural history, behavioral genetics, and primatology, and she used it to shed light on a variety of topics such as sex-role development and adolescent delinquency. In my first e-mail to her I asked, "Have you thought of writing a book?"

The thesis of *The Nurture Assumption*—that in the formation of an adult, genes matter and peers matter, but parents don't matter—raises issues about children and parents that could not be more profound. It calls into question the standard social science model of the child as a bundle of reflexes and a blank cortex waiting to be programmed by benevolent parents—which, when you think about it, is pretty improbable on biological grounds. Like other living things, children are products of evolution and must be active players in their own struggle to survive and eventually to reproduce. This has important implications, thoroughly explored herein.

For one thing, the biological interests of the parent and the child are not identical. So even if children acquiesce to their parents' rewards, punishments, examples, and naggings for the time being—because they are smaller and have no choice—they should not allow their personalities to be permanently shaped by these tactics.

Moreover, *Homo sapiens* is a species that lives in groups, and a group is like any other aspect of an organism's environmental niche: it has a texture of causes and effects that the organism had better adapt to. Prospering in a group means taking advantage of the fact that many heads are better than one and sharing in its accumulated discoveries. It means figuring out local norms that may seem utterly arbitrary but that are adaptive because they are shared (familiar examples include paper currency and driving on the right side of the road). It means striving to benefit from one's association with other people, rather than allowing oneself to be exploited or dominated. And since each group develops a commonality of interests that puts it into conflict with other groups, it means taking part in group-against-group competitions.

Today, children win or lose by their ability to prosper in this milieu; in the past they lived or died by it. It makes sense that they should take their calories and protection from their parents, because their parents are the only ones willing to provide them, but that they should get their information from the best sources they can find, which might not be their parents. The child will have to compete for mates, and before that for the status necessary to find and keep them, in groups other than the family—groups that play by different rules. Children and parents may even find themselves in partly competing groups. Nature surely did not design children to be putty in their parents' hands.

Equally unlikely is the idea that a baby's attachment to its mother sets the pattern for its later commerce with the world—another dogma dissolved in these pages. Relationships with parents, with siblings, with peers, and with strangers could not be more different, and the trillion-synapse human brain is hardly short of the computational power it would take to keep each one in a separate mental account. The attachment hypothesis owes its popularity to a tired notion bequeathed to us by Freud and the behaviorists: the baby's mind as a small blank slate that will retain forever the first few inscriptions written on it.

The Nurture Assumption is truly rare. Though its thesis is at first counterintuitive, one gets a sense of *real* children and parents walking through these pages, not compliant little humanoids that no one actually meets in real life. Among its other treats are a devastating methodological critique of much research in child development, an eye-opening analysis of why schools fail, an explanation of why female doctors and lawyers have children who insist that women are supposed to be housewives, and an uncommonly wise answer to the inevitable question: So you're saying it doesn't matter how I treat my child?

Being among the first to read this electrifying book has been one of the high points of my career as a psychologist. One seldom sees a work that is at once scholarly, revolutionary, insightful, and wonderfully clear and witty. But don't be misled by all the fun. *The Nurture Assumption* is a work of serious, original science. I predict it will come to be seen as a turning point in the history of psychology.

Steven Pinker
Cambridge, Massachusetts
May 1998

PREFACE

This book has two purposes: first, to dissuade you of the notion that a child's personality—what used to be called "character"—is shaped or modified by the child's parents; and second, to give you an alternative view of how the child's personality is shaped. My arguments against the old notion and in favor of the new one were originally sketched out in a 1995 article I wrote for the journal *Psychological Review.* The article began with these words:

> Do parents have any important long-term effects on the development of their child's personality? This article examines the evidence and concludes that the answer is no.

It was a challenge—a slap in the face, really—to traditional psychology. I expected people to be taken aback, maybe even angry. But what many readers focused on instead was the strange lack of a university affiliation—of any affiliation—under the author's name; the puzzling absence, in the acknowledgments footnote, of thanks to granting agencies for supporting my research. I was not a professor, not even a graduate student. Hardly anyone had ever heard of me, and here I was publishing an article in psychology's most distinguished academic journal —a journal that accepts maybe 15 percent of the manuscripts submitted to it.

I meant my readers to get mad but instead they got curious. They sent me e-mail. Members of the academic world wrote to me, inquiring politely (or not so politely) who I was and who were my mentors. My who-the-hell-are-you? mail, I call it. Here is my favorite example of the genre, from a professor at Cornell University:

> Your article constitutes a major contribution to personality and developmental psychology—which only makes me even more curious about you. Are you an academic? A clinician? An unemployed steel worker who has an interesting hobby of writing seminal scientific articles?

Of those choices, I told him, I'd have to pick (c): an unemployed steel worker. In fact, I said, I was an unemployed writer of college textbooks. I explained that I had no Ph.D.—I'd been kicked out of Harvard's Department of Psychology with only a master's degree. I had been stuck at home for many years due to chronic health problems; I had no mentors, no students. I became a writer of textbooks because that is something one can do at home. I was an *unemployed* writer of textbooks because I'd quit that job.

I never heard from him again. But others who got the same explanation did write back, and some of them have become my friends and colleagues. As yet I have met none of them in person; my link to the academic world is entirely through e-mail and postal mail.

In 1997, my article in the *Psychological Review* was chosen to receive a prize given by the American Psychological Association to "an outstanding recent article in psychology." The prize is named the George A. Miller Award, after a prominent psychologist and former president of the APA. It was proof that the gods do have a sense of humor. Thirty-seven years earlier I had received a letter from Harvard's Department of Psychology; they had decided, the letter said, that they didn't want to give me a Ph.D. because they didn't think I'd amount to much. The letter was signed by the Acting Chairman of the Department, George A. Miller.

In the years between my two encounters with the name of George A. Miller, I married one of my fellow graduate students (I am married to him still) and we reared two daughters; they appear, from time to time, in the pages of this book. I was in good health when I married and remained so for about fifteen years, but I made no attempt to return to graduate school. I didn't do anything to prove that Harvard had been wrong about me because I assumed they were right.

Getting sick was what changed my mind. Perhaps it was the intimations of mortality (if you think you may die in a fortnight, it concentrates the mind wonderfully); perhaps it was simple boredom. Confined to bed for a period of time, I started doing the kind of work that my professors at Harvard would have approved of. Some of it even got published.

Fortunately, the metamorphosis came too late to permit me to go back to graduate school. And thus I escaped indoctrination. Whatever I learned about developmental psychology and social psychology, I learned on my own. I was an outsider looking in, and that has made all the difference. I did not buy into the assumptions of the academic establishment. I was not indebted to their granting agencies. And, once I had given up writing textbooks, I was not required to perpetuate the status quo by teaching the received gospel to a bunch of credulous college students. I gave up writing textbooks because one day it suddenly occurred to me that many of the things I had been telling those credulous college students were wrong.

"If possible," counseled a physician in the pages of the *Journal of the American Medical Association,* "the effectiveness of an effort should be determined by someone outside the effort who has nothing to gain by its perpetuation." In other words, if you want to know the truth about the emperor's clothes, don't ask the tailors.

Though I am not one of them, I am deeply in their debt, because the theory of child development presented in this book is based in large part on the research produced by the academic establishment. In particular, I am grateful to the many members of the academic world who, over the years, responded so generously to my requests for copies of their published articles.

Not having access to a university library is an inconvenience that can be overcome. Public libraries served me well by providing me with scores of books borrowed from university shelves. My thanks to Mary Balk of the Middletown (New Jersey) Library and Jane Eigenrauch of the Red Bank Library for the many books they obtained for me on interlibrary loan. Thanks, too, to the helpful people—especially Joan Friebely, Sabina Harris, and David G. Myers—who sent me additional reading material by mail.

Many people helped to keep me from feeling isolated. My first e-mail friends in the academic world, Neil Salkind and Judith Gibbons, made me realize that "shut in" does not necessarily mean "shut out." Daniel Wegner saw to it that the manuscript I submitted to the *Psychological Review* got fair treatment; his comments challenged me to think more deeply about some of the assertions I made in the first version of the article, which led to improvements not only in the article but in the theory as well. The advice and encouragement I received from Steven Pinker, from my literary agent, Katinka Matson of Brockman, Inc., from my first editor at the Free Press, Susan Arellano, and from my second editor, Liz Maguire, were of tremendous value. A million thanks to all these people, and to Airie Dekidjiev, the editor of the paperback edition.

Thanks also to Florence Metzger, who kept my house clean and gave me, as a bonus, her cheerfulness and kindness.

My colleagues, friends, and members of my family made generous contributions of their time and expertise to read and comment upon earlier drafts of this book. I am deeply grateful to them for their comments, which lifted my spirits and my prose and kept me from making some embarrassing errors. Susan Arellano, Joan Friebely, Charles S. Harris, Nomi Harris, David Lykken, David G. Myers, Steven Pinker, and Richard G. Rich read and made perspicacious comments on the entire manuscript. Anne-Marie Ambert, William Corsaro, Carolyn Edwards, Thomas Kindermann, and John Modell did the same for parts of the book in their areas of interest.

My daughters, my son-in-law, my brother, and—most of all—my husband provided me with the support a writer needs. They put up with me; they believed in me. They have my love and my eternal gratitude.

1 "NURTURE" IS NOT THE SAME AS "ENVIRONMENT"

Heredity and environment. They are the yin and yang, the Adam and Eve, the Mom and Pop of pop psychology. Even in high school I knew enough about the subject to inform my parents, when they yelled at me, that if they didn't like the way I was turning out they had no one to blame but themselves: they had provided both my heredity and my environment.

"Heredity and environment"—that's what we called them back then. Nowadays they are more often referred to as "nature and nurture." Powerful as they were under the names they were born with, they are yet more powerful under their alliterative aliases. Nature and nurture rule. Everyone knows it, no one questions it: nature and nurture are the movers and shapers. They made us what we are today and will determine what our children will be tomorrow.

In an article in the January 1998 issue of *Wired,* a science journalist muses about the day—twenty? fifty? a hundred years from now?—when parents will be able to shop for their children's genes as easily as today they shop for their jeans. "Genotype choice," the journalist calls it. Would you like a girl or a boy? Curly hair or straight? A whiz at math or a winner of spelling bees? "It would give parents a real power over the sort of people their children will turn out to be," he says. Then he adds, "But parents have that power already, to a large degree."

Parents already have power over the sort of people their children will turn out to be, says the journalist. He means, because parents provide the environment. The nurture.

No one questions it because it seems self-evident. The two things that determine what sort of people your children will turn out to be are nature —their genes—and nurture—the way you bring them up. That is what you believe and it also happens to be what the professor of psychology believes. A happy coincidence that is not to be taken for granted, because in most sciences the expert thinks one thing and the ordinary citizen— the one who used to be called "the man on the street"—thinks something else. But on this the professor and the person ahead of you on the checkout line agree: nature and nurture rule. Nature gives parents a baby; the end result depends on how they nurture it. Good nurturing can make up for many of nature's mistakes; lack of nurturing can trash nature's best efforts.

That is what I used to think too, before I changed my mind.

What I changed my mind about was nurture, not environment. This is not going to be one of those books that says everything is genetic; it isn't. The environment is just as important as the genes. The things children experience while they are growing up are just as important as the things they are born with. What I changed my mind about was whether "nurture" is really a synonym for "environment." Using it as a synonym for environment, I realized, is begging the question.

"Nurture" is not a neutral word: it carries baggage. Its literal meaning is "to take care of" or "to rear"; it comes from the same Latin root that gave us *nourish* and *nurse* (in the sense of "breast-feed"). The use of "nurture" as a synonym for "environment" is based on the assumption that what influences children's development, apart from their genes, is the way their parents bring them up. I call this the *nurture assumption.* Only after rearing two children of my own and coauthoring three editions of a college textbook on child development did I begin to question this assumption. Only recently did I come to the conclusion that it is wrong.

It is difficult to disprove assumptions because they are, by definition, things that do not require proof. My first job is to show that the nurture assumption is nothing more than that: simply an assumption. My second is to convince you that it is an unwarranted assumption. My third is to give you something to put in its place. What I will offer is a viewpoint as powerful as the one it replaces—a new way of explaining why children turn out the way they do. A new answer to the basic question of why we are the way we are. My answer is based on a consideration of what kind of mind the child is equipped with, which requires, in turn, a consideration of the evolutionary history of our species. I will ask you to accompany me on visits to other times and other societies. Even chimpanzee societies.

Beyond a Reasonable Doubt?

How can I question something for which there is so much evidence? You can see it with your own eyes: parents do have effects on their kids. The child who has been beaten looks cowed in the presence of her parents. The child whose parents have been wimpy runs rampant over them. The child whose parents failed to teach morality behaves immorally. The child whose parents don't think he will accomplish much doesn't accomplish much.

For those doubting Thomases who have to see it in print, there are books full of evidence—thousands of books. Books written by clinical psychologists like Susan Forward, who describes the devastating and long-lasting effects of "toxic parents"—overcritical, overbearing, underloving, or unpredictable people who undermine their children's self-esteem and autonomy or give them too much autonomy too soon. Dr. Forward has seen the damage such parents wreak on their children. Her patients are in terrible shape psychologically and it is all their parents' fault. They won't get better until they admit, to Dr. Forward and themselves, that it is all their parents' fault.

But perhaps you are among those doubting Thomases who don't consider the opinions of clinical psychologists, formed on the basis of conversations with a self-selected sample of troubled patients, to be evidence. All right, then, there is evidence of a more scientific sort: evidence obtained in carefully designed studies of ordinary parents and their children—parents and children whose psychological well-being varies over a wider range than you could find in Dr. Forward's waiting room.

In her book *It Takes a Village,* First Lady Hillary Rodham Clinton has summarized some of the findings from the carefully designed studies carried out by developmental psychologists. Parents who care for their babies in a loving, responsive way tend to have babies who are securely attached to them and who develop into self-confident, friendly children. Parents who talk to their children, listen to them, and read to them tend to have bright children who do well in school. Parents who provide firm —but not rigid—limits for their children have children who are less likely to get into trouble. Parents who treat their children harshly tend to have children who are aggressive or anxious, or both. Parents who behave in an honest, kind, and conscientious manner with their children are likely to have children who also behave in an honest, kind, and conscientious manner. And parents who fail to provide their children with a home that contains both a mother and a father have children who are more likely to fail in some way in their own adult lives.

These statements, and others of a similar sort, are not airy speculation. There is a tremendous amount of research to back them up. The textbooks I wrote for undergraduates taking college courses in child development were based on the evidence produced by that research. The professors who teach the courses believe the evidence. So do the journalists who occasionally report the results of a study in a newspaper or magazine article. The pediatricians who give advice to parents base much of their advice on it. Other advice-givers who write books and newspaper articles also take the evidence at face value. The studies done by developmental psychologists have an influence that ripples outward and permeates our culture.

During the years I was writing textbooks, I believed the evidence too. But then I looked at it more closely and to my considerable surprise it fell apart in my hands. The evidence developmental psychologists use to support the nurture assumption is not what it appears to be: it does not prove what it appears to prove. And there is a rising tide of evidence *against* the nurture assumption.

The nurture assumption is not a truism; it is not even a universally acknowledged truth. It is a product of our culture—a cherished cultural myth. In the remainder of this chapter I will tell you where it came from and how I came to question it.

The Heredity and Environment of the Nurture Assumption

Francis Galton—Charles Darwin's cousin—is the one who usually gets the credit for coining the phrase "nature and nurture." Galton probably got the idea from Shakespeare, but Shakespeare didn't originate it either: thirty years before he juxtaposed the two words in *The Tempest,* a British educator named Richard Mulcaster wrote, "Nature makes the boy toward, nurture sees him forward." Three hundred years later, Galton turned the pairing of the words into a catchphrase. It caught on like a clever advertising slogan and became part of our language.

But the true father of the nurture assumption was Sigmund Freud. It was Freud who constructed, pretty much out of whole cloth, an elaborate scenario in which all the psychological ills of adults could be traced back to things that happened to them when they were quite young and in which their parents were heavily implicated. According to Freudian theory, two parents of opposite sexes cause untold anguish in the young child, simply by being there. The anguish is unavoidable and universal; even the most conscientious parents cannot prevent it, though they can easily make it worse. *All* little boys have to go through the Oedipal crisis; *all* little girls go through the reduced-for-quick-sale female version. The mother (but not the father) is also held responsible for two earlier crises: weaning and toilet training.

Freudian theory was quite popular in the first half of the twentieth century; it even worked its way into Dr. Spock's famous book on baby and child care:

> Parents can help children through this romantic but jealous stage by gently keeping it clear that the parents do belong to each other, that a boy can't ever have his mother to himself or a girl have a father to herself.

Not surprisingly, it was psychiatrists and clinical psychologists (the kind who see patients and try to help them with their emotional problems) who were most influenced by Freud's writings. However, Freudian theory also had an impact on academic psychologists, the kind who do research and publish the results in professional journals. A few tried to find experimental evidence for various aspects of Freudian theory; these efforts were largely unsuccessful. A greater number were content to drop Freudian buzzwords into their lectures and research papers.

Others reacted by going to the opposite extreme, dumping out the baby with the bathwater. Behaviorism, a school of psychology that was popular in American universities in the 1940s and '50s, was in part a reaction to Freudian theory. The behaviorists rejected almost everything in Freud's philosophy: the sex and the violence, the id and the superego, even the conscious mind itself. Curiously, though, they accepted the basic premise of Freudian theory: that what happens in early childhood—a time when parents are bound to be involved in whatever is going on—is crucial. They threw out the script of Freud's psychodrama but retained its cast of characters. The parents still get leading roles, but they no longer play the parts of sex objects and scissor-wielders. Instead, the behaviorists' script turns them into conditioners of responses or dispensers of rewards and punishments.

John B. Watson, the first prominent behaviorist, noticed that real-life parents aren't very systematic in the way they condition their children's responses and offered to demonstrate how to do the job properly. The demonstration would involve rearing twelve young humans under carefully controlled laboratory conditions:

> Give me a dozen healthy infants, well-formed, and my own specified world to bring them up in and I'll guarantee to take any one at random and train him to become any type of specialist I might select—doctor, lawyer, artist, merchant-chief, and, yes, even beggar-man and thief, regardless of his talents, penchants, tendencies, abilities, vocations, and race of his ancestors.

Fortunately for the dozen babies, no one took Watson up on his proposal. To this day, there are probably some aging behaviorists who think he could have pulled it off, if only he had had the funding. But in fact it was an empty boast—Watson wouldn't have had the foggiest idea of how to fulfill his guarantee. In his book *Psychological Care of Infant and Child* he had lots of recommendations to parents on how to keep their children from being "spoiled" and how to make them fearless and self-reliant (you leave them alone and avoid showing them affection), but there were no suggestions on how to raise children's IQs by twenty points, which would seem to be an important step toward getting them into medical or law school, in preparation for the first two occupations on Watson's list. Nor were there any guidelines for how to make them choose medicine over law, or vice versa. When it got right down to it, the only thing John Watson had succeeded in doing was to produce conditioned fear of furry animals in an infant named Albert, by making a loud noise whenever little Albert reached for a rabbit. Although this training no doubt discouraged Albert from growing up with the idea of becoming a veterinarian, he still had plenty of other career options to choose from.

A more promising behavioristic approach was that of B. F. Skinner, who talked about reinforcing responses rather than conditioning them. This was a far more useful method because it didn't have to make do with responses the child was born with—it could create new responses, by reinforcing (with rewards such as food or praise) closer and closer approximations to the desired behavior. In theory, one could produce a doctor by rewarding a kid for bandaging a friend's wounds, a lawyer by rewarding the kid for threatening to sue the manufacturer of the bike the friend fell off. But what about the third occupation on Watson's list, artist? Research done in the 1970s showed that you can get children to paint lots of pictures simply by rewarding them with candy or gold stars for doing so. But the rewards had a curious effect: as soon as they were discontinued, the children stopped painting pictures. They painted *fewer* pictures, once they were no longer being rewarded, than children who had never gotten any rewards for putting felt-tip pen to paper. Although subsequent studies have shown that it is possible to administer rewards without these negative aftereffects, the results are difficult to predict because they depend on subtle variations in the nature and timing of the reward and on the personality of the reward recipient.

Genius is said to be 99 percent perspiration, 1 percent inspiration. Behaviorism focuses on the perspiration and forgets about the inspiration. Tom Sawyer was a better psychologist than B. F. Skinner: by letting his friends reward *him* for the privilege of whitewashing the fence, he not only got them to do the work, he got them to *like* it.

I don't think Watson really wanted a dozen healthy infants to experiment with. I think his request was just a vainglorious way of expressing the basic belief of behaviorism: that children are malleable and that it is their environment, not innate qualities such as talent or temperament, that determines their destiny. The extremist statements were made for their publicity value: Watson was promoting himself for the position of Lord High Environmentalist.

The Art and Science of Studying Children

As an academic specialty, the study of how immature humans develop into adults had a rather late beginning—around 1890. The early developmentalists were interested in children but didn't pay much attention to their parents. If you look at a developmental psych book written before Freudian theory and behaviorism became popular, you will find little or nothing about parental influences on the development of the child's personality. Florence Goodenough's successful textbook, *Developmental Psychology,* first published in 1934, has no chapter on parent–child relationships. In her discussion of the causes of juvenile delinquency, Goodenough does talk about the effects of a "bad environment," but she is referring to those parts of a city where the dwellings are "run down and dilapidated" and where there are "many saloons, poolrooms, and gambling-houses."

At about the same time, Winthrop and Luella Kellogg reported the results of their experiment in primate-rearing. They reared a chimpanzee named Gua in their home, side by side with their infant son Donald, and treated them as much alike as possible. The word *environment* crops up frequently in the Kelloggs' book, but they used it only to distinguish "a civilized environment" or "a human environment" from the jungle or zoo in which Gua would otherwise have been reared. Fine distinctions between one civilized home and another had not yet been pinned to the term *environment.*

Perhaps the most influential of the early developmentalists was Arnold Gesell. For Gesell as for Goodenough, parents were a taken-for-granted part of the child's environment, anonymous and interchangeable. Children of a given age were pretty much interchangeable as well. Gesell spoke of "your four-year-old" or "your seven-year-old" and gave instructions on how to take care of them, much as a book about cars might have told you how to take care of "your Ford" or "your Studebaker." The home was like a garage where the children came home at night and where the anonymous attendants washed them, waxed them, and filled their tanks.

The modern variety of developmental psychology was born in the 1950s, when researchers stopped looking for ways that four-year-olds are similar to other four-year-olds and began to study the ways that they differ from one another. That led to the idea—and it was a novel idea at the time—of tracing the differences among the children to differences in the way their parents reared them. The harbinger of this kind of research was a study whose dual ancestry in Freudian psychology and behaviorism was clearly visible. It was designed to test how rewards and punishments dispensed by parents, including their methods of weaning and toilet training, affect their child's personality. In particular, the researchers were interested in aspects of the child's personality that pertained to Freudian concepts such as the development of the superego. One of the researchers was Eleanor Maccoby, now retired from Stanford University after a long and distinguished career. In a recent article, Maccoby described the outcome of this early study:

> The results of this body of work were in many respects disappointing. In a study of nearly 400 families, few connections were found between parental child-rearing practices (as reported by parents in detailed interviews) and independent assessments of children's personality characteristics—so few, indeed, that virtually nothing was published relating the two sets of data. The major yield of the study was a book on child-rearing practices as seen from the perspective of mothers. This book was mainly descriptive and included only very limited tests of the theories that had led to the study.

This inauspicious beginning did not discourage further efforts along the same lines—on the contrary, it was followed by a deluge of research that has continued to this day. Although the explicit links to Freudian theory and behaviorism were soon dropped, two ideas were retained: the behaviorists' belief that parents influence their children's development by the rewards and punishments they dole out, and the Freudians' belief that parents can mess up their children very badly and often do so.

That parents influence the development of their children was now being taken for granted. The goal of the later generations of researchers was not to find out *whether* parents influence their children's development but to discover *how* they influence it. The procedure became standardized: you look at how the parent rears the child, you look at how the child is turning out, you do that for a fair number of parents and children, and then, by putting together all the data and looking for overall trends, you try to show that some aspect of the parent's child-rearing method has had an effect on some characteristic of the child. Your hope is to find a relationship between the parents' behavior and the children's characteristics that is "statistically significant"—or, to put it in nontechnical terms, publishable.

Although the study described by Eleanor Maccoby failed to find results that were statistically significant, many of the thousands of subsequent studies, cut to the same pattern, were more successful. They did yield significant results and they were published in professional journals such as *Child Development* and *Developmental Psychology;* they became part of the mountain of evidence used to support the nurture assumption. Of the others—the ones that did not yield significant results—we know very little; most of them probably ended up in landfills. The only reason we know that the first study of this type found "few connections" between the parents' child-rearing practices and the children's personalities is that Dr. Maccoby admitted it in print—thirty-five years later.

Turning the Wild Baby into a Solid Citizen

Developmentalists who specialize in doing the kind of research I just described are called socialization researchers. Socialization is the process by which a wild baby is turned into a domesticated creature, ready to take its place in the society in which it was reared. Individuals who have been socialized can speak the language spoken by the other members of their society; they behave appropriately, possess the requisite skills, and hold the prevailing beliefs. According to the nurture assumption, socialization is something that parents do to children. Socialization researchers study how the parents do it and how well they do it, as judged by how well the children turn out.

Socialization researchers believe in the nurture assumption. As I said at the beginning, I used to believe in it too. On the basis of that belief, I coauthored three editions of a textbook on child development. I had begun work (without a coauthor this time) on a new development textbook when something happened to make me abandon the project. For years I had been feeling a vague discomfort about the quality of the data in socialization research. For years I had avoided thinking about observations that didn't fit neatly into the story my publishers expected me to tell my readers. One day I suddenly found I no longer believed that story.

Here are three of the observations that bothered me.

First observation. When I was a graduate student I lived in a rooming house in Cambridge, Massachusetts. It was owned by a Russian couple who, along with their three children, occupied the ground floor of the house. The parents spoke Russian to each other and to their children; their English was poor and they spoke it with a thick Russian accent. But the children, who ranged in age from five to nine, spoke perfectly acceptable English with no accent at all—that is, they had the same Boston–Cambridge accent as the other kids in the neighborhood. They *looked* like the other kids in the neighborhood, too. There was something foreign-looking about the parents—I wasn't sure if it was their clothing, their gestures, their facial expressions, or what. But the children didn't look foreign: they looked like ordinary American kids.

It puzzled me. Obviously, babies don't learn to speak on their own; obviously, they learn their language from their parents. But the language those children spoke was not the language they had learned from their parents. Even the five-year-old was a more competent speaker of English than her mother.

Second observation. This one has to do with children reared in England. It came to my attention—thanks to my weakness for British mystery novels—that generations of upper-class British males were reared in a way that doesn't make sense in terms of the nurture assumption. The son of wealthy British parents spent most of his first eight years in the company of a nanny, a governess, and perhaps a sibling or two. He spent little time with his mother and even less with his father, whose attitude toward children was typically that they should not be heard and, if possible, not be seen either. At the age of eight the boy was sent off to a boarding school and he remained at school for the next ten years, coming home only for "holidays" (vacations). And yet, when he emerged from Eton or Harrow, he was ready to take his place in the world of British gentlemen. He did not talk and act like his nanny or his governess, or even like his teachers at Eton or Harrow. In his upper-class accent and his upper-class demeanor, he bore a close resemblance to his father—a father who had had virtually nothing to do with bringing him up.

Third observation. Many developmental psychologists assume that children learn how they are expected to behave by observing and imitating their parents, particularly the parent of the same sex. This assumption, too, is a legacy from Freudian theory. Freud believed that the resolution of the Oedipal or Electra complex leads to identification with the same-sex parent and, consequently, to the formation of the superego. Little children who have not yet made it through the Sturm und Drang of the Oedipal period cannot be expected to behave properly because they have not yet acquired a superego.

Selma Fraiberg, a child psychologist whose books were popular in the 1950s, accepted the Freudian story of socialization. She used the following anecdote to illustrate how children behave during the iffy period when they've learned what they're not supposed to do but can't quite keep themselves from doing it.

> Thirty-month-old Julia finds herself alone in the kitchen while her mother is on the telephone. A bowl of eggs is on the table. An urge is experienced by Julia to make scrambled eggs. . . . When Julia's mother returns to the kitchen, she finds her daughter cheerfully plopping eggs on the linoleum and scolding herself sharply for each plop, "NoNoNo. Mustn't dood it. NoNoNo. *Mustn't* dood it!"

Fraiberg attributed Julia's lapse to the fact that she had not yet acquired a superego, presumably because she had not yet identified with her mother. But look closely at what Julia was doing when her mother came back and caught her egg-handed: she was making scrambled eggs and she was yelling "NoNoNo." Julia was imitating her mother! And yet Mother was not pleased.

The fact is that children cannot learn how to behave by imitating their parents, because most of the things they see their parents doing—making messes, bossing other people around, driving cars, lighting matches, coming and going as they please, and lots of other things that look like fun to people who are not allowed to do them—are prohibited to children. From the child's point of view, socialization in the early years consists mainly of learning that *you're not supposed to behave like your parents.*

In case you are wondering whether imitating the same-sex parent might work better in a less complex society, the answer is no. In preindustrial societies, the distinction between acceptable adult behavior and acceptable child behavior tends to be even greater than in our own. In village societies in the Polynesian islands, for instance, children are expected to be restrained and submissive with adults and to speak only when spoken to. The adults do not behave this way, either when interacting with their children or when interacting with other adults. Although Polynesian children may learn the art of weaving or fishing by watching their parents, they cannot learn the rules of social behavior that way. In most societies, children who behave like grownups are considered impertinent.

According to the nurture assumption, it is the parents who transmit cultural knowledge (including language) to their children and who prepare them for full membership in the society in which they will spend their adult lives. But the daughter of immigrant parents does not learn the local language and customs from her parents, the son of wealthy British parents sees his parents too rarely to make such a theory plausible, and children in many different cultures are likely to get into trouble if they behave too much like their parents. And yet, all these children somehow do learn to behave the way their society expects them to.

The nurture assumption is based on a particular model of family life: that of a typical middle-class North American or European family. Socialization researchers do not, as a rule, look at families in which the parents cannot speak the local language; they do not study children who go to boarding schools or who are reared by governesses and nannies. Although anthropologists and cross-cultural psychologists have done many studies of child-rearing methods in other societies, socialization researchers seldom check to see whether their theories are applicable to children growing up in these other societies.

Of course, some things are true in every society. In every society, babies are born helpless and ignorant and need older people to take care of them. In every society, babies must learn the local language and customs and form working relationships with the other members of their household. They must learn that the world has rules and that they cannot do whatever they feel like doing. This learning has to begin very early, at a time when they are still completely dependent on their adult caregivers.

There is no question that the adult caregivers play an important role in the baby's life. It is from these older people that babies learn their first language, have their first experiences in forming and maintaining relationships, and get their first lessons in following rules. But the socialization researchers go on to draw other conclusions: that what children learn in the early years about relationships and rules sets the pattern for later relationships and later rule-following, and hence determines the entire course of their lives.

I used to think so too. I still believe that children need to learn about relationships and rules in their early years; it is also important that they acquire a language. But I no longer believe that this early learning, which in our society generally takes place within the home, sets the pattern for what is to follow. Although the learning itself serves a purpose, the *content* of what children learn may be irrelevant to the world outside their home. They may cast it off when they step outside as easily as the dorky sweater their mother made them wear.

Environment is everything around a person - not just parents and home life

2 THE NATURE (AND NURTURE) OF THE EVIDENCE

From the beginning, academic psychology has been split by a great divide. On one side are those who believe in nature or who are most interested in things that are inherited. On the other are those who believe in nurture (*sic*) or who are most interested in things that are acquired through experience. Nowhere are the two sides farther apart than in developmental psychology. The socialization researchers dwell on the nurture side. The nature side is the land of the behavioral geneticists.

Socialization researchers and behavioral geneticists earn a living by teaching college undergraduates and graduate students and by doing research. Their status depends on the success of their research and the quantity and quality of their publications. They are specialists: neither side spends much time reading what the other has written, partly because they know they won't agree with it, partly because they don't have time. In general, academicians read mostly the publications in their own area and perhaps a few closely related areas.

My situation is different. I do not teach at a university and am not required to carry out a program of research in a specialized area. A writer of textbooks is supposed to present a balanced view, so in the years I spent writing and revising one textbook and preparing to write another, I read books and journal articles written from a variety of viewpoints. That gave me a perspective that most academic psychologists don't have—a bird's-eye view of the entire field. Sometimes things can be seen from a middle distance that aren't visible from close up.

In this chapter and the next, I will tell you what I learned from my bird's-eye survey of socialization research and behavioral genetics. I will tell you what the researchers found, what they said about what they found, and what was wrong with what they said about what they found.

If you are not one of them, you might wonder why you should care about what a bunch of college professors said. The reason is that their research and the way they interpret it provide the background for almost every bit of child-rearing advice you read in newspapers and parenting magazines or get from your pediatrician. Almost every bit of child-rearing advice Hillary Rodham Clinton gives her readers in *It Takes a Village* was based on the research produced by those college professors. Yes, Hillary did her homework.

The nurture assumption—the notion that parents are the most important part of the child's environment and can determine, to a large extent, how the child turns out—is a product of academic psychology. Though it has permeated our culture, it is not folklore. In fact (as you will see in Chapter 5), folks didn't use to believe it.

The Effects of Eating Broccoli

Socialization research is the scientific study of the effects of the environment—in particular, the effects of the parents' child-rearing methods or their behavior toward their children—on the children's psychological development. It is a science because it uses some of the methods of science, but it is not, by and large, an experimental science. To do an experiment it is necessary to vary one thing and observe the effects on something else. Since socialization researchers do not, as a rule, have any control over the way parents rear their children, they cannot do experiments. Instead, they take advantage of existing variations in parental behavior. They let things vary naturally and, by systematically collecting data, try to find out what things vary together. In other words, they do correlational studies.

You are probably familiar with another kind of correlational study, from a field called epidemiology. Epidemiologists study the environmental factors that make people sick or healthy. The methods they use in gathering and analyzing data are similar to those used in socialization research and are plagued by similar problems. I will take a short detour through epidemiology because the parallels between the two fields are informative.

Let's say we are epidemiologists and we decide to do a study on the relationship between broccoli consumption and health. Our method will be straightforward: we will ask a large number of middle-aged people how much broccoli they consume and then, five years later, check to see how many of them are still alive. We are using *being alive* as a simple measure of good health; on the whole, living people are healthier than dead ones.

Five years later we find the relationship between broccoli eating and survival shown in the table below. (Please note that these results are entirely fictional—I made them up.)

	Percentage Still Alive After Five Years		
	All Subjects	*Women*	*Men*
Broccoli lovers (eat it at least once a week)	99	99	99
Broccoli tolerators (eat it about once a month)	98	99	97
Broccoli shunners (wouldn't touch it with a ten-foot pole)	97	99	95

We feed these results into a computer. The computer tells us that broccoli eating did not have a significant effect on the longevity of all our subjects (there isn't much difference between 99, 98, and 97 percent), or on the longevity of women. But if we consider just the men, the relationship between broccoli eating and longevity was "statistically significant." That means it is unlikely—though not impossible—that the difference we found was simply a fluke, a lucky coincidence. It also means we can write up our results, publish them, and apply for a grant to study the relationship between *cauliflower* consumption and health.

Our study appears in an epidemiological journal. A newspaper reporter happens to read it. The next day there's a headline in the paper: EATING BROCCOLI MAKES MEN LIVE LONGER, STUDY SHOWS.

But does it? Does the study show that eating broccoli *caused* the male subjects to live longer? Men who eat broccoli may also eat a lot of carrots and brussels sprouts. They may eat less meat or less ice cream than the broccoli shunners. Perhaps they are more likely to exercise, more likely to buckle their seatbelts, less likely to smoke. Any of these other lifestyle factors, or all of them together, may be responsible for the longer lives of the broccoli eaters. Eating broccoli might have nothing to do with it. Eating broccoli might even have been *shortening* our subjects' lives, but this effect was outweighed by the beneficial effects of all the other things the broccoli eaters were doing.

Another complication is that broccoli eating may be related to marital status: married men may eat more broccoli than single men. It is a well-known fact that, on the average, married men live longer than single men. So perhaps it was being married that made the broccoli eaters live longer, not the broccoli. On the other hand, maybe it is eating broccoli that makes married men live longer.

Clearly, it is difficult to conclude anything at all from a correlation between eating broccoli and living longer. Equally clearly, people do draw conclusions from such correlations all the time. Even if we scrupulously point out in our journal article that there are other possible interpretations of our results, our caveats are unlikely to make it into the newspaper article—or, for that matter, into the minds of the other epidemiologists who read our journal article.

You see, epidemiologists do not do research solely for the purpose of getting grants from the Cauliflower Council: they have a higher purpose. Their purpose is to show that the lifestyle decisions people make today will determine whether they're still alive tomorrow. Researchers in this field find it difficult to keep an open mind because they start out with a preconception: the idea that there are "good" lifestyles and "bad" lifestyles, and that people who practice a good lifestyle will have better health than those with a bad lifestyle. We all know the rules for a good lifestyle: eat lots of veggies, avoid fatty foods, exercise daily, don't smoke, et cetera. Epidemiologists measure the goodness of their subjects' lifestyles and the goodness of their health; their goal is to show that a better lifestyle leads to better health.

Socialization researchers also start out with a preconception: the idea that there are good child-rearing styles and bad child-rearing styles, and that parents who use a good child-rearing style will have better children than those who use a bad child-rearing style. Just as we all know the rules for a healthy lifestyle, we all know the rules for a good child-rearing style: give children plenty of love and approval, set limits and enforce them firmly but fairly, don't use physical punishment or make belittling remarks, be consistent, et cetera. We also have a pretty clear idea of what we are looking for in a child: a "good" child is cheerful and cooperative, is reasonably obedient but not to the point of being a robot, is neither too reckless nor too timid, does well in school, has lots of friends, and doesn't hit people without good cause.

In both kinds of studies, the researchers collect data on the goodness of the style (life- or child-rearing) and the goodness of the presumed outcome (health or child). In both kinds of studies, the goal is to show that if you do the right thing you will obtain the desired result. In both kinds of studies, the results come in the form of correlations, and correlations are intrinsically ambiguous.

With apologies to the epidemiologists—my critique of their work is not meant to imply that you can give up eating broccoli and go back to a life of sloth and self-indulgence—I will return now to socialization research. Let's say we decide to do a correlational study on the environmental factors that increase children's intelligence. We hypothesize that parents who provide an intellectually stimulating environment for their children have smarter children, and we set about collecting data to "test" (translation: "try to prove") our hypothesis. We'll need a measure of how stimulating the home environment is, plus a measure of the children's intelligence. For a simple measure of the environment we'll use the number of children's books the home contains, and for a simple measure of intelligence, the children's IQ scores. (These measures are only rough estimates of the qualities we are really interested in, but they are convenient to use because they don't have to be converted into numbers—they already are numbers.)

What we are trying to do is to explain the variation in the children's IQ scores—the fact that some children score high, some score low, and some have average IQs—in terms of another variable, the number of books in the house. If our hypothesis is correct we will find that children who live in homes with lots of books have high IQs, children who live in homes with no books have low IQs, and children who live in homes with an average number of books have average IQs. In other words, we hope to find a positive correlation between IQ and books.

If the correlation were perfect (a correlation of 1.00), we would be able to predict each child's IQ precisely, just by knowing the number of books in his or her home. Since correlations in real life are never perfect, we will be content with a correlation of .70 or .50 or even .30. The higher the correlation, the more accurately we can predict the children's IQs by knowing how many books they have in their homes. Also, the higher the correlation, the more likely it is to be statistically significant. But even a low correlation can be statistically significant if the number of subjects is large enough. I recently came across a paper that reported a significant correlation of .19, based on 374 subjects. It was a correlation between how often children were hostile or uncooperative with their parents and how often these same children were hostile or uncooperative with their peers. A correlation of .19, even if it is significant in the statistical sense, is all but useless. With a correlation that low, knowing one variable tells you virtually nothing about the other. Knowing how obnoxious a given child was with his parents would tell you virtually nothing about how obnoxious he was with his peers.

It is unusual for a socialization study to have as many as 374 subjects. On the other hand, most socialization studies gather a good deal more data from their subjects than we did in our IQ-and-books study: there are usually several measurements of the home environment and several measurements of each child. It's a bit more work but well worth the trouble. If we collect, say, five different measurements of each home and five different measurements of the child's intelligence, we can pair them up in twenty-five ways, yielding twenty-five possible correlations. Just by chance alone, it is likely that one or two of them will be statistically significant. What, none of them are? Never fear, all is not lost: we can split up the data and look again, just as we did in our broccoli study. Looking separately at girls and boys immediately doubles the number of correlations, giving us fifty possibilities for success instead of just twenty-five. Looking separately at fathers and mothers is also worth a try.* "Divide and conquer" is my name for this method. It works like buying lottery tickets: buy twice as many and you have twice as many chances to win.

Although the divide-and-conquer technique often produces publishable results, writing them up can be a real challenge. Here is an actual report of a socialization study, as it appeared in print:

> *Mothers'* total expressiveness, mothers' positive expressiveness, and mothers' negative expressiveness were all positively correlated with *girls'* peer acceptance, but not with *boys'* peer acceptance. Conversely, *fathers'* total expressiveness and fathers' negative expressiveness were positively correlated with *boys'* acceptance, but not with girls' acceptance. Fathers' positive expressiveness was not related to boys' acceptance, but was related to girls' acceptance.
>
> Parents' emotional expressiveness was also significantly correlated with peer and teacher behavior measures. Greater maternal total expressiveness was associated, for boys, with greater prosocial behavior and less disruptiveness. A congruent pattern of results emerged in relation to maternal positive and negative expressiveness. A different pattern emerged in relation to

* The significant correlations are easy to spot, nicely marked with asterisks by the statistics software. It's a technique known as "psychoastronomy": looking for the stars.

paternal emotional expressiveness. Greater paternal total expressiveness was associated, for boys, with less aggression, less shyness, and more prosocial behavior. For girls, greater paternal total expressiveness was associated with less aggression, more prosocial behavior, and less disruptiveness. A congruent pattern of results emerged in relation to paternal positive and negative expressiveness, with one exception: a positive correlation between fathers' negative expressiveness and girls' shyness.

These findings reveal connections between parental emotional expressiveness within the family context and children's social competence.

The proliferation of this sort of report led two prominent developmentalists, in a long and thorough review of socialization research, to wonder "whether the number of significant correlations exceeds the number to be expected by chance." If a correlation is significant just by chance in one study, it is not likely to be significant in the next. Complex patterns of results, such as those I just quoted, generally do not hold up from one study to another.

And yet, I do not believe that the results of socialization studies are all attributable to chance, luck, clever data analyses, and the failure to report negative results. There are two kinds of correlations that turn up often enough to convince me that they are real. They are not strong correlations —strong correlations are hardly ever found in this kind of research—but they show up as consistent trends in study after study. Here is my summary of these trends:

Generalization 1: Parents who do a good job of managing their lives, and who get along well with others, tend to have children who are also good at managing their lives and getting along with others. Parents who have problems managing their lives, their homes, or their personal relationships tend to have children who also have problems.

Generalization 2: Children who are treated with affection and respect tend to do better at managing their lives and their personal relationships than children who are treated harshly.

That noise is a chorus of socialization researchers saying "Yes!" They like these generalizations; they take them as proof of their convictions. To them it is obvious that the children of pleasant and competent people grow up to be pleasant and competent people *because* of what they learned at home and how they were treated by their parents. To them it is obvious that children turn out better if they're treated better, and that they turn out better *because* of how they were treated.

This is not just what socialization researchers believe—it is what nearly everyone believes. But I challenge you to keep an open mind and to examine the rest of the evidence.

The Effects of the Genes

A foxhound does not behave like a poodle; the two breeds have different personalities. Someone who believed in nurture might point out that the foxhound was reared in a kennel with dozens of other dogs, whereas the poodle was reared in a city apartment and sleeps in its owner's bed. Someone who believed in nature would scoff and say, "You can't make a foxhound into a poodle by rearing it in an apartment and spoiling it rotten." This experiment could be done: you could rear a pack of poodles in a kennel, give each foxhound a doting owner and a lease on an apartment, and observe the results. What you'd find is that the nature and nurture advocates were both right: you *can't* make a foxhound into a poodle, but a foxhound reared in an apartment will behave differently from one reared in a kennel.

That experiment involves separating the effects of heredity (the genes that determine whether a puppy is born a foxhound or a poodle) from the effects of environment. The problem with socialization studies of the type I've described is that the effects of heredity and environment are not separated; nor are they separable. Every (or nearly every) parent-and-child pair who participate in a socialization study are biological relatives; they are as much alike in terms of their DNA as two poodles from the same litter. Not only do the parents provide the child's genes; they also provide the child's environment. The kind of environment they provide—the kind of parents they are—is, in part, a function of their genes. There is no way to distinguish the effects of the genes they provide from the effects of the environment they provide. Socialization researchers are trying to figure out what makes foxhounds different from poodles without switching the puppies around.

Although we can't switch human babies around for the sake of science, sometimes they are switched around for other reasons. An adopted child has two sets of parents: one provides the child's genes, the other provides the environment. Studying adopted children is one of the methods used by researchers in the field of behavioral genetics. The stated purpose of their research is to separate the effects of heredity from those of the environment. Like the socialization researchers, behavioral geneticists also have a hidden agenda: to show that heredity is a force to be reckoned with. To show that John Watson was wrong and that infants are not malleable pieces of clay, capable of being pushed and pulled into any shape whatever by the environment.

In the early days of behavioral genetics, studies of adopted children were designed to find out whether these children were more similar to their biological parents (who provided their genes) or to their adoptive parents (who provided their environment). The characteristic that got the most attention was IQ. In biological families, children's IQs tend to be correlated with those of their parents—parents with above-average IQs tend to have children with above-average IQs. The purpose of the early studies was to determine whether this correlation was due primarily to heredity or to the stimulating environment that intelligent parents were presumed to provide. If the IQs of adopted children were more like those of their biological parents, then heredity would be declared the victor. If they were more like those of their adoptive parents, then environment would win the game.

Although this technique makes a certain amount of sense when the characteristic being studied is IQ, it makes no sense at all for studying personality characteristics, which is what I am mainly interested in. It is reasonable to assume that being reared by smart parents increases a child's IQ, but it is not reasonable to assume, for example, that being reared by bossy parents makes a child bossier. Maybe being reared by bossy parents makes a child meek and passive. Another problem is that parents and children belong to different generations; they grow up in different times. Cultural changes in the society add to the differences between parents and children and make it harder to detect the similarities.

To get around these problems, modern behavioral geneticists look for correlations between people of the same generation. Instead of comparing children to their adoptive or biological *parents,* they compare them to their adoptive or biological *siblings.* They look at pairs of adoptive siblings (two unrelated children reared in the same home) or pairs of biological siblings, preferably identical and fraternal twins. This gives the researchers three levels of genetic similarity: the reared-together adopted children are biologically unrelated, the fraternal twins (like ordinary siblings) share about 50 percent of their genes, and the identical twins have identical genes. So genetic similarity varies but environmental similarity is held more or less constant, because each pair was reared in the same home by the same parents. Doing the opposite experiment—varying environmental similarity while holding genetic similarity constant—is also possible, but it involves locating identical twins reared apart. Identical twins reared apart are harder to find than poodles at a foxhunt.

Rounding up subjects for a behavioral genetic study is not an easy job. Almost anyone is eligible to take part in a socialization study, but for the typical behavioral genetic study, only adoptees and twins need apply. Moreover, the behavioral geneticists must examine at least two children in each family, whereas the socialization researchers make do with just one. The extra effort is worth it, though, because it provides the researchers with the tweezers they need to disentangle the effects of heredity and environment. Effects due to heredity show up as similarities that are greater in identical twins than in fraternal twins, and greater in fraternal twins than in adoptive siblings. Thus, the effects of heredity can be gauged by measuring the degree to which people who share genes are more similar than people who do not share genes. The effects of the home environment can be gauged by measuring the degree to which people who grew up in the same home are more similar than people who grew up in different homes.

A large number of human characteristics have now been studied with behavioral genetic methods. The results are clear and consistent: overall, heredity accounts for roughly 50 percent of the variation in the samples of people that have been tested, environmental influences for the other 50 percent. People differ from one another in many ways: some are more impulsive, others more cautious; some are more agreeable, others more argumentative. About half of the variation in impulsiveness can be attributed to people's genes, the other half to their experiences. The same is true for agreeableness. The same is true for most other psychological characteristics.

It seems like an unremarkable finding, pretty much what you might expect. But in the 1970s, when these results first started appearing in psychology journals, American psychology was still under the lingering influence of behaviorism, with its anti-heredity bias. The political climate of the country was also anti-heredity; the existence of inborn differences was felt to be incompatible with the ideal of human equality. The heredity–environment issue got mixed up with politics and feelings ran high. Behavioral genetics was an unpopular field in those days. But an interest in the workings of heredity is not a symptom of a particular political position—it can afflict even a flaming liberal. Over time, due partly to advances in molecular biology, the study of the effects of genes became more acceptable in academic circles. Behavioral geneticists have gradually become more numerous.

Nonetheless, they are still overwhelmingly outnumbered by socialization researchers. Perhaps that is why most socialization researchers find it easy to ignore the results of behavioral genetic studies. The behavioral geneticists, on the other hand, do not ignore the work of socialization researchers. They have pointed out time and again that the failure to control for the effects of heredity makes the results of most socialization studies uninterpretable. And they are right.

Generalization 1 said that pleasant, competent parents tend to have pleasant, competent children. Another way of stating this is that children tend to resemble their parents. Parents who do a good job of managing their lives, and whose relationships with other people (including their children) are cordial, tend to have children with similar characteristics. Is this because of the way the children were reared, or because of the competence and cordiality genes they inherited from their competent, cordial parents? There's no way of telling. The 50-50 result (50 percent heredity, 50 percent environment) that the behavioral geneticists obtained doesn't mean that half of the correlation between parents and children is due to genes and the other half is due to environmental influences. The 50-50 result means only that 50 percent of the variation among the children, in some particular characteristic such as cordiality, can be traced to differences in their genes. It doesn't say anything about how much of the correlation between the children's cordiality and the parents' cordiality—the similarity between them—is due to heredity. In fact, correlations between children and parents are usually well below .50. A correlation between children and parents is usually low enough that the genes they share could account for *all of it.*

Not clear? Let me try again, using an example from another species—a vegetable species this time. Plant some corn, pick an ear from each plant, taste it, and judge its sweetness. Notice that some plants produce sweeter corn than others. Save a cob from each plant to use for seed and next year plant the seeds. You will find that the seeds from the plants that produced sweeter corn grow into plants that also produce, on the average, sweeter corn—in other words, there will be a correlation between the parent corn's sweetness and the offspring corn's sweetness. That correlation is entirely due to heredity: the genes the offspring received from the parent account for 100 percent of the resemblance between them. But the genes account for only about half the variation in the sweetness of the offspring corn, because other factors—environmental factors such as soil quality, water, and sunshine—will also play a role. So it is possible for heredity to account for 100 percent of the resemblance between parent and offspring even though it accounts for only 50 percent of the variation among the offspring.

The environment has effects, both on children and on corn. In our own species, differences in environment account for about half of the variations in personality characteristics. The socialization researchers are correct in believing that environmental factors have effects on children. They are wrong, however, in believing that their research tells them what these factors are. Their research does not demonstrate what they want it to demonstrate, because they have failed to consider the effects of heredity. They have failed to allow for the fact that children and their parents resemble each other for genetic reasons.

Generalization 1 is true. On the average, pleasant, competent parents tend to have pleasant, competent kids. But that doesn't prove that parents have any influence—other than genetic—on how their children turn out.

A Two-Way Street

In a typical socialization study, the researchers start by rounding up a group of subjects: a number of children of about the same age (often recruited from a nursery school or elementary school classroom) and their parents. They then proceed to collect data on the parents' child-rearing methods—perhaps by interviewing them or having them fill out a questionnaire, perhaps by observing their behavior while they interact with their child. However it is measured, a parent's child-rearing method is assessed in respect to only one child, since only one child per family participates in this kind of study. This procedure would be all right if parents had a uniform child-rearing style—if "child-rearing style" were a more-or-less stable characteristic of a person, like eye color or IQ. But parents do not have a single, fixed child-rearing style. The way a parent acts toward a particular child depends on the child's age, physical appearance, current behavior, past behavior, intelligence, and state of health. Parents tailor their child-rearing style to the individual child. Child-rearing is not something a parent does to a child: it is something the parent and the child do together.

Not long ago I was in my front yard with my dog. A mother and her two children—a girl of about five and a boy of about seven—walked by in the street. My dog, who is trained not to go into the street, ran to the curb and started barking at them. The two children reacted in very different ways. The girl veered straight toward the dog, asking, "Can I pet him?" despite the fact that the dog was behaving in an unfriendly manner. Her mother said quickly, "No, Audrey, I don't think the dog wants you to pet him." Meanwhile, the boy had retreated to the other side of the street and was standing there looking scared, unwilling to walk past the barking dog even though the entire width of the street was between them. "Come on, Mark," his mother said, "the dog won't hurt you." (I was holding the dog's collar by then.) It took a minute before Mark got up the courage to rejoin his mother, who was waiting for him with her impatience carefully concealed under a good deal of genuine sympathy. As the three went down the street I could hear Audrey making fun of Mark. I didn't catch her words but the tone was unmistakable.

I was sorry for Mark but I identified more strongly with his mother: I, too, reared a pair of very different children. My older daughter hardly ever wanted to do anything that her father and I didn't want her to do. My younger daughter often did. Raising the first was easy; raising the second was, um, interesting.

My Uncle Ben, who had no children of his own, was fond of his grandnieces and often gave me advice on how to rear them. I remember a conversation I had with him when my daughters were about eight and twelve. I was complaining to him about the behavior of my younger child and Uncle Ben (who knew I hadn't had these problems with my older one) asked, "Well, do you treat them both the same?"

Do I treat them both the same? I didn't know what to say. How can you treat two children both the same when they *aren't* the same—when they do different things and say different things, have different abilities and different personalities? Could the mother of Mark and Audrey treat them both the same? What would that mean? Telling Audrey "The dog won't hurt you" (that was what she said to Mark) instead of "I don't think the dog wants you to pet him"?

If Mark and his mother took part in a socialization study, the researchers would probably get the impression that Mark's mother was overprotective. If Audrey and her mother took part in a socialization study, the researchers might see Audrey's mother as a sensible setter of limits. Each team of researchers would see her with only one of her children; each would get a different picture of what kind of parent she was. I would have been pegged as a permissive parent with my first child, a bossy one with my second.

The relationship between a parent and a child, like any other relationship between two individuals, is a two-way street—an ongoing transaction in which each party plays a role. When two people interact, what each one says or does is, in part, a reaction to what the other has just said or done, and to what was said or done in the past.

Even young babies make an active contribution to the parent–child relationship. By the time they are two months old, most babies are looking their parents in the eye and smiling at them. It is remarkably rewarding, being smiled at by a baby. A normal baby pays back her parents for all the trouble she causes them by showing that she is delighted to see them.

Some babies—notably those with the disorder called autism—don't do this. Autistic babies don't look their parents in the eye, don't smile at them, don't seem glad to see them. It is difficult to feel enthusiastic about a baby who isn't enthusiastic about you. It is difficult to interact with a child who won't look at you. The late Bruno Bettelheim, who for many years ran an institution for autistic children, claimed that autism was caused by the mother's coldness, her lack of feeling for the child. One of these mothers later attacked Bettelheim in print, calling him a "vile individual" who had "brought ostracism and suffering to entire families." Bettelheim was not only cruel: he was wrong. Autism is caused by a defect in the brain; autistic children are born that way. The mothers' apparent coldness was not the cause of their children's behavioral abnormalities—it was a *reaction* to it.

John Watson assumed that if two children are different, it must be because they are treated differently by their parents—an assumption shared by my childless Uncle Ben. But, as most parents realize shortly after the birth of their second child, children come into this world already different from each other. Their parents treat them differently *because* of their different characteristics. A fearful child is reassured; a bold one is cautioned. A smiley baby is kissed and played with; an unresponsive one is fed, diapered, and put in its crib. The effects the socialization researchers are interested in are parent-to-child effects: the parent has an effect on the child. There are also effects that go in the opposite direction: the child has an effect on the parent. Child-to-parent effects, I call them.

Generalization 2 said that children who are hugged are more likely to be nice, children who are beaten are more likely to be unpleasant. Turn that statement around and you get one that is equally plausible: nice children are more likely to be hugged, unpleasant children are more likely to be beaten. Do the hugs cause the children's niceness, or is the children's niceness the reason why they are hugged, or are both true? Do beatings make children unpleasant, or are parents more likely to lose their temper with unpleasant children, or are both true? In the standard socialization study, there is no way to distinguish these alternative explanations, no way to tell the causes from the effects. Thus, Generalization 2 does not prove what it appears to prove.

Parallel Universes

Castor and Pollux, Romulus and Remus—twins have fascinated their audiences for a very long time. For behavioral geneticists, they are an essential component of the research program. It isn't even necessary to find twins reared apart: the vast majority of the twins who participate in behavioral genetic studies were reared in the same home by their own biological parents. The technique hinges on a contrast between the two kinds of twins, identical versus fraternal. By comparing the similarity of the identical twin pairs with that of the fraternal twin pairs, the researchers can judge whether or not (and to what degree) a particular characteristic of the twins is under the control of the genes. Say, for example, that the characteristic being investigated is the tendency to be physically active or inactive. If identical twins are fairly similar in activity level (both twins are always in motion, or both twins are couch potatoes) and fraternal twins are noticeably *less* alike, this is taken as evidence for a genetic influence on that trait.

Socialization researchers have objected to this method because they believe it rests upon a shaky assumption: that the environments of the reared-together fraternal twins are as similar as the environments of the reared-together identical twins. If identical twins actually have more similar environments than same-sex fraternal twins, the greater similarity of identical twins could be due to the greater similarity of their environments, rather than (or in addition to) their more similar genes.

Do identical twins have more similar environments than fraternal twins? Matching outfits and matching toys are not at issue here: the question is whether identicals are treated more alike in terms of how much affection and discipline they receive. Are they given the same number of hugs? The same number of spankings?

The evidence suggests that parents do tend to treat identical twins more alike than they treat fraternal twins. When adolescent twins were asked how much affection or rejection they had received from their parents, identical twins were more likely than fraternal twins to give matching reports. If one identical twin said that her parents made her feel loved, the other was likely to say the same thing. But if one *fraternal* twin reported that her parents made her feel loved, the other might say either that she also felt loved or that she felt rejected. Parents may give their identical twins different outfits and different toys, but nonetheless they seem to love them both the same (or not love them both the same). Whereas with fraternal twins—who often differ considerably in appearance and behavior—they might find one a good deal more lovable than the other. So it is probably true that identical twins tend to have more similar environments than fraternal twins.

In fact, identical twins may have more similar environments than fraternal twins even if they grow up in different homes. Adult identical twins who had been separated early in life and reared apart give surprisingly similar descriptions of their childhoods; they agree fairly well on the amount of affection they received from their adoptive parents. Although it is possible that the similar reports are due to the similar way their memories work—cheerful twins have happy memories of childhood whereas gloomy ones tend to remember the slings and arrows—I don't think that is the whole story. I think reared-apart identical twins really do receive a similar amount of affection from their adoptive parents. One reason for this is that identical twins look alike: if one is cute, the other is cute. If one is homely, the other is homely. Researchers have found that children's cuteness or homeliness has a measurable effect on how their parents treat them. A study showed that a mother is, on the average, more attentive to her baby if the baby is cute than if the baby is homely. (The cuteness of the babies was rated by independent judges—a panel of undergraduates at the University of Texas.) Although all the babies in this study were well cared for, the cute babies were looked at more, played with more, and given more affection than the homely ones. In their report, the researchers quoted a letter written by Queen Victoria to one of her married daughters. According to the queen, who was something of an expert on babies (she had borne nine of them), "An ugly baby is a very nasty object."

Most ugly babies get better looking over time, but think about the ones who don't. People aren't as nice to homely children as they are to pretty ones. If they do something wrong, they are punished more harshly than the pretty ones. If they don't do anything wrong, people are quicker to think that they did. Homely children and pretty children have different experiences. They grow up in different environments.

Children's experiences are not, of course, determined solely by their looks. Other qualities also influence how people react to them. A timid child like Mark is treated differently from a bold one like Audrey. But timidity in a child has a substantial genetic component, so if Mark had an identical twin on the other side of the world, the twin might well be timid too. They would have different mothers, but the chances are good that both mothers would react the same way: they would be sympathetic and a bit impatient. Their fathers might be a bit less sympathetic, a bit more impatient. Outside the home, Mark and his separated twin would probably experience similar treatment from their peers: teasing and bullying. Recess is not much fun for timid boys.

To the extent that children's experiences are a function of built-in characteristics such as timidity or good looks, identical twins are more likely than fraternal twins to have similar experiences. The socialization researchers are right about this. The trouble is, as you will see in the next chapter, the trick is not to explain why identical twins are so much alike —whether it's because of their identical genes or their similar experiences. The trick is to explain why they are not *more* alike. Even identical twins reared in the same home are far from being identical in personality.

The Effects of the Effects of the Genes

Genes contain the instructions for producing a physical body and a physical brain. They determine the shape of the facial features and the structure and chemistry of the brain. These physical consequences of heredity are the straightforward consequences of carrying out the instructions in the genes; I call them direct genetic effects. Timidity can be a direct genetic effect; some babies are born with hypersensitive nervous systems. Being born beautiful is a direct genetic effect.

Direct genetic effects have consequences of their own, which I call *indirect* genetic effects—the effects of the effects of the genes. A child's timidity causes his mother to reassure him, his sister to make fun of him, and his peers to pick on him. A child's beauty causes her parents to dote on her and wins her a wide circle of admiring friends. These are indirect genetic effects. Identical twins lead similar lives because of indirect genetic effects.

Socialization researchers who protest the behavioral geneticists' use of twin data are correct in saying that behavioral genetic methods lump together the effects of the environmental similarities with the effects of the genes. In fact, behavioral genetic methods cannot distinguish the effects of genes from the effects of the effects of genes—they cannot distinguish between direct and indirect genetic effects. What they call "heritability" is actually a combination of direct and indirect genetic effects.

It would be nice to be able to distinguish them, but given that we cannot do so with the methods currently available, I am content to have indirect genetic effects attributed to "heredity" rather than "environment." Although technically they are part of the child's environment, they are consequences of the child's genes. However, I agree with the socialization researchers that the behavioral geneticists have not handled this problem well. They can be faulted not for lumping direct and indirect effects together, but for failing to state clearly that this is what they are doing.

Let me state it clearly here and now. Behavioral genetic studies are designed to distinguish the effects of the genes from the effects of the environment. The researchers look at one characteristic at a time, dividing up the variation in that characteristic—the differences among their subjects—into two parts, the part that's due to genes and the part that's due to the environment. The result, for the majority of the psychological characteristics that have been studied, is that roughly half of the variation is attributable to the subjects' genes, the other half to their environment. But the half attributed to heredity *includes indirect effects,* the environmental consequences of the effects of the genes. That means that the other half of the variation must be due to "pure" environmental influences—influences that are not, either directly or indirectly, a function of the genes.

Half of the variation gives socialization researchers plenty to work with. However, their job isn't just to prove that the environment as a whole has effects on children: it is to prove that the particular aspects of the environment they are interested in—namely, the parents' child-rearing methods—have effects on children. In my judgment they have not proved it. Yes, competent parents tend to have competent children, but that could be due to heredity. Yes, well-treated children tend to be nicer than those who are treated harshly, but that could be due to child-to-parent effects.

Socialization researchers do not like the idea that some of the effects they report may be due to inherited similarities between children and their biological parents; they seldom mention this possibility in their published articles. But the idea that children have effects on their parents—that the relationship is two-way—has gradually won acceptance. Almost every article that reports a correlation between parents' behavior and children's behavior now includes, somewhere near the end, a disclaimer that admits that the direction of cause and effect is unclear, that the reported correlation could be due to the child's effect on the parents, rather than (or in addition to) the parents' effect on the child. The disclaimer serves the same purpose as the warning on the cigarette pack: the rules say it has to be there, but no one pays any attention to it.

My impression is that socialization researchers believe that child-to-parent effects do occur but that such effects are found mainly in other people's data. They interpret their own ambiguous results in terms of the nurture assumption because the nurture assumption itself is never questioned. Their research is not designed to test the hypothesis that the environment provided by parents has lasting effects on the child's behavior and personality, because that is not considered to be a testable hypothesis: it is a given.

Questioning the nurture assumption is what I'm here for. In this chapter I have told you about some of the things that are wrong with the evidence used to support it. In the next I will tell you about the evidence against it.

Parents treat children diff. b/c each child is diff. Parents - Children react to eachother.

3 NATURE, NURTURE, AND NONE OF THE ABOVE

Tales of the eerie resemblances between identical twins separated early in life and reared in different homes have made their way into the popular press and the popular imagination. There was the story of the two Jims —both bit their nails, enjoyed woodworking, drove the same model Chevrolet, smoked Salems, and drank Miller Lite; they named their sons James Alan and James Allan. There was the story in my local newspaper, accompanied by a photo of two men with the same face, both wearing fire helmets—reunited because both had become volunteer firefighters. There was the story of Jack Yufe and Oskar Stöhr, one reared in Trinidad by his Jewish father, the other in Germany by his Catholic grandmother. When reunited, they were both wearing rectangular wire-frame glasses, short mustaches, and blue two-pocket shirts with epaulets; both were in the habit of reading magazines back to front and flushing toilets before using them; both liked to startle people by sneezing in elevators. And there was the story of Amy and Beth, adopted into different homes—Amy a rejected child, Beth doted upon—both girls suffering from the same unusual combination of cognitive and personality deficits.

These true stories of reared-apart identical twins are a testimony to the power of the genes. They suggest that genes can cause striking similarities in personality characteristics, even in the face of substantial differences in rearing environments. They imply that genes can control behavior in subtle, intricate ways that cannot be explained in terms of our current understanding of genetic mechanisms and brain neurophysiology.

But the flip side of the coin is seldom mentioned. The flip side of the coin is that identical twins reared in the *same* home are not nearly as alike as you would expect them to be. Given how similar the reared-apart twins are, you probably think that the reared-together ones must be as alike as two copies of your annual Christmas letter. In fact, they are no more alike than identical twins separated in infancy and reared in different homes. Though they have many little quirks in common, there are also many little differences between them.

They are no more alike than the ones reared in different homes! Here are two people who not only have exactly the same genes but who also grew up in the same home at the same time with the same parents, and yet they do not have the same personality. One might be friendly (or shy), the other more (or less) so. One might look before she leaps, the other might not leap at all. One might disagree with you but hold his peace while the other tells you you're full of crap. I am talking about *identical twins.* These people are so alike in appearance that you have trouble telling them apart, but give them a personality test and they will check off different answers. The correlation of personality traits (as estimated by scores on personality tests and in various other ways) is only about .50 for identical twins reared in the same home.

Growing Up in the Same Home Does Not Make Children More Alike

At the University of Minnesota, a group of behavioral geneticists are running an ongoing research project called the Minnesota Study of Twins Reared Apart. When reared-apart adult twins are located, they are awarded all-expense-paid trips to Minneapolis and treated to a solid week of psychological testing; one wonders whether second prize is *two* solid weeks of psychological testing. As it happens, very few of the twins turn down the offer. The chance to meet one's wombmate, possibly for the first time since the umbilical cords were cut, is irresistible.

Among the twins who came to Minneapolis to be tested were a pair known as the Giggle Twins. Although these women had been reared in separate homes, and both twins described their adoptive parents as dour and undemonstrative, both were inordinately prone to laughter. In fact, neither had ever met anyone who laughed as much as she did until the day she was reunited with her identical twin.

Observing the Giggle Twins, it is easy to jump to the conclusion that laughter is genetic. But they are just one set of twins, and what I've told you about them is an anecdote, not data. Also, the adoptive homes in which these twins were reared actually sound a lot alike. Perhaps both twins laugh so much in adulthood because neither of them got enough laughter during childhood. In truth, there is no way to determine with certainty whether these twins are both gigglers because of their identical genes or because they both happened to have had experiences which produced this effect in them. Although any *differences* between them have to be environmental—they can't be genetic since they both have the same genes—the *similarities* can be genetic, environmental, or a combination of the two.

But what cannot be done for the Giggle Twins themselves can be done for the trait they are noted for. Give behavioral geneticists a few dozen pairs of twins or siblings (biological or adoptive, reared together or apart) and they can tell you whether the tendency to laugh a lot—I'll call this trait "risibility"—is genetic, environmental, or a combination of the two. The methodology of behavioral genetics is based on a variation of the old question, Are adopted children more like their adoptive parents or more like their biological parents? Substituting "siblings" for "parents" eliminates the complications of trying to compare people of widely different ages, but otherwise the idea is the same. The method rests on two basic premises: that people who share genes should be more alike than people who don't, and that people who shared a childhood environment should be more alike than people who didn't.

From these two premises, we can generate predictions. If risibility is entirely genetic, we would expect to find that identical twins are very similar in risibility (though not exactly alike, since even a single individual varies from day to day in readiness to laugh) and that it doesn't make any difference whether they were reared together or apart. If risibility is entirely environmental, we would expect to find that reared-together identical twins, fraternal twins, and adoptive siblings are all equally alike in risibility and that pairs reared in different homes are not at all alike. Finally, if risibility is due to a combination of heredity and environment—certainly the best bet—we'd expect to find that people who share genes are somewhat alike, people who were reared in the same home are somewhat alike, and people who share *both* genes and a rearing environment are the most alike.

Sounds logical? Guess again. If risibility follows the pattern of the other traits that have been studied so far, what we would actually find is None Of The Above.

The unexpected results started appearing in the mid-1970s. By the late 1970s, enough data had been collected to make it look like there was something wrong with the basic premises of behavioral genetics. Not the genetic premise—that was okay. People who share genes *are* more alike in personality than people who don't share genes. It was the premise about sharing an environment that didn't seem to be working properly. Study after study was showing that pairs of people who grew up in the same home were *not* noticeably more alike in personality than pairs who grew up in two different homes. And yet the results didn't fit the entirely-genetic prediction either, because genetic relatives weren't alike enough—the correlations were too low. Something other than genes was exerting an effect on the subjects' personalities, but it didn't seem to be the home in which they were reared. Or if it was the home, it was working in an inexplicable manner. It wasn't making siblings *more* alike, it was making them *less* alike.

Perhaps you are wondering why these results were unexpected. Why *should* children reared in the same home be alike? If your parents were dour and undemonstrative, don't you feel you could have gone either way—either be just like them or just the opposite? Can't you imagine a family with sourpuss parents and two children who went in opposite directions: one a sourpuss like the parents, the other a barrel of fun?

The problem is that researchers who study child development—including behavioral geneticists—would like to believe that parents' attitudes, personalities, and child-rearing practices have predictable effects on their children. Epidemiologists try to predict what effects certain eating habits and lifestyles will have on a person's physical health and longevity; developmentalists try to predict what effects parents' behaviors and child-rearing styles will have on their child's mental health and personality.

Parents vary in their attitudes toward children and their ideas about family life. In some families humor is considered a virtue and laughter its reward; kids are permitted to interrupt or make impertinent remarks if they're funny enough. I grew up in a family like that. In high school I had a friend named Eleanor whose family was considerably more intellectual than mine (mine wasn't intellectual at all). One evening she had dinner at my house and afterwards she told me she wished she had been born into my family instead of hers. Dinner at the Riches' was lively, with everyone talking at once and lots of wisecracking and laughter. Eleanor's parents were straitlaced and proper; dinner at her house, she said, was boring. Don't you think that someone who grew up in my family should score higher on a test of risibility than someone who grew up in Eleanor's? Don't you think that two people who grew up in my family should be more alike in risibility than one who grew up in my family and one who grew up in Eleanor's?

If you believe that children can "go either way"—that they can either turn out like their parents or, with equal ease, go in the opposite direction —then what you are saying is that parents have no predictable effects on their children. If you are espousing a milder version of that view—that *most* children are influenced by their parents but occasionally you find one rebelling and going in the opposite direction—then we would expect to find some overall tendency for siblings to be similar, since the majority do not rebel. Because children are different to begin with—one sibling may be born an Abbott, the other a Costello—we wouldn't expect them to react in exactly the same way to the parents' attitudes and behaviors. Nonetheless, on the average, people reared in a family that encourages joke-telling and laughter should be higher in risibility than people reared in a family of the we-are-not-amused variety.

But that is not what the behavioral geneticists found. They looked at a wide variety of personality traits (though not, as far as I know, risibility) and the results were about the same for all of them. The data showed that growing up in the same home, being reared by the same parents, had little or no effect on the adult personalities of siblings. Reared-together siblings are alike in personality only to the degree that they are alike genetically. The genes they share can entirely account for any resemblances between them; there are no leftover similarities for the shared environment to explain. For some psychological characteristics, notably intelligence, there is evidence of a transient effect of the home environment during childhood—the IQ scores of preadolescent adoptive siblings show a modest correlation. But by late adolescence all nongenetic resemblances have faded away. For IQ as for personality, the correlation between adult adoptees reared in the same home hovers around zero.

Research results in psychology often prove to be evanescent. Interesting effects that show up in one study often fail to show up in the next. But results in behavioral genetics are what statisticians call "robust." Study after study shows the same thing: almost all the similarities between adult siblings can be attributed to their shared genes. There are very few similarities that can be attributed to the home in which they both grew up.

Growing up in the same home does not make siblings alike. If there really are "toxic parents," they aren't toxic to all their children. Or they aren't toxic in the same way. Or, if they are toxic in the same way, each child reacts to the toxicity differently, even if they are identical twins. What does it mean if the presumed effects of toxic parents are discernible on only one of their children—the one who ends up in the office of the clinical psychologist—and the others are fine?

Scylla or Charybdis

By and large, socialization researchers ignored the unsettling results being reported by behavioral geneticists. Of the few who took notice, the most prominent was Eleanor Maccoby, the Stanford professor mentioned in Chapter 1 (the one who admitted, many years later, that the first socialization study didn't pan out).

In 1983, Maccoby and her colleague John Martin published a long and penetrating review of the field of socialization research. They talked about research methods, results, and theories. They talked about the effects of parents on children and also about the effects of children on parents. After eighty densely printed pages of this, they summed up their impressions of the field in a few brisk paragraphs. They pointed out that the correlations found between the parents' behavior and the children's characteristics were neither strong nor consistent. They wondered, in view of the large number of measurements made, whether the correlations that did turn up might have occurred by chance. And they drew their readers' attention to the puzzling findings coming from the field of behavioral genetics: that adopted children growing up in the same home are not at all alike in personality, and that even for biological siblings the correlations are very low.

From the weakness of the trends found in socialization studies and the unsettling results emerging from behavioral genetic studies, Maccoby and Martin drew the following conclusions:

> These findings imply strongly that there is very little impact of the physical environment that parents provide for children and very little impact of parental characteristics that must be essentially the same for all children in a family: for example, education, or the quality of the relationship between the spouses. Indeed, the implications are either that parental behaviors have no effect, or that the only effective aspects of parenting must vary greatly from one child to the other within the same family.

Either that parents have no effect or that they have different effects on each of their children—those were the only two alternatives that Maccoby and Martin offered. Neither was much to the liking of socialization researchers. It was like telling epidemiologists that either broccoli and exercise have no effects on health or else they make some people healthier and others sicker. Agreed, broccoli and exercise probably do have different effects on different people, but at least in epidemiology there are overall trends—eating veggies and getting regular exercise appear to be good for *most* people. In socialization research, according to Maccoby and Martin, it wasn't even clear that there were overall trends.

I want to examine their statement a little more closely, because it is of central importance. "These findings," they said—by which they meant the weak and inconsistent trends found by the socialization researchers, plus the lower-than-expected correlations between reared-together siblings found by behavioral geneticists—"imply strongly that there is very little impact of the physical environment that parents provide for children and very little impact of parental characteristics that must be essentially the same for all children in a family." In other words, most of the things that were believed to have important effects on children turn out not to have important effects on them. If the parents work or don't work, read or don't read, drink or don't drink, fight or don't fight, stay married or don't stay married—all these "must be essentially the same for all children in a family," and therefore all appear to have "very little impact" on the children. Similarly, if the physical environment of the home is an apartment or a farmhouse, spacious or crowded, messy or tidy, full of art supplies and tofu or full of auto parts and Twinkies—all these, too, "must be essentially the same for all children in a family," and therefore appear to have "very little impact."

With a stroke of the pen, Maccoby and Martin had crossed out most of the things that socialization researchers had been making a living on for decades. With a second stroke, they threatened to cross out the rest. Take your pick, they said: either the home and the parents have *no* effects, or else the only things that have effects are those that differ for each child in the family. The first alternative would mean that the nurture assumption is wrong; the second offered the only hope of rescuing it.

No one chose the first alternative. No one. The developmentalists who paid attention to what was going on in the field as a whole, rather than in their own little corner of it, rallied around Maccoby and Martin's second alternative. The rest ignored their warning that the sky is falling and kept on with their plowing.

Maccoby and Martin's second alternative says that "the only effective aspects of parenting must vary greatly from one child to the other within the same family." In other words, the parents and the home still matter, but each child inhabits, in effect, a different environment within the home. Developmentalists who take this approach speak of "within-family environmental differences," meaning experiences that children who grow up in the same family do not share. For example, the parents might prefer one child to the other, so the preferred child grows up with loving parents while the other grows up with indifferent or rejecting parents. Or the parents might be strict with one child, lenient with the other. Or they might label one "the athlete" and the other "the brain." Within-family environmental differences might also result from the interactions of the children themselves. One grows up with a bossy older sister, the other with a pesky younger brother. The home is depicted, not as a single homogeneous environment, but as a bunch of little microenvironments, each inhabited by one child.

It's a perfectly reasonable idea. There is no question that such microenvironments exist, no question that each child in the family does have different experiences within the same home and different relationships with the other people who live in it. Everyone knows that parents don't treat their children exactly alike, even if they try to. Mom always loved you best, so naturally you turned out better.

But immediately we run into problems, because that path leads directly to an endless loop of causes and effects. How do we know Mom didn't love you best because you were better to begin with? Are you smart because you were labeled "the brain" or were you labeled "the brain" because you were smart? If parents treat each of their children differently, are they *responding* to the differences among their children or are they *causing* them?

In order to get out of this loop, we need to show that parents are not simply reacting to characteristics their children already had—characteristics they were born with. We need to find a reason why a parent might behave differently toward two children that cannot be attributed to genetic differences between them. Then—and this is the tricky part—we need evidence that these differences in parental treatment *actually have effects on the children.* We need evidence of parent-to-child effects, because if all we've got are child-to-parent effects we haven't shown that parents have any influence whatever on how their children turn out.

Birth Order

There is one thing I can think of that makes parents act differently to different children and that can't be explained in terms of the characteristics the children were born with: birth order. The firstborn and secondborn have equal chances in the lottery in which genes are handed out, but once they are born they find themselves in very different microenvironments. They have different experiences in the home, and these experiences can be predicted with some accuracy on the basis of which one was born first. The firstborn has the parents' full attention for at least a year and then suddenly is "dethroned" and has to compete with a rival; the secondborn has competition right from the start. The firstborn is reared by nervous, inexperienced parents; the secondborn by parents who know (or think they know) what they're doing. Parents give firstborns more responsibility, more blame, and less independence.

If children's personalities are affected by how their parents treat them, and if parents treat firstborns differently from laterborns, then the order in which they were born should leave traces on children's personalities—traces that should still be detectable after they grow up. The traces are called birth order effects. They are a favorite topic among writers of pop psychology. Here, for example, is John Bradshaw, the guru of "dysfunctional families," expounding on the distinctive personality characteristics of firstborns, secondborns, and thirdborns:

> A first child will make decisions and hold values consistent with or in exact opposition to the father. . . . They are other-oriented and socially aware. . . . First children often have trouble developing high self-esteem. . . . Second children naturally relate to the emotional maintenance needs of the system. . . . They will pick up "hidden agendas" immediately but not be able to express clearly what they feel. Because of this, second children often seem naive and puzzled. . . . The third child hooks into the relationship needs of the system. . . . Appears very uninvolved but is actually very involved. Feels very ambivalent and has trouble making choices.

The problem for academic psychologists is that they can't go around making statements like these unless there is some evidence to back them up. They would have to be able to show that, on the average, firstborns really do have more self-esteem problems than second- or thirdborns, and that thirdborns really do feel more ambivalent than their older siblings. Scores on a personality test would serve the purpose, if it could be shown that firstborns, secondborns, and thirdborns differed systematically from each other in the responses they gave.

For more than fifty years, academic psychologists of all persuasions have been looking for these systematic differences—looking for convincing evidence that birth order has effects on personality. Both behavioral geneticists and socialization researchers would love to find such evidence. For behavioral geneticists, it would provide a way to reconcile their unsettling results with their assumptions (yes, behavioral geneticists, too, believe in the power of nurture). For socialization researchers, the potential payoff is obvious: proof that what goes on in the home has important and lasting effects.

Piles and piles of birth order data have been collected over the years, much of it in the form of scores on personality tests. Thousands of subjects have indicated, at the top of the page, their position in the family they grew up in, and, in the spaces below, whether they have confidence in their abilities or have trouble expressing their feelings or hate having to make choices. Hundreds of researchers have collected these pages and analyzed the data they contain. Sad to say, the enterprise has been a waste of time and paper. In 1990, Judy Dunn and Robert Plomin—she's the world's leading expert on sibling relationships, he's the world's leading expert on behavioral genetics—looked hard and (I suspect) longingly at birth order data. This is what they concluded:

> When differences in parents' behavior to their different children are discussed, often the first issue that comes to mind is the birth order of the children. It is frequently assumed that parents systematically treat their firstborn child differently from laterborn children. . . . In an important sense such differences are not relevant. This is because individual differences in personality and psychopathology in the general population—the differences in outcome that we are trying to explain—are *not* clearly linked to the birth order of the individuals. Although this evidence goes against many widely held and cherished beliefs, the judgment of those who have looked carefully at a large number of studies is that birth order plays only a bit-part in the drama of sibling differences. . . . If there are no systematic differences in personality according to birth order, then any differences in parental behavior that are associated with birth order cannot be very significant for later developmental outcome.

Dunn and Plomin referred to "those who have looked carefully at a large number of studies." Foremost among those careful lookers are the indefatigable Swiss researchers Cécile Ernst and Jules Angst—that's right, Ernst and Angst, I am not making them up.

In their herculean review of birth order research, Ernst and Angst examined all the studies they could find on personality and birth order—studies published anywhere in the world between 1946 and 1980. The data consisted of direct observations of the subjects' behavior; ratings by their parents, siblings, or teachers; and scores on various personality tests. By putting together all these results, Ernst and Angst expected to verify the hypothesis that "Personality varies with birth order: there is a 'firstborn personality.' "

They did not verify it. What Ernst and Angst found, first of all, was that most of the studies that purported to show birth order effects were irredeemably flawed. In most cases the researchers had failed to take into account differences in family size and socioeconomic status, variables that are themselves correlated and that can bias the results. Ernst and Angst eliminated the flawed studies, put together what they had left, and what did they find? No consistent birth order effects on personality. The majority of studies yielded no significant effects. When effects did occur they were often restricted to some subset of subjects—girls but not boys, small families but not large ones—with no rhyme or reason to the patterns.

Just to be sure they hadn't overlooked anything, Ernst and Angst did a study of their own. It was a huge study by the standards of social science: they gave personality tests to 7,582 college-age residents of Zurich. Twelve different aspects of personality were measured: sociability, extraversion, aggressiveness, excitability, nervousness, neuroticism, depression, inhibition, calmness, masculinity, dominance, and openness. (Nope, they didn't measure risibility.)

The results offer no comfort to believers in the efficacy of the family environment. Among subjects coming from two-child families, there were no significant differences between the first- and the secondborn in any of the measured personality traits. Among subjects coming from families of three or more, there was one small difference, probably a fluke: the lastborn scored slightly lower on masculinity. (When so many variables are measured, a significant difference is likely to turn up just by chance.)

Ernst and Angst summed up the outcome of their efforts this way: "An environmental variable"—birth order—"that is considered highly relevant is thus disaffirmed as a predictor for personality and behavior. This may signify that most of our opinions in the field of dynamic psychology will have to be revised."

But the belief in birth order effects isn't killed so easily: it's one of those things that can be knocked down repeatedly and pops right back up again, time after time. The most recent attempt to revive the idea comes from historian of science Frank Sulloway. In his book *Born to Rebel,* Sulloway claims that innovations in scientific, religious, and political thought are generally supported by laterborns and opposed by firstborns. This is because laterborns have more of the quality he calls "openness to experience." The innovative thoughts themselves, I notice, are not necessarily the products of laterborns: Galileo, Newton, Einstein, Luther, Freud, and Mao Tse-tung were all firstborns. But when it comes to accepting the new ideas of others, it appears (from the data presented in Sulloway's book) that firstborns tend to drag their heels. From early childhood, says Sulloway, they are heavily invested in the status quo. Unless they get on poorly with their parents, or have other reasons which he enumerates, firstborns have no motivation to rebel. They have no wish to upset an applecart from which they already get more than their share of the apples. Whatever is being given out, most notably parental attention, they get first shot at it. All they have to do to maintain their favored position is to say Yes Mommy, Yes Daddy. Since the brown-nose slot has already been filled, younger siblings must search for another role to play in the family. Thus, laterborns are the ones who rebel. As adults, laterborns are more likely to espouse what Sulloway calls "heterodox" (as opposed to orthodox) views.

Perhaps I am biased against Frank Sulloway's theory because I myself am a firstborn with heterodox views. Sulloway, himself a laterborn, is very hard on firstborns: they are depicted in his book as selfish, intolerant, jealous, close-minded, aggressive, and domineering. Cain, as he points out more than once, was a firstborn. Sulloway clearly identifies with Abel.

Stuck with the role of the domineering aggressor, I've tried to make the best of it. My critique of *Born to Rebel* is in the back of this book, in Appendix 1. Sulloway reexamined the studies reviewed by Ernst and Angst and came up with different results, results that support his theory. But I found the reanalysis unconvincing. And Sulloway doesn't mention the fact that Ernst and Angst did a study of their own—a carefully done study that was larger than any of the ones they reviewed—and found no birth order effects of interest. In particular, they found no difference between firstborns and laterborns in openness.

Birth order effects are like those things you think you see out of the corner of your eye but that disappear when you look at them closely. They do keep turning up but only because people keep looking for them and keep analyzing and reanalyzing their data until they find them. They turn up more often in older or smaller studies than in newer or larger ones. They turn up most often when the subjects' personalities are judged by their parents or siblings—a finding to which I will return in the next chapter.

Parental love and attention are not distributed evenly; Sulloway got that right. In his book he cites the finding that two-thirds of mothers with two children admitted to researchers that they favored one child over the other. What he doesn't mention is that a large majority of these non-impartial mothers said it was their *younger* child who got more attention and affection. This result was backed up by a later study in which both mothers and fathers were interviewed. About half admitted that they gave more love to one child than the other. Of these parents, 87 percent of the mothers and 85 percent of the fathers favored the younger child.

Contrary to Sulloway's notions and contrary, perhaps, to his childhood memories, it is the younger child, not the older one, who more often gets the lion's share of the parents' affection and attention. This is true the world over. In places where traditional methods of child-rearing are still used (I will describe them in Chapter 5), babies are cosseted and three-year-olds are dethroned without warning or apology when a younger sibling is born. Your elder brother may inherit the kingdom, the mansion, or the family farm, but that doesn't mean that Mom always loved him best. Well, maybe she did love him best, but it wasn't because he was born first.

I will have more to say about Sulloway's theory in the next chapter. Right now the topic is birth order and on this I will let those plainspoken Swiss researchers, Ernst and Angst, have the last word. In italics (theirs).

> Birth order research seems very simple, since position in a sibship and sibship size are easily defined. The computer is fed some ordinal numbers, and then it is easy to find a plausible post hoc explanation for any significant difference in the related variables. If, for example, lastborn children report more anxiety than other birth ranks, it is because for many years they were the weakest in the family. If firstborns are found to be the most timid, it is because of incoherent treatment by an inexperienced mother. If, on the other hand, middle children show the greatest anxiety, it is because they have been neglected by their parents, being neither the first- nor the last-born. With some imagination it is even possible to find explanations for greatest anxiety in a second girl of four, and so on, ad infinitum. *This kind of research is a sheer waste of time and money.*

Parenting Styles

Behavioral geneticists accepted Ernst and Angst's advice and gave up on birth order. But they gave up reluctantly, because it would have been an ideal way out of their dilemma. They already knew that parents' behavior can vary—that parents act differently toward different children. What they needed was a way of showing that these variations in parenting are not simply a response to the children's preexisting characteristics (child-to-parent effects) but that they actually have measurable effects (parent-to-child effects) on the children's personalities. Birth order effects could have done that. If differential parental behavior such as favoring one child over another really does have an influence on the children's personalities, the consequences should have shown up in birth order studies, because more often than not the parents favor the younger child. But most studies—especially the larger, newer, more carefully done ones—find no differences between the adult personalities of firstborns and laterborns. The only logical conclusion to be drawn from these results is that microenvironmental differences such as parental favoritism have no consistent effects on the child's personality. No effects that are still detectable in adulthood.

Maccoby and Martin's first alternative was that parents have no effects on their children. Their second was that the aspects of parenting that do have effects must vary from one child to another within the family. Birth order effects were the one kind of evidence that could have provided support for the second alternative. The failure to find convincing evidence of birth order effects left it twisting in the wind.

In the years since Maccoby and Martin offered their Scylla-or-Charybdis choice, no tempting third alternative has turned up. Behavioral genetic studies continue to show that the family home has few, if any, lasting effects on the people who grew up in it. If there are any long-term effects, they must be different for each sibling and unpredictable, because they do not show up in studies in which data from a number of people are combined. Of course, if we look at one particular person, it's easy to come up with a story about how the home environment (the critical, demanding mother, the ineffectual father) shaped the child's personality and produced the messed-up grownup we see today. That kind of post hoc speculation—unprovable, undisprovable—is the stock-in-trade of biographers.

Like the behavioral geneticists (and unlike biographers), socialization researchers have continued to turn out data. Many of them are still doing the same kind of studies they did before Maccoby and Martin—studies designed to find differences in parental child-rearing methods and to link these differences to the children's social, emotional, and intellectual functioning. These researchers are still looking for the effects of differences *between* families, not microenvironmental differences *within* families. I think it's necessary to examine this research a little more closely, since it is featured in every textbook of developmental psychology, including, alas, my own.

In 1967, developmentalist Diana Baumrind defined three contrasting styles of parenting. She named them Authoritarian, Permissive, and Authoritative, but I have always found those terms confusing so I will call them Too Hard, Too Soft, and Just Right.

Too Hard parents are bossy and inflexible: they lay down rules and enforce them strictly, with physical punishment if necessary. These are the shut-your-mouth-and-do-what-you're-told type of people. Too Soft parents are just the opposite: they don't *tell* their children to do things, they *ask* them. Rules? What rules? The important thing, they believe, is to give children lots of love.

The third choice is Just Right. You already know what these parents are like—I described them in the previous chapter when I was talking about broccoli eaters. Just Right parents give their children love and approval but they set limits and enforce them. They persuade their children to behave properly by reasoning with them, rather than by using physical punishment. Rules are not set in stone; these parents take their children's opinions and desires into account. In short, Just Right parents are exactly what end-of-the-twentieth-century middle-class Americans of European descent think that parents *ought* to be.

Baumrind and her followers have produced dozens of studies, all claiming to show the same thing: that the children of Just Right parents turn out better. The words are more convincing than the numbers, however. If you look closely at the data and the statistics, you'll see a lot of the kind of creative data analysis I described in the previous chapter. You take a lot of measurements of the parents and a lot of measurements of the children, so the chances are good that you'll get some significant correlations. If perchance you don't, you resort to the divide-and-conquer method. You look at boys and girls separately. You look at fathers and mothers separately. You look at white and nonwhite families separately. Often, the benevolent effects of Just Right parenting are different for girls and boys, different for fathers and mothers. Often, the benevolent effects of Just Right parenting are found only for white kids.

But I am quibbling. Looked at as a whole, these studies do show a modest but reasonably consistent tendency for good parents to have good kids. The children of Just Right parents tend to get along better with other kids and other adults and to make better grades in school. They get into less trouble in their teens. In general, they manage their lives in a competent fashion—slightly more competently, on the average, than the children of Too Hard or Too Soft parents.

The trouble with these findings is that they conflict with the behavioral genetic data. Remember that the style-of-parenting researchers are looking for differences *between* families—ways in which the Smith family is different from the Joneses. They typically look at only one child per family—one Smith, one Jones. The behavioral geneticists, on the other hand, look at *two* children per family, and what do they find? They find that it makes little or no difference whether a kid grows up in the Smiths' house or the Joneses'. The two Smith kids are similar in personality only if they are biological siblings. If they are adopted children it doesn't matter whether they both live in the Smiths' house or one of them lives with the Joneses—in either case they are not similar at all.

The implications of the behavioral genetic findings are unavoidable. Either the parents' child-rearing style has no effects on the children's personalities (Maccoby and Martin's first alternative), or the parents do not have a consistent child-rearing style (I'll call this alternative 2a), or they have a consistent style but it has different effects on each child (alternative 2b). Not one of these alternatives is compatible with the views of the style-of-parenting researchers, not even 2b. If being a Just Right parent makes some children better and others worse, what's the point of studying child-rearing styles?

I do not believe that parents have a consistent child-rearing style, unless they happen to have consistent children. I had two very different children—one of them is adopted but the same thing can happen with biological siblings—and used two very different child-rearing styles. My husband and I seldom had hard-and-fast rules with our first child; generally we didn't need them. With our second child we had all sorts of rules and none of them worked. Reason with her? Give me a break. Often we ended up taking the shut-your-mouth-and-do-what-you're-told route. That didn't work either. In the end we pretty much gave up. Somehow we all made it through her teens.

If parents adjust their child-rearing style to fit the child's characteristics, then Baumrind and her colleagues might be measuring child-to-parent effects rather than parent-to-child effects. It's not that good parenting produces good children, it's that good children produce good parenting. If parents *don't* adjust their child-rearing style to fit the child, then Baumrind and her colleagues might be measuring genetic effects rather than environmental effects. It's not that good *parenting* produces good children, it's that good *parents* produce good children.

Here's what I think. Middle-class Americans of European descent try to use the Just Right parenting style, because that is the style approved by their culture. If they don't use it, it's because they have problems or their kid does. If they have problems, it could be because they have disadvantageous personality characteristics that they can pass on to their kid genetically. If the kid has problems—a difficult temperament, for instance—the Just Right parenting style might not work and the parents might end up switching to the Too Hard method. So among Americans of European descent, parents who use a Too Hard child-rearing style are more likely to be the ones with problem kids. This is exactly what the style-of-parenting researchers find.

In other ethnic groups—notably Americans of Asian or African descent—cultural norms differ. Chinese Americans, for example, tend to use the Too Hard parenting style—the style Baumrind called Authoritarian—not because their kids are difficult, but because that's the style favored by their culture. Among Asian and African Americans, therefore, parents who use a Too Hard child-rearing style should *not* be more likely to have problem kids. Again, this is exactly what the researchers find.

What they find, in fact, is that Asian-American parents are the most likely of all American parents to use the Too Hard style and the least likely to use the Just Right style, and yet in many ways Asian-American children are the most competent and successful of all American children. Although this finding contradicts their theory, the style-of-parenting researchers continue on undaunted.

And it isn't just them—other developmentalists do the same thing. Data that conflict with the nurture assumption are ignored, ambiguous data are interpreted as confirmation of the nurture assumption.

Other Between-Family Differences

Differences between families are often a function of parental characteristics that are partly genetic, which means that many of the results reported by socialization researchers can be due to genetic transmission of traits from parents to children. When parents have trouble managing their own lives or getting along with others, their children are subject to a kind of double jeopardy, because they are at risk of inheriting disadvantageous genes and also of having a lousy home life. If such children do not turn out well, their problems are usually attributed to their lousy home life, but the true cause could be their disadvantageous genes. In most cases it's impossible to tell.

Let us look, therefore, at a few between-family differences that do not depend on advantageous or disadvantageous characteristics of the parents. Parents make some kinds of lifestyle decisions that are unrelated to how successful or unsuccessful they are at managing their lives.

For example, a classic question in developmental psychology is whether the children of mothers with paying jobs differ in personality or behavior from those whose mothers stay at home. A generation ago, mothers stayed at home unless their husbands couldn't make a decent living, and back then most developmentalists believed that the children of working mothers were at risk of psychological dysfunction. But now that working mothers are found in all walks of life, children whose mothers have jobs are found to be virtually indistinguishable from the minority whose mothers stay home. A developmentalist who was asked to write a review on the effects of maternal employment on children said that "few consistent differences emerge" and ended up writing mostly about the effects on the parents.

A related issue concerns the effects of day care. When only families with problems put their kids into day-care centers, institutional care was thought to be bad for young children. Now day-care centers are used by well-off families as well as the not-so-well-off, and it no longer seems to matter whether babies or preschoolers spend most of their daylight hours there or at home. In a 1997 review, a developmentalist asked the question, "Do infants suffer long-term detriments from early nonmaternal care?" Recent studies, she concluded, "have demonstrated that the answer is 'no.' " Even the variation in quality among day-care centers makes less difference than you might think: "The surprising conclusion from the research literature is that variation in quality of care, measured by experts, proves to have little or no impact on most children's development."

Researchers have also looked at the effects of homes that vary in family composition and lifestyle. There are still many families that consist of a mother and father and kids, but an increasing number have less conventional arrangements. When the unconventional arrangement is inadvertent—the result of a failed marriage or a failure to marry—there is an increased risk that the kids will experience failures in their own lives (I discuss the plight of the children of divorce and single parenthood in Chapter 13). But when the unconventional arrangement results from a consciously made lifestyle decision, no differences in child outcome have been found. Researchers in California have been studying a sample of unconventional families since the mid-1970s. Some of the parents are hippies and live on communes; others have "open marriages"; still others are single mothers of the Murphy Brown variety. The children are as bright, as healthy, and as well adjusted as children who live in more conventional families.

Another kind of unconventional arrangement involves children being reared by lesbian or gay parents. Here again, no important differences have emerged: children with two parents of the same gender are as well adjusted as children with one of each kind. There appears to be nothing unusual about their sex-role development: the girls are as feminine as other girls, the boys as masculine as other boys. Researchers have so far found no increased tendency for the children of homosexual parents to become homosexuals themselves, but it's too soon to make long-term predictions. Evidence from genetic studies suggests that genes may play a role in sexual orientation, and if this is the case we would expect homosexuality to occur with greater frequency among the biological offspring of homosexuals. Psychologists no longer consider this to be a sign of maladjustment.

Many of the children in conventional families are "accidents": more than 50 percent of the pregnancies in the United States are unintended. But there are other families—a growing number of them—whose children were conceived at great cost and difficulty with the aid of modern reproductive technology. These children owe their existence to techniques such as in vitro fertilization. According to a recent study, their parents provide a superior variety of parenting. But the children themselves are no different from anyone else's: "No group differences were found for any of the measures of children's emotions, behavior, or relationships with parents."

A recent study looked at three kinds of unconventional families at once—those without fathers, those with lesbian mothers, and those created through modern reproductive techniques—by examining children conceived through donor insemination. Some of the mothers were lesbians, others were heterosexuals; some were single, others had partners. The children of these mothers were well adjusted and well behaved—in fact, their adjustment and behavior was above average—and the researchers found no differences among them based on family composition. The ones without fathers were doing as well as the ones with fathers.

Among the many family differences that have an impact on a child's life at home, surely one of the most important is the presence or absence of siblings. The only child leads a very different life from the child with siblings. Her relationship with her parents is likely to be far more intense. She gets all the worry, responsibility, and blame heaped on the oldest child, plus all the attention and affection heaped on the youngest. In the past, when most families had at least two children and deviations from this pattern were usually a sign that something had gone wrong, the only child had a bad reputation. But people are marrying later now and having fewer children. Studies done in the past fifteen years have found no consistent differences between only children and children with one or two siblings. Minor differences do turn up, but sometimes they favor the only child, sometimes the child with siblings.

Searching for the Key

Children who grow up in different families are likely to have very different home environments. Some have siblings; others do not. Some have two parents of opposite sexes who are married to each other; others do not. Some are cared for exclusively by their mothers and fathers; others are not. These major differences between families have no predictable effects on the children reared in them—a finding that agrees with behavioral genetic data. Less obvious differences between families—namely, the parents' child-rearing style—are claimed to have predictable effects, but, as Maccoby and Martin pointed out, the reported effects are weak and can be accounted for in other ways.

That leads us back to Maccoby and Martin's second alternative, that the only aspects of parenting that do have effects are those that differ for each child in the family. But if major differences *between* homes have no predictable effects, why should we expect the smaller differences *within* the home to have predictable effects? Does it make any sense to say that what matters is whether Mom loved you best, if it doesn't matter whether Mom was home or at work, married or single, gay or straight?

The idea that each child grows up in a unique microenvironment within the home was supposed to be a way out of the bind that behavioral geneticists found themselves in. Heredity can't account for everything: their work showed that only half the variation in personality traits could be ascribed to genetic differences between individuals. The other half, therefore, had to be due to the environment—which they, like everybody else, assumed meant "nurture." Only one behavioral geneticist, David Rowe of the University of Arizona, pointed out that parents aren't the be-all and end-all of the child's life and that the child has environments other than the home—environments that might be more important. The others went on searching inside the home, like people looking for a lost key: "It's *got* to be in here *somewhere!*"

Perhaps you too are thinking, "It's *got* to be in there *somewhere!*" Everybody *knows* that parents make a difference! Fifty thousand psychologists couldn't possibly be wrong! What about all the evidence that dysfunctional families produce dysfunctional kids? But genes matter too, and children can inherit from their parents the traits that caused or contributed to the family's dysfunction. (I'll take a closer look at dysfunctional families in Chapter 13. It's not just genes.)

It's not just genes. You believe in the power of the home environment because you've seen the evidence with your own eyes. Parents who don't know the first thing about parenting and their terrible kids. The explosive temper of the child who's been rewarded for throwing tantrums. The low self-esteem of the child whose parents are constantly berating her. The nervousness of the child whose parents are inconsistent. And the vast differences in personality between people who grew up in different cultures. My job is not an easy one. I have to find alternative explanations for all the things you've observed that make you so certain that parents have lasting effects on their children.

Thomas Bouchard, a behavioral geneticist at the University of Minnesota, is one of the researchers working on the Minnesota Study of Twins Reared Apart. In 1994 he admitted in the journal *Science* that how the childhood environment influences adult personality "remains largely a mystery." Perhaps a greater mystery is why psychologists have remained fixated for so long on the notion that people's personalities are formed by some combination of nature and nurture. Nature—the DNA we get from our parents—has been shown to have effects but it can't be the whole story. Nurture—all the other things our parents do to us—has not been shown to have effects despite heroic efforts on its behalf.

It is time to look for another alternative, None Of The Above.

4 SEPARATE WORLDS

Folktales passed down to us from earlier times often feature a hero or heroine who was treated badly at home but who eventually left home and became a great success. Consider the story of Cinderella. In the book I had as a child, the story began as follows:

> There was once a man who married for his second wife a woman who was both vain and selfish. This woman had two daughters who were as vain and selfish as she was. The man had a daughter of his own, however, who was sweet and kind and not vain at all.

The sweet, kind daughter was, of course, Cinderella. Unlike the Disney movie, this version depicts the (unnamed) stepsisters as beautiful. It was only their personalities that were ugly. In this respect, they closely resembled their mother. Cinderella presumably inherited her sweet nature from *her* mother, who was dead. Dead mothers were not a rarity in the old days; as many families were broken by death as are broken today by divorce.

In a fairy tale, events are compressed. Cinderella must have suffered years of abuse from her stepmother and stepsisters. She had no recourse: her father was unwilling or unable to stand up for her, and there were no laws or agencies in those days to protect children against mistreatment. She must have learned early on that it was best to remain as inconspicuous as possible, to do what she was told, and to accept verbal and physical insults without protest. And then—then came the ball, and the fairy godmother, and the prince.

The folk who gave us this tale ask us to accept the following premises:

that Cinderella was able to go to the ball and not be recognized by her stepsisters, that despite years of degradation she was able to charm and hold the attention of a sophisticated guy like the prince, that the prince didn't recognize her when he saw her again in her own home dressed in her workaday clothing, and that he never doubted that Cinderella would be able to fulfill the duties of a princess and, ultimately, of a queen.

Preposterous? Maybe not. The whole thing works if you accept one simple idea: that children develop different selves, different personas, in different environments. Cinderella learned when she was still quite small that it was best to act meek when her stepmother was around, and to look unattractive in order to avoid arousing her jealousy. But from time to time, like all children who are not kept under lock and key, she would slip out of the cottage in search of playmates. (They *couldn't* keep her locked in the cottage—there was no indoor plumbing.)

Outside the cottage things were different. Outside the cottage no one insulted Cinderella or treated her like a slave, and she discovered that she could win friends (including the kindly neighbor whom she would later refer to as "my fairy godmother") by looking pretty. Her stepsisters didn't recognize her at the ball not just because she was dressed differently: her whole demeanor was different—her facial expressions, her posture, the way she walked and talked. They had never seen her outside-the-cottage persona. And the prince, of course, had never seen her *inside*-the-cottage persona, so he didn't recognize her when he called at the cottage in search of the girl who dropped the shoe. She was quite charming at the ball, though admittedly lacking in sophistication. But that, he figured, could be easily remedied.*

The Two Faces of Cinderella?

Perhaps it sounds like I am describing someone with a "split personality," like the protagonist of *The Three Faces of Eve*. But what made Eve abnormal was not the fact that she had more than one personality, or even that the alternate personalities were very different. Eve's problem was that her personalities appeared and disappeared unpredictably and didn't have access to each other's memories.

Having more than one personality is not abnormal. William James, brother of the novelist Henry James, was the first psychologist to point

* I have no comment on the part of the story that says "and they lived happily ever after." It is, after all, a fairy tale.

this out. Over a hundred years ago, William described multiple personalities in normal adolescents and adults—that is, in normal *male* adolescents and adults.

> Properly speaking, *a man has as many social selves as there are individuals who recognize him* and carry an image of him in their mind. . . . But as the individuals who carry the images fall naturally into classes, we may practically say that he has as many different social selves as there are distinct *groups* of persons about whose opinion he cares. He generally shows a different side of himself to each of these different groups. Many a youth who is demure enough before his parents and teachers, swears and swaggers like a pirate among his 'tough' young friends. We do not show ourselves to our children as to our club-companions, to our customers as to the laborers we employ, to our own masters and employers as to our intimate friends. From this there results what practically is a division of the man into several selves; and this may be a discordant splitting, as where one is afraid to let one set of his acquaintances know him as he is elsewhere; or it may be a perfectly harmonious division of labor, as where one tender to his children is stern to the soldiers or prisoners under his command.

In other words, to put James's observations into current terminology, people behave differently in different social contexts. Contemporary personality theorists do not dispute this. What they argue about is whether there is any "real" personality under all these masks. If a man can be tender in one context and stern in another, which is he *really?* If three different men all are tender with their children and stern with their prisoners, isn't it the situation that determines personality and not the man?

The passage from William James comes from his book *The Principles of Psychology*—America's first psychology textbook, published in 1890 (I own a copy of it, too tattered to be valuable). Because psychology was just beginning, James had it pretty much to himself for a while, and he stuck his finger in every pie. He talked about personality, cognition, language, sensation and perception, and child development. James was the one who said—incorrectly, as it turned out—that the world of the newborn infant is "one great blooming, buzzing confusion."

Today, these fields of psychology are entirely separate, presided over by specialists who seldom read articles outside their own field once they've made it through graduate school. Arcane arguments about adult personality are unlikely to attract the interest of socialization researchers. The word "selves" is not in the vocabulary of most behavioral geneticists.

Which is a pity, because I think it's relevant. Indeed, I think James's observation that people behave differently in different social contexts, and the subsequent discussions about why this happens and whether there is a "real" personality underneath, contain important clues to one of the big puzzles of personality development.

Here is the puzzle. There is evidence (I told you about it in Chapters 2 and 3) that parents cannot modify the personality their child was born with, at least not in ways that can be detected after the child grows up. If that is true, how come everyone is so certain that parents do have important effects on the child's personality?

Different Places, Different Faces

Unlike Eve of the Three Faces, most people do not have multiple personalities that lack access to each other's memories. Normal people may behave differently in different social contexts, but they carry along their memories from one context to another. Nonetheless, if they learn something in one situation they do not necessarily make use of it in another.

In fact, there is a strong tendency *not* to transfer the knowledge or training to new situations. According to learning theorist Douglas Detterman, there is no convincing evidence that people spontaneously transfer what they learned in one situation to a new situation, unless the new situation closely resembles the old one. Detterman points out that undergeneralization may be more adaptive than overgeneralization. It is safer to assume that a new situation has new rules, and that one must determine what the new rules are, than to blithely forge ahead under the assumption that the old rules are still in effect.

At any rate, that is how babies appear to be constructed. Developmentalist Carolyn Rovee-Collier and her colleagues have done a series of experiments on the learning ability of young babies. The babies lie in a crib, looking up at a mobile hanging above them. A ribbon is tied to one of their ankles in such a way that when they kick that foot, the mobile jiggles. Six-month-old babies catch on to this very quickly: they are delighted to discover that they can control the mobile's movements by kicking their foot. Moreover, they will still remember the trick two weeks later. But if any detail of the experimental setup is changed—if a couple of the doodads hanging from the mobile are replaced with slightly different doodads, or if the liner surrounding the crib is changed to one of a slightly different pattern, or if the crib itself is placed in a different room—the babies will gaze up at the mobile cluelessly, as though they had never seen such a thing in their lives. Evidently babies are equipped with a learning mechanism that comes with a warning label: what you learn in one context will not necessarily work in another.

It is true: what you learn in one context will not necessarily work in another. A child who cries at home gets—if he's lucky—attention and sympathy. In nursery school, a child who cries too much is avoided by his peers; in grade school he is jeered at. A child who acts cute and babyish for her daddy evokes a different reaction from her classmates. Children who get laughs for their clever remarks at home wind up in the principal's office if they don't learn to hold their tongue in school. At home the squeaky wheel gets the grease; outside, the nail that sticks up gets hammered down. Or, as in Cinderella's case, vice versa.

Like Cinderella, most children have at least two distinct environments: the home and the world outside the home. Each has its own rules of behavior, its own punishments and payoffs. What made Cinderella's situation unusual was only that her two environments—and hence her two personalities—were unusually divergent. But children from ordinary middle-class American families also behave differently inside the home and outside of it. I remember when my children were in elementary school and my husband and I used to go to Back-to-School Night to meet their teachers. Year after year we would see parents talking to their child's teacher and coming away shaking their heads in disbelief. "Was she talking about *my* kid?" they would say, making it sound like a joke. But sometimes the teacher really seemed to be talking about a child who was a stranger to them. More often than not, this child was better behaved than the one they knew. "But he's so obstinate at home!" "At home she never shuts up for a minute!"

Children—even preschoolers—are remarkably good at switching from one personality to another. Perhaps they can do this more easily than older people. Have you ever listened to a couple of four-year-olds playing House?

Stephie (in her normal voice, to Caitlin): I'll be the mommy.
Stephie (in her unctuous mommy voice): All right, Baby, drink your bottle and be a good little baby.
Stephie (whispering): Pretend you don't like it.
Caitlin (in her baby voice): Don't want botta!
Stephie (in her unctuous mommy voice): Drink it, sweetheart. It's good for you!

Stephie plays three parts here: author/producer, stage director, and the starring role of Mommy. As she switches back and forth between them, she gives each one a different voice.

Context and Behavior

The "bottle" that Stephie was pretending to feed to Caitlin was a cylindrical wooden block. Developmentalists are interested in this kind of pretense because it appears to be an advanced, symbolic form of behavior, and yet it appears remarkably early—before the age of two. Much has been written about the environmental influences that make pretense appear earlier or later; not surprisingly, attention has focused on the role of the child's mother. Researchers have found that a toddler engages in more advanced types of fantasy when the mother joins in the fantasy with the child.

But there is a catch. Greta Fein and Mary Fryer, specialists in children's play, reviewed the research and concluded that, although young children do play at a more advanced level when they are playing with their mothers, "the hypothesis that mothers contribute to subsequent play sophistication receives no support." When the mother encourages the child to engage in elaborate fantasies, the child can do it; but later, when the child is playing alone or with a playmate, it makes no difference what kinds of games she played with her mother.

Other developmentalists attacked this conclusion. Fein and Fryer responded by saying that they "did not intend to disparage the importance of adult caregivers in the lives of young children" and that they hadn't previously realized "how deep is the belief" in the omnipotence of parents. But they stuck to their guns. The evidence indicates that mothers influence children's play only while the children are playing with the mothers. "When theories don't work," Fein and Fryer counseled, "chuck 'em or change 'em." My view precisely.

Learning to do things with Mommy is all well and good, but the infant does not automatically transfer this learning to other contexts. This is a wise policy, because what is learned with Mommy might turn out to be useless in other contexts—or worse than useless. Consider, for example, a baby I will call Andrew. Andrew's mother was suffering from postpartum depression, an affliction that is not uncommon in the first few months after childbirth. She was able to feed Andrew and change his diapers, but she didn't play with him or smile at him very much. By the time he was three months old, Andrew too was showing signs of depression. When he was with his mother he smiled infrequently and was less active than babies of that age usually are—his face was serious, his movements muted. Fortunately, Andrew didn't spend all his time with his mother: he spent part of it at a day nursery, and the caregiver at the nursery was not depressed. Watch Andrew with his nursery caregiver and you will see a different baby, smiley and active. The somber faces and muted movements common in the babies of depressed mothers are "specific to their interactions with their depressed mothers," according to researchers who studied babies like Andrew.

Different behaviors in different social contexts have also been noted in older infants, infants of walking age. Researchers have studied how toddlers behave at home (by asking their mothers to fill out questionnaires) and at day-care centers (by observing them there or by asking the caregivers at the center) and found that the two descriptions of the children's behavior do not agree. "There exists the possibility that the toddler's actual behavior differs systematically in the home and day-care settings," admitted one researcher.

Sisters and Brothers

Granted that what children learn from interacting with their mothers might not help them get along with their peers in nursery school, but surely what they learn from interacting with their *siblings* should be transferable? You would think so—I would have thought so too. But on second thought, children are probably better off starting from scratch with their peers. The child who dominates her younger brother at home may be the smallest one in her nursery school class; the dominated younger brother may turn out to be the largest and strongest in his. Here is what one team of researchers has to say on this topic:

> There was no evidence of individual differences in sibling interactions carrying over into peer interactions. . . . Even the second-born child, who has experienced years in a subordinate role with an older sibling, can step into a dominant role [with a peer].

And this from another:

> Few significant associations were found between measures of children's sibling relationships and characteristics of their peer relationships. . . . Children who were observed to be competitive and controlling to their siblings were reported by their mothers to have positive friendships. Children whose mothers reported that they had hostile sibling relationships received higher scores on friendship closeness. . . . Indeed, we should not expect competitive and controlling behaviour toward a younger sibling to be necessarily associated with negative and problematic behaviour with friends.

Unless they happen to have a twin, children's relationships with their siblings are unequal. In most cases the elder is the leader, the younger is the follower. The elder attempts to dominate, the younger to avoid domination. Peer relationships are different. Peers are more equal, and often more compatible, than siblings. Among American children, conflict and hostility erupt far more frequently among siblings than among peers.

Conflict between siblings is the theme of Frank Sulloway's book *Born to Rebel,* which I mentioned in the previous chapter. In Sulloway's view, siblings are born to be rivals, fighting to get their fair share—or, in the case of firstborns, more than their fair share—of family resources and parental love. Children do this, he says, by specializing in different things: if one niche in the family is filled, the next child must find some other way of winning parental attention and approval.

I do not disagree with that. Nor do I doubt that people often drag their sibling rivalries along with them to adulthood and sometimes to the grave. My Aunt Gladys and my Uncle Ben hated each other all their lives. What I doubt is that people drag the emotions and behaviors they acquire in their sibling relationships to their other relationships. With anyone other than her brother Ben, my Aunt Gladys was as sweet and kind as the Cinderella in my childhood storybook.

The patterns of behavior that are acquired in sibling relationships neither help us nor hinder us in our dealings with other people. They leave no permanent marks on our character. If they did, researchers would be able to see their effects on personality tests given to adults: firstborns and laterborns would have somewhat different personalities in adulthood. As I reported in the previous chapter (also see Appendix 1), birth order effects do not turn up in the majority of studies of adult personality. They do, however, turn up in the majority of studies of one particular kind: the kind in which subjects' personalities are judged by their parents or siblings. When parents are asked to describe their children, they are likely to say that their firstborn is more serious, methodical, responsible, and anxious than their laterborns. When a younger brother or sister is asked to describe the firstborn, a word that turns up is "bossy." What we're getting is a picture of the way the subject behaves at home.

At home there are birth order effects, no question about it, and I believe that is why it's so hard to shake people's faith in them. If you see people with their parents or their siblings, you do see the differences you expect to see. The oldest does seem more serious, responsible, and bossy. The youngest does behave in a more carefree fashion. But that's how they act when they're together. These patterns of behavior are not like albatrosses that we have to drag along with us wherever we go, all through our lives. We don't even drag them to nursery school.

Never Leave Home Without It

My favorite example of a failure to transfer behavior from one context to another involves picky eating—a common complaint among the parents of young children. You would think a picky eater in one setting would be a picky eater in another, wouldn't you? Yes, it has been studied, and no, that's not what the researchers found. One third of the children in a Swedish sample were picky eaters *either* at home or in school, but only 8 percent were picky in *both* places.

Ah, but what about that 8 percent? It is time to admit that I have been misleading you: the correlation between behavior at home and behavior outside the home may be low, but it's not zero. I mentioned another example in Chapter 2: the children who behaved obnoxiously with their parents but not with their peers, or vice versa. The correlation between obnoxious behavior in the two settings was only .19, which means that if you saw how a child behaved with her parents you would be unlikely to predict correctly how she would behave with her peers. Still, the correlation was not zero; in fact, it was statistically significant.

Significant, but surprisingly low. Surprising because, after all, it was the *same child* behaving in both contexts—the same child with the *same genes.* We know from behavioral genetic research that personality traits such as disagreeableness and aggressiveness have heritabilities of around 50 percent. That means a sizable portion of a child's personality (the exact percentage isn't important) is built in, innate, not acquired through experience. Children who have a built-in tendency to be disagreeable take this tendency with them wherever they go, from one social context to another. What they've learned may be tied to the context it was acquired in, but what they were born with they cannot leave behind. The child who is a picky eater both at home and at school may have food allergies or a delicate digestive system. Thus, the fact that some children are picky both at home and at school, and some children are obnoxious both with their parents and with their peers, could be due to direct genetic effects.

Indirect genetic effects—the effects of the effects of the genes—can also lead to a carryover of behavior from one context to another. Cinderella's case was unusual: her prettiness put her in danger whenever she was within striking distance of her stepmother. Only in the world outside the cottage was her prettiness an asset. Most pretty children find their prettiness an asset wherever they go. Most homely children learn that homeliness is a disadvantage in every social context. Perhaps some of the children who were obnoxious both with their parents and with their peers were physically unattractive children who had given up trying to get their way by being pleasant, because it didn't work with anyone. Or perhaps they were born with unpleasant dispositions, which made their dealings with all sorts of people problematic. A disagreeable temperament can lead to trouble both directly and indirectly: directly because it makes the child respond unfavorably to other people, indirectly because it makes other people respond unfavorably to the child.

Code-Switching

The carryover of behavior from one context to another due to genetic effects is a nuisance for me—it gets in the way of the point I am trying to make. I am trying to convince you that children learn separately, in each social context, how to behave in that context. But social behavior is complicated. It is determined partly by characteristics people are born with, partly by what they experience after they are born. The inborn part goes with them wherever they go and thus tends to blur the distinctions between social contexts. To solve this problem I will turn to a social behavior that's acquired entirely through experience: language.

Perhaps I'd better qualify that statement. Language is acquired through experience; yet it is also innate. It is one of the things that we inherit from our ancestors but that does not vary among normal members of our species, like lungs and eyes and the ability to walk erect. Every human baby born with a normal brain is equipped with the ability and desire to learn a language. The environment merely determines *which* language will be learned.

In North America and Europe, we take it for granted that we must teach our babies how to communicate with language; indeed, we consider that to be one of a parent's important jobs. We start the language-learning lessons early, talking to our babies the minute they're out of the womb, if not before. We encourage their coos and babbles and make a big deal out of their "mamas" and "dadas." We ask them questions and await their replies; if they don't reply we answer the questions ourselves. If they make a grammatical error we rephrase their poorly formed phrase into proper English (or proper whatever). We speak to them in short, clear phrases about things they're interested in.

Thus encouraged, not to say prodded, our babies start talking when they're barely a year old and are speaking in sentences when they're barely two. By the age of four they're competent speakers of English (or whatever).

Now I ask you to imagine a child who goes outside her home for the first time at the age of four and discovers—as Cinderella did—that out there everything is different. Only in this case, what's different is that everyone is speaking a language she can't understand, and no one can understand *her* language. Will she be surprised? Probably not, judging from the reaction of the babies who learned to jiggle the mobile by kicking one foot. Change the liner surrounding the crib and they're in a different world. They assume that the new world has new rules, yet to be learned.

Children of immigrant parents, like the kids of the Russian couple who ran the rooming house in Cambridge (described in Chapter 1), are in exactly that situation. They learn things at home—most conspicuously a language but other things as well—that prove to be useless outside the home. Unfazed, they learn the rules of their other world. They learn, if necessary, a new language.

Children have a great desire to communicate with other children, and this desire serves as a powerful incentive to learn the new language. A psycholinguist tells the story of a four-year-old boy from the United States, hospitalized in Montreal, trying to talk to the little girl in the next bed. When his repeated attempts to talk to her in English proved futile, he tried the only French words he knew, fleshed out with a few nonsense syllables: "Aga doodoo bubu petit garçon?" An Italian father living in Finland with his Swedish-speaking wife and son tells of the time he took his three-year-old son to a park and the boy wanted to play with some Finnish-speaking children. He ran up to them shouting the only words of Finnish he had learned: "Yksi, kaksi, kolme . . . yksi, kaksi, kolme"—Finnish for "One, two, three."

This fools-rush-in approach is practiced mainly by younger children; older ones are more likely to start off with a least-said-soonest-mended strategy. Researchers studied a seven-and-a-half-year-old boy—I'll call him Joseph—who moved with his parents from Poland to rural Missouri. In school, Joseph listened quietly for several months, watching the other children for clues to what the teacher was saying. With neighborhood friends he was more willing to risk making mistakes and he started practicing his English with them almost immediately. At first Joseph's speech sounded like that of a toddler—"I today school"—but within a few months he was speaking serviceable English and after two years he was using it like a native, with hardly a trace of an accent. The accent eventually went away entirely,* even though he continued to speak Polish at home.

It is common for immigrant children to use their first language at home and their second language outside the home. Give them a year in the new country and they are switching back and forth between their two languages as easily as I switch back and forth between programs on my computer. Step out of the house—click on English. Go back in the house—click on Polish. Psycholinguists call it code-switching.

Cinderella's alternate personas are an example of another kind of code-switching. Step out of the cottage—look pretty, act charming. Go back in—look homely, act humble. If she had also spoken one language in her home and another language outside it, as Joseph did, that would have been just another difference between life inside the cottage and life outside it. Mastering bilingualism is probably easier for a child than switching back and forth from looking pretty to looking homely.

Code-switching is sort of like having two separate storage tanks in the mind, each containing what was learned in a particular social context. According to Paul Kolers, a psycholinguist who studied bilingual adults, access to a given tank may require switching to the language used in that context. As an example, he mentioned a colleague of his who had moved from France to the United States at the age of twelve. This man does his arithmetic in French, his calculus in English. "Mental activities and information learned in one context are not necessarily available for use in another," Kolers explained. "They often have to be learned anew in the second context, although perhaps with less time and effort."

It is not only book-learning that is stored in separate tanks. "Many bilingual people," reported Kolers, "say that they think differently and respond with different emotions to the same experience in their two languages." If they use one language exclusively at home, the other exclusively outside the home, the home language becomes linked to the thoughts and emotions experienced at home, the other to the thoughts

* Psycholinguists sometimes make the claim that babies, before they are a year old, lose the ability to hear the difference between speech sounds that are not distinguished in their language. That can't be right, though. If babies really lost the ability to discriminate the sounds, children such as Joseph couldn't learn to speak a second language without an accent. More likely what happens is that the babies learn not to pay attention to differences that are irrelevant in their language. If later on those differences become relevant, they are able to turn their attention back on.

and emotions experienced outside the home. At home Cinderella thought of herself as worthless, outside her home she found that she could win friends and influence people. A bilingual Cinderella might still be scrubbing floors if the prince had addressed her in the language used in her cottage.

Personality theorists don't pay much attention to language. And yet, language, accent, and vocabulary are aspects of social behavior, just as "personality traits" such as agreeableness and aggressiveness are. Like other aspects of social behavior, the language a person uses is sensitive to context, and this is as true for monolingual speakers as it is for bilingual ones. William James said that a person "shows a different side of himself" in different social contexts and gave as his first example the youth who swears like a pirate when he's with his friends but is "demure enough before his parents and teachers." A high school student tells this anecdote about one of his classmates:

> A girl at my school was walking down the hall and remembered she forgot something.
>
> "Oh shoot!" she exclaimed.
>
> As she looked around and saw her friends she said, "I mean oh shit."

The girl's parents and teachers make similar adjustments in their verbal behavior. They do not use the same vocabulary or sentence structure when they're talking to a teenager as when they're talking to a two-year-old. They do not use the same vocabulary or sentence structure when they're talking to their automobile mechanic as when they're talking to their doctor.

Though it is a social behavior, language has the advantage of being free of the genetic complications that plague other kinds of social behaviors. The tendency to be agreeable or aggressive is partly genetic, but the tendency to speak Polish rather than English, or to use swear words with some people and not with others, is entirely environmental.

Language and Social Context

Code-switching is an extreme example; most children's mental tanks do quite a bit of leaking. After all, they carry their memories with them wherever they go, from one context to another. A child who comes out of his house at the age of four and finds that people out there *do* speak the language he learned at home doesn't have to learn it all over again, although he may be cautious at first about using it outside his home.

For most children, the home environment and the outside-the-home environment do not have steel walls between them. The parents come to school to watch their children perform in plays and for conferences with the teacher. The children reveal bits of their home lives in Show-and-Tell and "What I Did on My Summer Vacation." They invite their school friends to their homes for birthday parties.

When William James spoke of the "division of the man into several selves," he said there were two kinds of divisions: harmonious, as exemplified by the man who is tender with his children but stern with his prisoners, and discordant, "as where one is afraid to let one set of his acquaintances know him as he is elsewhere." Cinderella's division was discordant: she was afraid to let her stepmother see her as she was elsewhere. Some psychologists and psychiatrists believe that severe abuse in childhood can lead to multiple personality disorder, the three-faces-of-Eve phenomenon. The connections between the mental tanks are broken, or never get formed, and each personality accumulates its own memories and fails to share them with the others.

Most children do not risk a beating if they reveal bits of their outside-the-home behavior to their parents. But it is common for children to act as though some terrible punishment would ensue if they reveal bits of their home behavior outside the home. Philip Roth, in his novel *Portnoy's Complaint,* tells an anecdote that is almost certainly autobiographical. Here's Alexander Portnoy—the son of first-generation Jewish Americans who speak English liberally sprinkled with Yiddish words—describing an incident from his childhood:

> I was already the darling of the first grade, and in every schoolroom competition, expected to win hands down, when I was asked by the teacher one day to identify a picture of what I knew perfectly well my mother referred to as a "spatula." But for the life of me I could not think of the word in English. Stammering and flushing, I sank defeated into my seat, not nearly so stunned as my teacher but badly shaken up just the same . . . in a state resembling torment—in this particular instance over something as monumental as a kitchen utensil.

Alexander thought *spatula* was a Yiddish word—a home word, a family word—and he would rather be struck dead than use it in public. I had a similar experience in third or fourth grade when I used the word *pinky* to refer to my little finger. The girl I was talking to (not a close friend) asked, "What did you say?" and I was struck with panic. I had made a fatal error: *pinky* must be a home word! The girl asked again, "What did you say?" and I mumbled "Nothing." She became insistent and I became more and more embarrassed but refused to tell her what I had said. Years later I realized that she, too, must have been unsure of the status of the word *pinky* and was trying to find out if it was a legitimate outside-the-home word.

Joseph spoke Polish to his parents and English to his teachers, schoolmates, and friends. But sometimes his friends came over to his house to play with him and he spoke to them in English, and thus English crept into his home. Or perhaps, like Alexander Portnoy, he was embarrassed to use his home language outside the home, so when he went shopping with his parents he spoke to them in English. However it starts, the children of people who immigrate to English-speaking countries usually end up bringing English home with them, speaking English to their parents. Here's the son of Korean immigrants, describing how he communicated with his mother: "She would mostly speak to me in Korean, and I would answer her in English." Here's an anthropologist explaining why Jewish immigrants from Eastern Europe failed to transmit their language to their children: "They talked Yiddish to their children and the children answered in English." The same sort of thing happens, in a smaller way, in homes in which everyone speaks English: I have heard a great many native-born Americans complain that their children come home talking in the uncouth accents of their peers.

If immigrant parents insist that their children continue to address them in their native language—that is, the parents' native language—the children may do so, but their ability to communicate in that language will remain childish, while their ability to communicate in the outside-the-home language continues to grow. Here's a young Chinese-American woman, the child of immigrants, who went to Harvard:

> I had never discussed literature or philosophy with my parents. We talked about our health, the weather, that night's dinner—all in Cantonese, since they do not speak English. While at Harvard, I ran out of words to communicate with my parents. I literally did not have the Cantonese vocabulary to explain the classes I was taking or my field of concentration.

Many immigrant parents see their children losing the language and culture of their homeland and try very hard to prevent it. My local newspaper ran a story about a woman from West Bengal, India, who started a Bengali language school for her children and the children of other Bengali-speaking immigrants.

> Like many immigrants, Bagchi wants her children to understand their cultural background. To do that, she believes, they first must be fluent in Bengali, their parents' native tongue and one of 15 languages spoken in India. . . . But learning a language isn't easy if you study it for only a few hours each week. School, television and peer groups immerse children in English, and despite the best of efforts by both parents and children, it often is a challenge to become fluent in the parents' language. "They dream in English. They do not dream in Bengali," Bagchi said, describing Bengali children born in the United States.

They dream in English. It makes no difference whether the first language they learned from their parents was English or Bengali, English has become their "native language." Joseph spoke nothing but Polish for the first seven and a half years of his life, but if he remains in the United States his "native language" will not be Polish. As an adult he will think in English, dream in English, do his arithmetic *and* his calculus in English. He may forget his Polish entirely.

Parents do not have to teach their children the language of their community; in fact—hard as it may be for you to accept this—they do not have to teach their children any language at all. The language lessons we give our infants and toddlers are a peculiarity of our culture. In parts of the world where people still live in traditional ways, no lessons are given and parents generally do very little conversing with their babies and toddlers—they consider learning the language the child's job, not the parents'. According to psycholinguist Steven Pinker, mothers in many societies "do not speak to their prelinguistic children at all, except for occasional demands and rebukes. This is not unreasonable. After all, young children plainly can't understand a word you say. So why waste your breath in soliloquies?" Compared to American toddlers, the two-year-olds in these societies seem retarded in their language development, but the end result is the same: all the children eventually become competent speakers of their language.

You are thinking, Yes, but even though the mother doesn't speak to the baby, the baby hears her speaking to other people. True. But even this is unnecessary. There is an old story, told by the Greek historian Herodotus, of a king who wanted to find out what language children would speak if left to their own devices. He had a couple of babies reared in a lonely hut by a shepherd and gave instructions that no one was to talk to them or speak a word in their hearing. Two years later he visited the children and, the story goes, they ran up to him saying something that sounded like "bekos," which is the word for *bread* in an ancient language called Phrygian. The king concluded that Phrygian must have been the world's first language.

Would it shock you to learn that in the United States there are thousands of babies being reared like that? No, it is not an experiment. These are babies born to profoundly deaf couples. Most deaf people marry other deaf people, but more than 90 percent of the babies born to these couples have normal hearing. These babies miss out on some of the experiences we consider crucial to normal development. No one comes running when they scream in terror or in pain. No one encourages their coos and babbles or makes a big deal out of their "mamas" and "dadas." Nowadays most deaf parents use sign language to communicate with their hearing children, but there was a period when the use of sign language was frowned upon, and during that period some deaf parents didn't communicate with their young children at all, except in the most rudimentary ways. And yet, these children suffered no harm. Despite the fact that they didn't learn any language at all from their parents, they became fluent speakers of English. Don't ask them how they learned it; they can't remember and many of them consider the question offensive. I assume they learned the same way Joseph did.

Socialization researchers are unlikely to study families in which the parents speak Polish or Bengali, much less families in which the parents communicate only in sign. They don't worry about how and where children acquire their language because it is a constant: all the parents in their studies speak English and so do all the children, and the researchers assume that children must learn their language from their parents. They make the same assumption about other aspects of socialization. They are wrong about language and I believe they are wrong about other aspects of socialization. Bilingualism is simply the most conspicuous marker of context-specific socialization—socialization that is tied to a particular social context.

A Place for Everything and Everything in Its Place

As the spatula story suggests, children appear to be motivated to keep their two lives separate. Child abuse often goes undetected because children don't like to talk about it when they're outside their home. They don't want anyone to know that their home is different—that their stepmother beats them and makes them scrub the floor. Conversely, school-age children often fail to tell their parents if they've been victims of bullying on the playground. I was a social outcast for four years of my childhood—none of my classmates would talk to me—and my parents never knew about it.

But the motivation to keep the home life from leaking out is stronger than the motivation to keep the outside world from leaking in, and it is especially strong in those who have an inkling that their homes might be abnormal in some way. If their mother drinks, their parents throw things at each other, or their father is an invalid, kids don't want anyone to know about it. The child of immigrants might avoid inviting friends over to play. The kid whose parents are wealthier than their neighbors may be as anxious to keep it a secret as the kid whose parents are poorer: what they hate is being different from their peers.

In order to know what has to be concealed, children need some way of learning whether their homes are normal or abnormal. One way they can do this is by watching television; however, this works only if the families they see on television are not too obviously different from the families they see in their neighborhood. If the differences are too great, then children must base their concepts of normal family life on what they learn from their friends and classmates.

Getting information from friends and classmates can be tricky. Mutual efforts by a pair of children to find out about the other's family often fail because both children fear they have something to hide, which is what happened when I used the word *pinky* to my classmate. But children have a clever way of getting around this problem: they play House. In the game of House, children can cooperatively develop an idea of what a "normal" family is like and at the same time limit their risks, because, after all, it's just a game.

Have you ever listened to children play House or similar games of pretense? The families they depict are straight out of Ozzie and Harriet. Talk about stereotypes! A developmentalist recorded this announcement, made by a little boy playing the part of Daddy: "Okay, I'm all through with work, honey. I brought home a thousand dollars." The girl playing Mommy was quite pleased. But a little boy who wanted to cook dinner was told firmly by his playmate that "daddies don't cook." Another child, a girl, was heard to insist that girls had to be nurses—only boys could be doctors—even though her own mother was a physician.

Aside from being sexist, the parents depicted in the game of House are curiously benign. They may argue with each other and scold their "baby," but they seldom go further than that. It isn't that children eschew depictions of violence—on the contrary, as researchers Iona and Peter Opie observed, "In these playlets children are stolen to be eaten [and] mutilation is accepted almost as commonplace." But in the games of pretend violence, the villains are witches or monsters or robbers and the children themselves often pretend to be orphans, thus explaining why good ol' Mom and Dad aren't around to protect them. If their real parents neglect them or abuse them, that's the last thing they want their friends to know.*

Kids want desperately to be normal, and part of being normal is having normal parents. If their parents are different in some way—and they're bound to be different in *some* way—they want to conceal these embarrassing differences from their peers. The humor writer Dave Barry has captured the feeling:

> After canteen we'd stand outside the school, surrounded by our peers, waiting for our parents to pick us up; when my dad pulled up, wearing his poodle hat and driving his Nash Metropolitan—a comically tiny vehicle resembling those cars outside supermarkets that go up and down when you put in a quarter, except the Metropolitan looked sillier and had a smaller motor—I was mortified. I might as well have been getting picked up by a flying saucer piloted by some bizarre, multitentacled, stalk-eyed, slobber-mouthed alien being that had somehow got hold of a Russian hat. I was horrified at what my peers might think of my dad; it never occurred to me that my peers didn't even notice my dad, because they were too busy being mortified by THEIR parents.

Parents belong in the home; when they come out of the home it makes their children nervous. Aside from the embarrassment, it makes it harder for children to know which context they're in, which rules they're supposed to follow. They are not aware of this, of course; context almost always affects behavior at a level that is not normally accessible to the conscious mind. It isn't until adolescence or adulthood that people occasionally become aware of the way their behavior changes in various social contexts. Perhaps there are people you don't like to be with because you don't like the way you act when you're with them.

The youth described by William James was "demure enough before his parents and teachers" but behaved differently when he was with his friends. He acted the way his parents and teachers had taught him to act, but only in social contexts that included his parents or teachers. It's difficult to teach your dog not to sleep on the sofa when you're not

* This changes later on. Teenagers love to complain to each other about how their parents treat them.

around, because what you are actually teaching him is to stay off the sofa when you *are* around. When you're not at home, he never gets whacked for jumping up on the sofa.

Seventy years ago a pair of ahead-of-their-time developmentalists tested children's ability to resist temptation. They gave the children opportunities to cheat or steal in a variety of settings: at home, in the classroom, in athletic contests; alone or in the presence of peers. They discovered that children who were honest in one context were not necessarily honest in others. The child who was honest at home might lie or cheat in the classroom or on the athletic field.

When children or adolescents misbehave outside their homes, they are sometimes referred to as "unsocialized" and their parents are blamed. According to the nurture assumption, it is the parents' job to socialize the child. But if children fail to transfer things their parents taught them to other social contexts, it is not their parents' fault.

Will the Real Personality Please Stand Up?

Babies are born with certain characteristics, certain tendencies to behave one way or another. They may have a greater than average tendency to be physically active, or to seek the company of others, or to get angry. These inborn tendencies are built upon and modified by the environment—that is, by each of the child's environments, separately.

Personality has two components: an inborn component and an environmental component. The inborn part goes with you wherever you go; it influences, to some extent, your behavior in every context. The environmental component is specific to the context in which you acquired it. It includes not only the way you learned to behave in those contexts, but also the feelings you associate with those contexts. If your parents made you feel worthless, those feelings of worthlessness are associated with the social contexts in which your parents did that to you. The feelings of worthlessness will be associated with outside-the-home contexts only if the people you encountered outside your home also made you feel like that.

Stability of personality across social contexts depends in part on how different or similar a person's various contexts have been. Cinderella's two social contexts were unusually divergent, so there was more than the usual amount of variation in her personality. But someone who met her after the prince carried her off to the castle wouldn't know that. They would see only her outside-the-cottage personality.

The psychologists who study adult personality typically assess it by giving their subjects a self-report personality test—a standardized list of self-descriptive statements, each of which the subject must agree or disagree with. In most cases the subjects are college students and the test is administered in a classroom or laboratory at the college. Thus, what the test is measuring is the subject's college personality, along with any thoughts or emotions associated with that particular classroom or laboratory. If the test is given again months later, to measure consistency across time, it is again given in a classroom or laboratory—usually the same one. The subject may be in a better or worse mood this time, but basically it's the same personality, with the same associated thoughts and emotions, so the results are reasonably consistent.

Personality psychologist James Council gave college students a self-report test that was designed to measure their ability to become absorbed in imaginative activities. Then he tried hypnotizing them. The subjects who scored highest on absorption were the easiest to hypnotize, but *only if he tried hypnotizing them in the same room where they took the absorption test.* When the test was given in one room and the hypnotizing was done in another, there was no significant correlation between the two. In a second experiment, Council asked subjects to fill out a questionnaire that asked them about traumatic childhood experiences such as physical or sexual abuse. Then, immediately afterward, they took a personality test designed to look for signs of emotional problems. There was a significant correlation between reports of childhood trauma and signs of emotional problems. But when Council tried the same thing on a different batch of subjects, this time giving them the personality test *first,* the correlation disappeared. Taking the trauma test evoked unpleasant thoughts and emotions and associated them with the test-taking setting. The effects of those unpleasant thoughts and emotions could be detected on a personality test if it was given after the trauma test and in the same setting. Council believes that these "context effects" call into question "the validity of a great deal of personality research."

Let's say you wanted to demonstrate that childhood trauma leads to emotional problems in adulthood. One way you could do it is with the method Council used: remind your subjects of their trauma and then, immediately after and in the same room, have them take the personality test. But an even better way would be to bring them back to the place where they experienced the trauma and have them take the personality test *there.* What you will be demonstrating, however, is not the power of childhood trauma to mess up people's minds. You will be demonstrating the power of context.

When behavioral geneticists study adult personality, they give their subjects personality tests in classrooms or laboratories. They find that the homes in which these subjects grew up had little or no effect on their adult personalities. If behavioral geneticists want to find effects of the home environment, they should take their subjects back to the homes in which they grew up and give them the test there. But what they will be demonstrating is not the power of the childhood home to influence adult personality. They will be demonstrating the power of context.

If you never go home again, the personality you acquired there may be lost forever. After Cinderella married the prince she never returned to her stepmother's cottage. Her self-effacing cottage persona was left behind forever, along with her broom and her raggedy clothes.

Most people do go home again. And the moment they walk in the door and hear their mother's voice from the kitchen—"Is that you, dear?" —the old personality they thought they had outgrown comes back to haunt them. In the world outside they are dignified, successful women and men, but put them back at the family dinner table and pretty soon they are bickering and whining again, just like they did in the good old days. No wonder so many people hate going home for holidays.

Made of the Myth

One of the reasons you didn't believe me when I told you that the nurture assumption is a myth is that there's so much evidence to support it. You can *see* that parents have effects on their children! And socialization researchers have collected *mountains* of data to prove it!

Yes, but where did you see it, and where did they collect it? You are right that parents have effects on their children, but what evidence do you have that these effects persist when the parents are not around? The child who acts obnoxious in the presence of her parents may be demure enough before her classmates and teachers.

Much of the evidence used by socialization researchers to support their belief in the nurture assumption consists of observations of the child's behavior in the presence of the parents, or questionnaires about the child's behavior filled out by the mother. Researchers want to demonstrate effects of the home environment—for example, after a divorce—so they observe the children in the home, a home where a lot of unpleasant things have happened recently. Worse yet, they ask the parents—not exactly what you'd call neutral observers, especially after the turmoil of a divorce—to fill out a questionnaire about the child's behavior. Predictably, these methods often show that the children of divorce are in significantly worse shape than those whose parents remained married. If the observations are made outside the home, away from the parents, the differences between the offspring of divorced and nondivorced parents get much smaller or go away entirely. (However, some of the differences do persist—they can be detected even in adulthood. I'll come back to the children of divorce in Chapter 13.)

Context effects are a serious problem in developmental psychology. They produce correlations that don't mean what the researchers think they mean or what they want them to mean. The correlations may turn up in the laboratory as well as in the home. Older children and adolescents are often interviewed or asked to fill out questionnaires in a school classroom or laboratory. This is a method the style-of-parenting researchers often use: they give the kids a self-report personality test or a questionnaire about what kind of trouble they've gotten into lately, and another questionnaire asking them how their parents treat them. Now we have not only a context effect (because the kid fills out both questionnaires in the same setting) but also what might be called a "person effect"—the same person who's telling you that he smoked four joints this week and flunked a math test is also telling you what jerks his parents are. One team of researchers checked up on their subjects. They gave teenagers a questionnaire asking them about their parents' child-rearing methods and also had their parents fill out the same questionnaire. The correlation between the parents' reports and the kids' reports was only .07—in other words, there was no agreement at all. And yet socialization researchers accept at face value kids' descriptions (and parents' descriptions) of what goes on in their homes and use data of this sort to support their theories.

Socialization research has demonstrated one thing clearly and irrefutably: a parent's behavior toward a child affects how the child behaves in the presence of the parent or in contexts that are associated with the parent. I have no problem with that—I agree with it. The parent's behavior also affects the way the child *feels* about the parent. When a parent favors one child over another, not only does it cause hard feelings between the children—it also causes the unfavored child to harbor hard feelings against the parent. These feelings can last a lifetime.

There are hundreds of books that give advice to parents—books that tell you what you're doing wrong and how to do a better job of raising your kids. Find a good one and it may help to explain why your children behave the way they do when they're at home. My goal is to explain what makes them behave the way they do in the world outside the home—the world where they will spend the rest of their lives.

5 OTHER TIMES, OTHER PLACES

In the mid-1950s, a pair of American researchers were studying the child-rearing practices of the inhabitants of Khalapur, a rural village in a remote part of northern India. One day they asked a Khalapur mother what kind of man she hoped her young son would be when he grew up. The woman shrugged. "It is in his fate," she said, "no matter what I want."

At that time, and for many hundreds of years before that time, the future of a baby born to a farming family in rural India was almost entirely determined by its health and its sex: if it survived, a boy would become a farmer, a girl a farmer's wife. In Khalapur, the researchers observed, babies were not the "objects of anxiety" that they are in the United States. They were not objects of anxiety because Khalapur parents did not feel they could make mistakes in rearing their children that would jeopardize the children's chances of future success.

People's beliefs about how much (or whether) parents influence their children's development, as well as their views about what children are like and how they should be treated, vary over time and place. The fatalistic attitude of the Khalapur mother, which sounds oddly passive to us today, was once common in the Western world. According to Danish sociologist Lars Dencik, the idea that childhood events play an important role in determining one's "fate" is a relatively new one:

> The significance of childhood for a person's "fate" in life has become something of an ideological dogma in the modern epoch. It was considered to be rather the reverse a few generations ago: people became what they were precisely because of their "fate." Adult life was predestined by inherited

and other irreversible factors. Childhood was not the phase of a person's life to which we paid all that much attention, nor did it prompt the nagging anxiety which we see all around us today. On the contrary, children were liable to be neglected, abused, and ill-treated, without anyone thinking there were questions to be raised here and without anyone having a guilty conscience about it whatsoever. The guilty conscience, which accuses us of not paying sufficient attention to the interests of the child, and which nowadays so plagues parents and other caregivers, is in fact a very new and rather unique feeling in our modern epoch.

We feel obligated to pay attention to the interests of the child for two reasons: because children are now seen as individuals with rights of their own, including the right to be well treated; and because of the "ideological dogma" that Dencik refers to, which says that people's adult lives are determined in large part by their childhood experiences. Those who hold to this dogma are also likely to believe that a certain class of experiences —namely, those involving parents—are of particular importance in determining the future course of a child's life. This belief is, of course, the nurture assumption.

The nurture assumption is linked to a specific model of family life and child-rearing that is common, though not universal, in Western societies today. This model calls for the child to be reared in a nuclear family consisting of one mother, one father, and one or more siblings. The parents are the "primary caregivers" and they are expected to shower their children with love and attention and to sprinkle them, as needed, with discipline. All this precipitation takes place in the privacy of the home— a home that may be visited by friends and relatives but that is inhabited solely by the members of the nuclear family, the only permissible exception being a grandparent or two. In short, as family historian Tamara Hareven put it, "The modern family is privatized, nuclear, domestic, and child-centered."

The Brief History of Privacy

The child in late-twentieth-century North America or Europe has two lives that seldom overlap: a home life and an outside-the-home life. Home life is private, the other is public, and different behaviors are required in each. Displays of emotion that are acceptable at home are frowned on outside the home. Elementary school children are not supposed to cry in public, or have tantrums, or express affection. What would be a minor mishap at home—throwing up on the floor, say, or wetting one's pants —becomes a major disaster in school. Wearing the proper clothing, combing one's hair in an acceptable fashion, comporting oneself in the approved manner—all are much more important outside the home than within it.

Within the home, family members are permitted—indeed, expected —to be less formal and freer in expressing their emotions. But people's home lives are carried out behind closed doors, and no one really knows what goes on behind the closed doors of other people's houses. Children don't know how their friends' parents and siblings behave when there are no visitors around. They may not even know the intimate details of the lives of their own siblings. Modern families are small and modern houses are large, and parents like to give every child a room of her own or his own. Privacy is regarded as one of our basic, inalienable, even constitutionally protected rights.

But privacy is a modern concept. The distinction between "private life" and "public life" is a recent one. Even *home* is a modern concept. Three or four hundred years ago, houses were very different from the ones we live in today. There was no separate place of business: the house was the workplace as well as the place where people ate, slept, talked, fought, and made love.

Three hundred years ago, a Norwegian couple named Frederik and Marthe Brun lived in a small town near Oslo. A description of their home by historian Witold Rybczynski gives us a glimpse of what family life was like at that time in Europe. Frederik Brun was a bookbinder; he was fairly prosperous and his house was relatively large for its time and place—about the size of a small modern bungalow. It served as his workplace and shop, and provided living quarters for fifteen people: Frederik, Marthe, their eight children, three male employees, and two female household servants. Other people—relatives, neighbors, customers—wandered in and out. Frederik and Marthe did not have a bed of their own: they shared a big four-poster with their three youngest children. The bed was situated in the main room of the house, a large room on the ground floor which was also used for eating meals and entertaining guests. The older children, two boys and three girls, slept in two beds in a smaller room upstairs.

The Bruns didn't miss their privacy because they had never had any. Being alone was not a normal situation for our ancestors. Today we put our babies in a crib and leave the room and wonder why some of them scream in protest. What we should be wondering is why any of them tolerate it. That most babies come to accept being left by themselves is a testimony to the adaptability of our species. Until quite recently in evolutionary time our ancestors made their living by hunting and gathering, and a hunter-gatherer baby was probably never left alone unless it was being abandoned. There were predators to worry about, and open fires, and who knows what they would pick up and put in their mouths,* so babies were carried about until they could walk well and had sense enough to avoid the most obvious dangers. At night they slept with their mothers.

Even today, babies in most parts of the world sleep in the same room, often in the same bed, with their mothers. Some researchers studying child-rearing practices in a Mayan community in Guatemala told the Mayan mothers that in the United States babies are commonly put to bed in a separate room. The mothers were appalled.

> One mother responded, "But there's someone else with them there, isn't there?" When told that they are sometimes alone in the room the mother gasped and went on to express pity for the U.S. babies. Another mother responded with shock and disbelief, asked whether the babies do not mind, and added with feeling that it would be very painful for her to have to do that. The responses of the Mayan parents gave the impression that they regarded the practice of having infants and toddlers sleep in separate rooms as tantamount to child neglect.

When a Mayan child is bumped from his mother's bed to make way for a younger sibling, he will sleep with his father or his grandmother or an older sibling. Mayans regard it as a hardship to sleep alone.

To people reared in traditional cultures, the way North Americans rear their kids is "unnatural." We justify our methods by saying we want our children to be independent, and indeed our children do appear to be pretty independent. But there is no proof that putting them to bed by themselves is what makes them independent. They are put to bed by themselves *because* we believe children should be independent. Child-rearing practices are the product of a culture, not necessarily the baton with which the culture is passed from one generation to the next (I'll return to this point in Chapter 9).

* Ethologist Irenäus Eibl-Eibesfeldt described an incident he witnessed while studying a hunter-gatherer society in Africa. A nineteen-month-old baby, left in the care of his sister, "stuffed feces in his mouth while the sister was not watching him closely." The sister got a scolding.

Telling People How to Raise Their Kids

We like our children to be independent, and yet we want them closely bound to us by emotional ties. Love between parent and child has become a sacred thing, exalted in innumerable movies and TV commercials that show children running into their parents' open arms, or parents looking with moist-eyed tenderness at their children (who are likely to be sleeping or, in commercials, eating). Mother love, father love—surely they're not cultural artifacts! Surely they are universal!

True, most parents do feel affection for their offspring. But the intensely sentimental attitude toward children that we see today in our society is relatively recent. During much of human history, in many parts of the world, childhood was a period of hardship and danger rather than a time of security and fun. Children were considered to be the possessions of their parents, and their parents (or stepparents) could do whatever they wanted with them. Babies and children could be ignored, mistreated, sold, or abandoned, and many were.

A lot depended on where and when they happened to be born. The history of childhood is not a steady ascent: it has had its ups and downs. For European children, probably the worst time was during the period from the Middle Ages through the eighteenth century. Juliet Schor, a professor of economics at Harvard, has described what parenting practices were like during that period.

> For the most part, children were not "cared for" by their parents. The rich had little to do with their offspring until they were grown. Infants were given to wet-nurses, despite widespread evidence of neglect and markedly lower chances of survival. . . . In all social classes, infants and children were routinely left unattended for long periods of time. To make them less of a nuisance, babies were wrapped in swaddling clothes, their limbs completely immobilized, for the first months of their lives.

Things got better for European and American children during the nineteenth century. When men began to work at jobs that took them away from the home for much of the day, the home became a private place—a haven from the world—instead of a place of business. The family came to be seen as a unit held together by mutual affection rather than by economic considerations. Around the same time, general health improved and more children were surviving to adulthood. These changes, which occurred earlier in wealthy homes than in poorer ones, brought with them a heightened interest in children. Children began to be valued more for themselves and less for what they could contribute in the way of free labor.

With men working outside the home, women were increasingly seen as having the role of attending to the family's needs. In particular, they were given full responsibility for their children's well-being. This, too, was a change: throughout most of European history it was men who had the say in this domain, as in most others. As late as 1794, according to German sociologist Yvonne Schütze, Prussian common law gave the father the right to determine how long his wife would nurse her infant.

Nor did men butt out even after child-rearing became the woman's area of expertise. There is a long list of dead white guys who took it upon themselves (while still alive) to tell people how to rear their kids. The list goes way back; it includes a seventeenth-century Puritan minister who informed his American congregation that all children possess "a stubbornness and stoutness of mind arising from natural pride," which must "be broken and beaten down." It includes the French philosopher Jean-Jacques Rousseau, who had a very different message for his eighteenth-century audience: that all children are born good and will remain that way if they are not meddled with too much. Rousseau, by the way, had no children of his own—that is, he *reared* none of his own. The babies born to his long-time mistress were deposited one by one, with his full knowledge, at the door of a foundling home. They may have been born good but they were not born lucky.

According to Yvonne Schütze, it was Rousseau who aroused Europeans' interest in the child as a subject of philosophical speculation. It was Rousseau who gave them the idea that a rational upbringing should be based on the child's essential nature, which could be determined through abstract thought. Philosophers and physicians, teachers and preachers, thereafter vied with each other to translate their abstract thoughts into concrete suggestions. For a while the advice remained fairly liberal, but by the time it became common to publish pamphlets and handbooks of advice addressed directly to mothers, the tide had turned again. The advice given in the latter part of the 1800s and the early part of the 1900s tended to be harsh. And women—particularly women of the educated classes—read these pamphlets and handbooks and followed the advice.

For example, physicians during that period often warned against overfeeding children, and mothers took those warnings to heart. Sir Anthony Glyn, reminiscing about life in England for his generation and the one preceding it, told of the spartan meals furnished to British children in the early 1900s. In the United States, a popular book around the turn of the century was Luther Emmett Holt's *On the Care and Feeding of Children,* which likewise recommended restricting the diet of children. The mother of a later advice-giver, Dr. Spock, was an adherent of Dr. Holt's views. As a child, Benjamin Spock was forbidden to eat, among other things, bananas. Benjamin was said to be "skeletally thin" when he left home for Andover at the age of sixteen.

Another notion promulgated by physicians was the fear that children's bodies would become crooked unless special devices or treatments were used to keep them straight. A German woman who grew up in the 1800s described how this "epidemic fear" of crookedness infected her own mother and the mothers of her friends.

> The fact that our posture was straight, and there was nothing noticeably wrong with us, did not reassure our mothers at all . . . . This one and that one of my girl friends were given fabulous machines to wear in their families, and at night were strapped into orthopedic beds . . . . It was finally ascertained that while I had an impeccable skeleton, my right shoulder was in fact stronger than my left one, and that every day I should hang for a while from a kind of horizontal bar, lie daily for an hour on the hard floor on my back, and every fortnight have four to six leeches placed on the suspect shoulder.

The most prevalent fear was that of "spoiling" one's children. Mothers were expected to love their children but not to let them know how much they were loved, because too much affection and attention were thought to be bad for them. At that time, Yvonne Schütze explains, mother love was supposed to express itself "in the mother's restraining herself, contrary to any need of her own to show tenderness—there is no assumption of a need for tenderness on the part of the child." German mothers were warned not to pick up the baby when it cried, lest they turn it into "the tyrant of the house."

The school of hard advice reached its apogee not in Germany but in America, in a book written by John B. Watson—yes, the same John Watson who proposed that he be given a dozen healthy infants. Since no one gave them to him, he took to telling other people how to rear their young.

> There is a sensible way of treating children. Treat them as though they were young adults. Dress them, bathe them, with care and circumspection. Let your behaviour always be objective and kindly firm. Never hug and kiss

> them, never let them sit in your lap. If you must, kiss them once on the forehead when they say good night. Shake hands with them in the morning. Give them a pat on the head if they have made an extraordinarily good job of a difficult task. Try it out. In a week's time you will find how easy it is to be perfectly objective with your child and at the same time kindly. You will be utterly ashamed of the mawkish, sentimental way you have been handling it.

According to Schütze, Watson's was "the first attempt to scientifically supervise the *psychological* relationship between mother and child." Previous advice had concentrated on children's physical well-being, or on teaching them manners, or on giving them religion. Now mothers were responsible, not only for protecting their children against crookedness, digestive upsets, uncouthness, and atheism, but also for protecting them against fearfulness, bossiness, underachievement, and unhappiness. And, as though this added responsibility weren't enough, at around the same time mothers became blamable not only for what they did and didn't do for their children, but also—thanks a lot, Dr. Freud—for their unconscious feelings and motivations. "The mother of the second half of the twentieth century," says Schütze, "can carry out her duties until she drops from exhaustion, and yet she is culpable if she does not have the feeling of personal enrichment, or if she has even unconscious negative feelings."

The mother of the second half of the twentieth century, unlike the mother of the first half, is expected to love her child wholeheartedly and to demonstrate it uninhibitedly. If she does not, or if her love is marred by the slightest shadow of "unconscious negative feelings," something is likely to go seriously wrong with the child. The corollary is that if anything does go seriously wrong with the child, it must be the mother's fault.

Current advice-givers, some of whom are female,* tell parents that their children require "unconditional love." Marianne Neifert, who calls herself "Dr. Mom," rotates John Watson's recommendations by 180 degrees:

* If there is a tendency for female advice-givers to give softer-hearted advice than male advice-givers, it is a slight one. The advice given in 1937 by Hildegarde Hetzer, a professor of psychology in Germany, was almost as stern as Watson's. She inveighed against "disorderly" mothers who "are excessively emotional toward their children, shower them with affection, coddle and spoil them, and attach too much importance to them."

Make a point of giving daily nonverbal messages of love and acceptance, through eye contact, touching, and hugging. All children need physical expressions of your love, no matter how old they get.

Obviously, Dr. Watson and Dr. Neifert can't both be right. Do children need physical affection or don't they? Can't we answer such questions by scientific means, as Watson alleged?

The problem is that the scientists doing the research are products of the same culture that gave rise to Dr. Neifert. No, I am not going to argue that science is "socially constructed" and that we can't see reality directly or test it without biases introduced by the worldview of our culture. I personally believe that reality is real and that science is an excellent way of figuring out how it works. But child-rearing is not physics. The research that gets done and the interpretations that get put on it are without a doubt the product of our culturally conditioned views of childhood and parenting—views that change, sometimes dramatically, sometimes in less than a generation. Because childhood and parenting are intrinsically emotional topics, it may be impossible to test theories about them with the same dispassion used to test theories about neutrinos and quarks.

Consider, for instance, the research on something called "mother-infant bonding." Starting in 1970, physicians Marshall Klaus and John Kennell published a series of articles and books on the effects of close physical contact between mothers and newborns in the first hour or two after birth. Their claim was that mothers who were allowed skin-to-skin contact during a short period immediately after childbirth became "bonded" to their babies—in other words, they fell madly in love with them. In contrast, mothers whose babies were quickly whisked off to the hospital nursery, and who therefore missed the emotional experience provided by immediate physical contact, were less likely to give their babies the unqualified love they require and more likely to neglect or abuse them.

The notion of bonding caught on like wildfire. It revolutionized hospital procedures. Authorities who, a generation earlier, would have attributed children's problems to "spoiling" were now attributing them to insufficient contact between mother and infant in the first hours after birth. The idea quickly spread to other countries. Yvonne Schütze tells about meeting a German mother who insisted that her problems with her daughter were due to the fact that she had not been permitted to bond with the child immediately after giving birth to her—nine years earlier. A British pediatrician warned:

> A normal baby should be delivered straight into his mother's arms. . . . The infant should lie nude and unwashed in contact with his mother's breasts. . . . The parents and the new baby should then be left alone for the first hour. . . . Animal studies of the effects of short periods of separation of mother and offspring have shown disastrous consequences—rejection and even killing of the baby.

The history of research on bonding has been reviewed in detail by psychologist Diane Eyer, and I will not attempt to duplicate her efforts. According to Eyer,

> By the early 1980s, research on the bonding of mothers and their newborns had been dismissed by much of the scientific community as having been poorly conceived and executed. Yet many pediatricians and social workers still see postpartum maternal bonding as a way of preventing child abuse. While the emphasis on bonding immediately after childbirth seems to have subsided, the concept has continued to flourish ideologically; women's proximity to their infants (whether they desire it or not) is still seen as a formula for preventing later problems of the child.

Eyer is overly optimistic when she says that the emphasis on bonding immediately after childbirth seems to have subsided. My younger daughter (yes, the one who gave her parents such a hard time) gave birth to her first child—my first grandchild—in March 1996. She refused anesthesia during the latter part of labor because she wanted herself and her baby to be fully alert in the period immediately after delivery—she didn't want anything to get in the way of their bonding.

The birth of my granddaughter really brought home to me how times have changed. When I was taking care of my own babies, back in the sixties, I felt guilty about picking them up when they cried; I had been taught in graduate school, by B. F. Skinner himself, that doing so would "reinforce" their crying and make them cry more. I no longer believe that, so I was all prepared to assure my daughter that it wouldn't spoil Jennifer to pick her up whenever she cried. But that advice turned out to be unnecessary. Instead, I found myself reassuring my daughter that it wouldn't do the baby any harm if she was occasionally allowed to cry for a few minutes.

"Natural" Childbirth

The bonding research caught on like wildfire because it came at the right time: a time when ideology called for making family life more "natural"—a time, ironically enough, when women were rebelling against being told what to do by white male scientists and physicians. Klaus and Kennell are, I believe, white male physicians. However, their ideas about bonding were "natural" in a way, because they were based on an animal model—specifically, on goats. If a mother goat is separated from her newborn kid for a short time immediately after delivery, she will reject it when she is reunited with it. If she is allowed to spend some time with the kid and *then* it is removed for an hour or two, she will accept it when it is returned. This observation led Klaus and Kennell to hypothesize that there is a hormonally based "sensitive period" immediately after birth.

The catch is that not all mammals behave like goats. Even closely related species may differ in regard to the presence or absence of a postpartum sensitive period. Some species of deer will accept an unfamiliar fawn; others will not. But I don't think the popular concept of bonding is based on goats. More likely it is based on an idealized notion of the "natural" mother in the "primitive" society—the noble savage, the hunter-gatherer who squats down and delivers her baby without any fuss in the forest or field, cuts the umbilical cord with her teeth, wipes the baby's face with a handful of leaves, puts it to her breast, and goes back to gathering roots and berries.

Don't believe it. Childbirth isn't like that. First of all, it is painful and difficult for women in all societies, and for women in preindustrial societies it is downright risky. In sub-Saharan Africa today, the chance that a woman will die as a consequence of pregnancy or childbirth is one in thirteen.

Second, because it is difficult and risky, it is rare for women to give birth alone. (The only exceptions are one or two societies in which experienced mothers sometimes go off by themselves to give birth and are admired for their staunchness; however, this does not happen with a first birth.) Traditionally, a woman in labor is assisted by one or, more commonly, several older women who give her encouragement during labor and catch the baby when it is born. Giving birth is not normally a solitary activity for human females and probably never has been. Nor is it common for the mother to be alone with her newborn after childbirth.

As for the practice of putting the baby immediately to the mother's breast, this is done in some traditional societies but not in all. Here is a description of childbirth among the Efe, a short-statured people (formerly called Pygmies) who dwell in the Ituri forest of the Republic of Congo (formerly called Zaire):

> The primary birth attendant squats in front of the [laboring] woman, ready to birth the infant. . . . Once born, the infant is placed on the banana and palm leaf mat. . . . The infant is then bathed in cold water to induce crying. . . . After the umbilical cord is cut [usually by the primary birth attendant], the infant is briefly brought outside for viewing by the men of the camp. When returned to the hut, the newborn is passed among the women who may suckle the infant whether or not they are lactating. Mothers do not hold their infants immediately after birth because of the belief that harm will come to the infant if first held by the mother. As a result, it is common for the newborn to spend several hours in the presence of female camp members before being given to the mother.

What is "unnatural" about our own birthing practices is not the treatment of the baby, which varies widely from one time and place to another, but the presence of the father at the delivery. Childbirth is traditionally an event attended only by women. But in our society the father is there, because of the belief that the father should witness—that he should *want* to witness—the "miracle of birth."

"Natural" Child-Rearing

For more than three hundred years, self-proclaimed experts in Europe and North America have been telling women how to rear their children. This advice has not fallen on deaf ears; indeed, it is clear that women—particularly educated women—have taken it very much to heart. When doctors warned against crookedness, mothers allowed their children to be strapped day and night into infernal devices. When doctors warned against overfeeding, children went hungry in the midst of plenty. The question arises: Would mothers have done these things in the absence of the warnings from eminent physicians? If there were no books or pamphlets to tell them how to rear their children, wouldn't they have reared them as nature intended them to?

But how did nature intend us to rear our children? Preliterate cultures have a wide variety of child-rearing practices, ranging from benign to, well, not so benign. Here, for example, is a description of how babies used to be fed among the Nyansongo people of Kenya:

> Traditionally, Nyansongo infants were fed [millet] gruel from birth, or a few days afterward, as a supplement to mother's milk. The gruel was administered by force feeding: cupping her hand against the infant's lower lip, the mother poured gruel into it and held his nose so that he would have to suck in the gruel in order to inhale.

Although such practices vary from one culture to another and from one generation to another within cultures—Nyansongo babies are no longer force-fed in this way—it is nonetheless possible to discern some commonalities. I will give you my impressions of childhood in traditional tribal and small village societies, based on my reading of the anthropological literature.

Childhood in a Traditional Society

Although childbirth is an exciting event everywhere, it is not always a welcome one. Sometimes the first decision that has to be made is not what to name the baby but whether to keep it. If the previous child has not yet been weaned, or if times are hard, or if the new baby appears to have something wrong with it, the mother may decide to abandon the newborn. Generally, such decisions are made right away, before anyone has a chance to get attached to the infant. And they are not made dispassionately but with sadness and regret.

Once the decision has been made to keep the baby, he* is likely to be very well taken care of. He is nursed whenever he whimpers, usually several times an hour, and is never left alone. By day his mother carries him around in a sling on her hip or back; by night he sleeps by her side. His father may also sleep with them, but this is not always the case. In some societies men have separate sleeping quarters, and in many they are permitted to have more than one wife. (Most men, however, can afford only one.)

When the baby is awake he is the center of attention. Little girls—his sisters and his cousins and his aunts—vie with each other to hold him. Adult men, especially his father, stop for a kitcheecoo. All over the world, everybody loves a baby. Well, everybody but the sibling whose place he usurped in their mother's arms.

His own place is not likely to be usurped for at least two years, because frequent breast-feeding and a low-calorie diet make it unlikely that his

* When talking about children with their mothers, to avoid confusion I use the masculine pronoun for the child.

mother will conceive again any sooner than that. Generally, children are breast-fed in these societies for two and a half or three years. After they have some teeth they are also given solid foods, prechewed by the mother if necessary.

Breast-feeding ends, usually abruptly, when the mother becomes aware that she is pregnant again. If the child doesn't like it—and he seldom does—he will be cajoled, ignored, laughed at, or beaten when he protests, depending on where and when he had the luck to be born.

Upon the arrival of the new baby, the older child, now around three, loses his place in his mother's arms once and for all and the new baby becomes the center of attention. In our society, children are carefully prepared in advance for this "dethronement," and parents, who feel guilty about it, pretend a greater interest in the older child than they may actually feel. We don't want the older one to resent the younger. In traditional societies, the older one seldom has such a gentle introduction to siblinghood. The dethronement is real and is likely to come with no warning: the child is presented with a fait accompli and must make the best of it. Naturally, he feels resentful toward the baby—he may even try to hit or scratch it. This demonstration of sibling rivalry is treated gently in some societies: the mother just pushes the older one's hand away. In others, the older one may be beaten if he even looks the wrong way at the baby, because it is believed that the child's murderous desires, whether acted upon or not, could harm the baby.

When the two-and-a-half- or three-year-old is deposed from his mother's arms, typically he is given over to the care of an older sibling. In many cases this is the one just ahead of him in age, the very child his own birth displaced—a child who may be no more than five or six herself. The older sibling schleps the younger one along with her when she goes out to play with the other children in the neighborhood. The children with whom she plays—the members of the neighborhood play group—are her siblings, half-siblings, cousins, and younger aunts and uncles. Houses in most traditional societies are arranged in clusters, and within a cluster everyone is related.

Although he can walk now, the little sibling who is brought along to the children's play group is still, to all intents and purposes, a baby. During his tenure in his mother's arms he had an active social life and attentive care for his physical needs, but he was taught virtually nothing. Parents in traditional societies do not believe that babies have sense or that they understand what is said to them; therefore, they generally do not talk to them. Nor do they attempt to teach them to talk. Consequently, the child acquires very little language before the age of two and a half or three—much less than a North American child of that age. Developmentalist James Youniss has pointed out how odd it is by middle-class American standards that the parents in many societies seem to lose interest in their children just when they begin to acquire language.

The two-and-a-half-year-old is at first unable to take an active part in the play. Depending on what game the older ones are playing, he might be allowed to participate as a sort of living doll or be left to watch or whine on the sidelines. He becomes a full participant at three or three and a half. According to the German ethologist Irenäus Eibl-Eibesfeldt:

> Three-year-old children are able to join in a play group, and it is in such play groups that children are truly raised. The older ones explain the rules of play and will admonish those who do not adhere to them, such as by taking something away from another or otherwise being aggressive. . . . Initially the older children behave very tolerantly toward the younger ones, although eventually they place definite limitations on behavior. By playing together in the children's group the members learn what aggravates others and which rules they must obey. This occurs in most cultures in which people live in small communities.

Boys, in particular, spend most of their time with their peers and a minimum of time at home. In a small rural village on the island of Okinawa, a mother complained to researchers that her five-year-old son came home during the day only to shovel rice down his throat and rush out again, because his friends were waiting for him. In African villages where older boys are given the responsibility of watching the cattle, the younger boys tag along and a boring job becomes an opportunity for play, out of sight of adults.

I am talking here about societies that use agriculture or animal husbandry to provide a more or less stable supply of food and that therefore have a greater population density than hunter-gatherers. In such societies there are always enough children to form a play group, and usually enough to enable the play group to split into two—a boys' group and a girls' group—or into three—older boys, older girls, and a mixed-age, mixed-sex group composed of young children and the even younger children they are babysitting. The splitting into sex- and age-segregated groups occurs spontaneously, wherever there are enough children in one place to make it possible.

Girls tend to play closer to home than boys and are more likely to have younger children to care for, because mothers in most societies—probably all societies—prefer girls as babysitters. But boys are pressed into service if no girls are available, and they take the job very seriously. In one of Jane Goodall's books about chimpanzees there is a photo of an African man with a badly mutilated face, the result of an injury he suffered when he was a child. He had been taking care of his baby brother when a big male chimpanzee came out of the forest and seized the baby.* The boy was only six, but he chased after the formidable animal. The chimpanzee dropped the baby and attacked the boy. The baby survived.

Along with the responsibility for the younger sibling's welfare comes the right to dominate. Older siblings are given complete authority to control and discipline younger ones, and there is no point in a child's tattling to his parents about how big brother or sister is treating him, because unless he can show impressive wounds, his complaints will be ignored. In traditional societies it is considered natural for older children to dominate younger ones—this happens automatically, around the world, wherever adults do not intervene. Adults don't intervene in these societies unless things really seem to be getting out of hand, and that is surprisingly rare. Sometimes the older child will tease the younger one, or be overly punitive toward him, but in general siblings get along remarkably well. Children share food with their younger brothers or sisters without being told and defend them from teasing by others.

Parents in our society try so hard to get their children to love each other and what they get is constant squabbling. Parents in traditional societies make no effort to get their children to love each other and it happens as a matter of course. There are two reasons, I believe, for this difference.

First, in traditional societies the children don't have as much to fight over. The custom of giving all the attention to the baby is hard on the child who has just been deposed from his mother's arms, but it means that all the children in the family—all but the baby—are in the same boat. They do not compete for their parents' attention because it doesn't work. Nor do they compete for toys, because there aren't any toys to speak of. Children in these societies play with things like sticks and pebbles and leaves, and there are plenty of those to go around. American children do a great deal of squabbling over objects that don't exist in traditional societies.

Second, American parents don't realize, or don't accept, that it is natural for older children to dominate younger ones. Because parents think

* Wild chimpanzees hunt and kill baby monkeys and, on rare occasions, baby humans.

their children should be equal, they try to keep the older one from dominating the younger one, and as a consequence the older one ends up resenting the younger one. Only by putting their might on the younger one's side can parents prevent domination by the elder, and this makes it look, to the elder child, like the parents are favoring the younger one. In fact, as I mentioned in Chapter 3, parents usually do favor the younger one, but for some reason they expect the older one not to notice it.

In developed societies, sibling rivalry is considered an inevitable part of family life. But the kind of sibling rivalry we are used to seeing—the kind that goes on at least until the kids leave for college and sometimes much longer than that—is not universal. In traditional societies sibling rivalries tend to be short-lived; they are over as soon as both siblings are past infancy and have stopped competing for their mother's attention. Relationships between brothers tend to be close and enduring. Your brother is your closest ally. He is the one who will stand by your side in defense of your village.

Discipline and Training

Parents in traditional societies don't worry about what the experts say and they don't worry about the long-term effects of their child-rearing practices. Never having read any B. F. Skinner, they use punishment, rather than positive reinforcement, to make their children behave. Parents give little or no praise in these societies. When a child does something wrong, they hit her (physical punishment is widespread in all societies, including our own), or make fun of her, or scare her with threats of ghosts, evil strangers, or wild animals. Often, no explanation is given for the punishment, and what is punished is the outcome of the child's behavior—a broken bowl, for instance—rather than her good or bad intentions.

Children in our society have to listen to a lot of long-winded explanations of how to do something or what's wrong with the way they're doing it. Verbal explanations and verbal feedback are much less common in preliterate societies. Among the Zinacantecos of Mexico, girls learn weaving by watching older women weave. North Americans do not take well to this method of instruction. A college student from the United States described her experiences with a Zinacanteco "teacher":

> When I began taking back-strap loom weaving from Tonik, an older Zinacanteco woman, I became increasingly restless, when after two months of what I termed observation and she termed learning, I had not touched the

loom. Many times she would verbally call my attention to an obscure technical point, or when she would finish a certain step she would say, "You have seen me do it. Now you have learned." I wanted to shout back, "No, I haven't! Because I have not tried it myself." However, it was she who decided when I was ready to touch the loom, and my initial clumsiness brought about comments such as, "Cabeza de pollo!" (chicken head) "You have not watched me! You have not learned!"

Most of what children need to know in order to live in a preliterate society is learned by imitation. They watch their parents or older siblings performing a chore and attempt to imitate them. If they do it wrong, they are laughed at when they're little, reprimanded or punished when they're older. When they get it right, they are rewarded by having the chore assigned to them.

Child-Rearing With and Without Guilt

Child-rearing is easier when it's done without guilt and without having to think about what long-term effects your actions might have on your child's fragile little psyche. It's easier on the *parent,* that is. From the child's point of view, it's six of one, half a dozen of the other. People in preliterate societies do awful things to their children and so do people in literate societies. In both cases parents think they are rearing their children as nature intended them to; in both cases they are actually rearing their children according to the rules set down by the culture or subculture to which they belong. In our culture one of the rules is: Listen to the experts.

One of my most painful memories of motherhood concerns something that happened when my older daughter was three. It was her first day of nursery school. She was a quiet, somewhat timid child with no experience of being away from home unaccompanied by a parent. I brought her into the nursery school classroom and, after a while, she got interested in what the other children were doing and wandered away from me. Almost at once, a teacher came over to me and asked me to leave. "She'll be all right," the teacher said. So I left, and they closed the door behind me. Then I heard my child throw herself against the door, hammering on it and screaming. I heard the teacher talking to her but the hammering and screaming continued. I wanted to go back in but the teacher had told me not to, so I didn't. I stood there listening to my child's outraged screams, suffering as much as she was.

My daughter did okay in nursery school, but I never forgot how I listened to the teacher—a woman only a little older than myself—instead of yielding to my very strong desire to go back in, pick up my child, hold her until she stopped crying, and remain there until she was ready to have me leave. I listened to the teacher because she was an authority and she made me feel that she knew more than I did about what was best for my child.

In our society, we listen to the experts. Today these experts tell us that children need lots of attention and lots of love. When our kids do something wrong we're supposed to reason with them, not whack them. We're supposed to warn them against dangers such as drugs and sex, and, just in case our warnings slip their mind, we're supposed to keep careful track of where they are and what they're doing. If they go wrong in spite of all our efforts, we must have failed to carry out one or more of these instructions, or carried them out in an insufficiently conscientious way.

Parents in North America and Europe—particularly educated, financially well-off parents—read the experts' advice and do their best to follow it. These same parents also participate, and allow their children to participate, in the research designed to prove that the advice is correct. And this whole precarious, circular structure rests on a set of assumptions about children and parents that are peculiar to our culture and our time. A set of assumptions written in sand.

6 HUMAN NATURE

The word *nature,* when contrasted with *nurture,* has two distinguishable meanings. The first is used when the question is, Why do people vary? If, for example, one child has a larger vocabulary and is generally more verbal than other children her age, we can ask whether her superior verbal ability is due to "nature" or to "nurture": Did she inherit it from her father the deviser of crossword puzzles and her mother the professor of English, or is it a consequence of growing up in a verbally stimulating environment?

The second meaning concerns the *similarities* among us: Why are people all the same? For example, all human children who are born with a normal brain—and many who are not—learn to communicate in a language. We can ask whether this propensity to acquire a language is due to "nature" or to "nurture": Is it a built-in characteristic of our species, or is it the result of experiences that normal children invariably have while they are growing up?

Nowadays "nature and nurture" are usually used to account for the differences among us. But in the early days of developmental psychology, attention more often focused on the similarities. Back in the 1930s, developmentalists generally did not make fine distinctions between one child's environment and another's, and then use those distinctions to account for the ways that the first child differed from the second. They were interested in studying the universals of human development, such as language acquisition. If young humans acquire a language and young apes do not (this was long before anyone thought of teaching an ape sign language), is it because language is part of human nature and not part of ape nature? Or is it because humans grow up in a human environment and apes grow up in an ape environment?

What the early developmentalists really wanted to know was whether children would acquire the abilities we consider characteristically human if they were not reared in a human environment. But even in those days, when researchers could get away with experiments that today would get them fired before their lips had time to form the word "tenure," it wasn't easy to get a dozen healthy infants to experiment with.* So Winthrop Kellogg, a professor of psychology at Indiana University, thought up a more modest experiment: he proposed to rear an ape in a human environment. With the cooperation of his wife Luella, he would rear a child and a chimpanzee together, treating them both like human children, to see whether a chimpanzee brought up under such conditions would develop human capabilities.

The experiment and its results are described in a book published in 1933, *The Ape and the Child.* Luella's name is listed, right after her husband's, on the title page. But it was Winthrop who was the professor of psychology, and it was his career the experiment advanced. I wonder how he convinced Luella to go along with it. I wonder if she knew what she was getting into. Did she realize that Gua, the chimpanzee, would not be the only subject in the experiment, and that the other would be her infant son Donald?

Donald of the Apes

Donald was ten months old, and Gua seven and a half months, when she came to live with the Kelloggs in 1931. Right from the start she was treated like a human baby—that is, the way human babies were treated in the 1930s. The Kelloggs put clothes on her, and the stiff shoes that babies wore in those days. She wasn't caged or tied up, which meant that she had to be watched every second except when she was asleep (but then, the same was true of Donald). She was potty trained. Her teeth were

* It wasn't possible to *keep* a dozen healthy infants, but it was often possible to *borrow* them for experimental purposes. In the late 1930s, developmental psychologist Myrtle McGraw managed to borrow a total of forty-two babies for the purpose of determining whether humans have an innate ability to swim. Her method was straightforward: drop a baby into a tub of water and let go. She found that newborns have a reflex that prevents them from breathing water into their lungs, but they shortly lose this ability. The older babies she tested would struggle desperately to keep their heads above water, fail to do so, and come up spluttering and coughing.

brushed. She was fed the same foods as Donald and had the same naptimes and bathtimes. There is a photograph in the Kelloggs' book of Gua and Donald sitting side by side, dressed identically in footed pajamas of the kind my mother used to call "Dr. Dentons." Donald is frowning; Gua's lips are curved upward in a modest smile. They are holding hands.

Aside from the difference in temperament recorded in that revealing photo, the two were remarkably well matched. Chimpanzees develop more rapidly in infancy than humans, but Donald was two and a half months older and that helped to even things up. They played together like siblings, chasing each other around the furniture, roughhousing and giggling. Donald had a walker, a big heavy thing, and one of his favorite sports, according to his parents, was "to rush at the ape in this rumbling Juggernaut and laugh as she scurried to keep from being run over, often without success." But Gua didn't hold grudges and she enjoyed rough-and-tumble play. In fact, the two got along better than most siblings. If one of them cried, the other would offer pats or hugs of consolation. If Gua got up from her nap before Donald, she "could hardly be kept from the door of his room."

Gua was more fun than a barrel full of Donalds. When the Kelloggs tickled her or swung her around, she would laugh just like a human baby. If they tried swinging Donald, he would cry. Gua was more affectionate (expressing her affection with hugs and kisses) and more cooperative. While being dressed, the ape—but not the boy—would push her arms into open sleeves and bend her head to allow her bib to be tied on. If she did something wrong and was scolded for it, she would utter plaintive "oo-oo" cries and throw herself into the scolder's arms, offering a "kiss of reconciliation" and uttering an audible sigh of relief when she was permitted to bestow it.

In mastering the challenges of civilized life, Gua often caught on a little faster than the stolid Donald. She was ahead in obeying spoken commands, learning to feed herself with a spoon, and giving a warning signal when she needed to use the potty (unfortunately, though, her potty training never became completely reliable). The ape equalled or exceeded the child in most of the tests that Dr. Kellogg devised: she was as adept as Donald at figuring out how to use a hoe-shaped implement to pull a piece of apple toward her, and learned more quickly to use a chair to reach a cookie suspended from the ceiling. When the chair was moved to a new starting point, so that it had to be pushed in a different direction to reach the cookie, Donald continued to push it in the same direction as before, whereas Gua kept her eye on the cookie and claimed the prize.

There was one thing, however, in which the boy was clearly superior: Donald was the better imitator. Does that surprise you? According to Frans de Waal, a Dutch primatologist who spent several years observing the chimpanzees and their human visitors at a Netherlands zoo, "Contrary to general belief, humans imitate apes more than the reverse."

This was clearly the case with Donald and Gua. "It was Gua, in fact, who was almost always the aggressor or leader in finding new toys to play with and new methods of play; while the human was inclined to take up the role of the imitator or follower." Thus, Donald picked up Gua's annoying habit of biting the wall. He also picked up a fair amount of chimpanzee language—the food bark, for instance. How did Luella Kellogg feel, I wonder, when her fourteen-month-old son ran to her with an orange in his hands, grunting "uhuh, uhuh, uhuh"?

The average American child can produce more than fifty words at nineteen months and is starting to put them together to form phrases. At nineteen months, Donald could speak only three English words.* At this point the experiment was terminated and Gua went back to the zoo.

The Kelloggs had tried to train an ape to be a human. Instead, it seemed that Gua was training their son to be an ape. Their experiment tells us more about human nature than about the nature of the chimpanzee, but it also tells us that there is remarkably little difference between them—at least in the first nineteen months. In this chapter I will look at some of the differences between chimpanzee nature and human nature that appear after the age of nineteen months, and at some of the similarities that remain.

I said at the beginning of the book that my answer to what makes children turn out the way they do—the theory I will offer you as a replacement for the nurture assumption—is based on a consideration of what kind of mind the child is equipped with, which requires, in turn, a consideration of the evolutionary history of our species. This is where we consider that history. We will be going, for business and for pleasure, on a trip through evolutionary time. Along the way I will say some highly speculative things—much more speculative than anything I say in the other chapters of this book. But heck, if other writers can speculate on

* If you are thinking, as I did when I read the Kelloggs's book, that maybe Donald simply had the bad luck to be born on the wrong slope of the bell curve, forget it. According to Ludy T. Benjamin, a historian of psychology, Donald went on to graduate from Harvard Medical School.

the evolutionary history of our species, why shouldn't I? Be assured: the speculations are not what my theory rests upon.

Mindreaders

Would Donald have learned to speak English if Gua hadn't gone back to the zoo? Yes, of course he would have. In Chapter 4, I described children whose parents are recent immigrants to the United States or are profoundly deaf. These children do not speak English at home: they acquire it outside their home. The same thing would have happened to Donald. If he didn't learn English in order to communicate with his parents, he would have learned it in order to communicate with the other kids in the neighborhood. When his social world broadened to include playmates other than Gua, he would have discovered that in the world outside his home, no one speaks Chimpanzee.

But language is only one of the things that distinguish humans from apes. There are other differences, equally important and interesting, that are just beginning to develop at nineteen months. For the past few years, psychologists who study the cognitive capabilities of human children have been all excited about something they call the "theory of mind."

According to these researchers, children have a theory of mind by the time they are four years old. That is, they know they have a mind and they believe other people do too. Their own mind is furnished with thoughts and beliefs and they assume that other people, too, have thoughts and beliefs. They also know that thoughts and beliefs are not necessarily true—that it is possible to hold mistaken beliefs. They understand, in fact, that it is within their power to give incorrect information to someone else and thereby cause that person to have a mistaken belief. This understanding is what enables them, for the first time, to tell intentional lies.

The sophistication of the theory of mind continues to develop as children grow up. We adults understand that people's behavior is determined by their thoughts and feelings about things, rather than by the things themselves, and that to predict what someone will do you have to know what they're thinking and feeling. Some of us are experts at figuring out what other people are thinking and feeling, but even amateurs are remarkably good at it, because people usually make no attempt to conceal the contents of their minds from others. In fact, they talk about their thoughts and feelings all the time. One of the things that language does is to give us a direct phone line into other people's brains, making it a great deal easier to figure out what they are thinking. On the other hand, if someone does wish to mislead us, language makes it a great deal easier for them to do so.

The theory of mind doesn't start with phone lines, though. It starts with windows—those windows of the soul, the eyes. Our ability to read minds begins to develop in early infancy, when we first look into our parents' eyes. Babies begin to make eye contact with their parents when they are about six weeks old. A normal baby can tell very early—so early that this ability must be innate—when someone is looking at him. He shows this by smiling when his mother looks at him and by turning away his face if she continues to look at him too long. Prolonged eye contact makes babies uncomfortable.

By the end of the first year, the baby can also tell where someone is looking when they're *not* looking at him. Watching his mother's face when she's looking at an unfamiliar object helps the baby decide whether to approach the object or avoid it. If she looks worried he will probably avoid it. Watching his mother's face while she's talking to an unfamiliar person helps the baby to decide whether the stranger is a friend or a foe. If the stranger looks too intently at the baby before he has had a chance to make up his mind, the baby will probably turn away his face. If the stranger tries to pick him up at this point, the baby is likely to resist and cry out in fear.

By the middle of the second year, the toddler is glancing at his mother to see what she's looking at when she tells him a word; he assumes that the word applies to the object she is looking at. When he points at something, he checks to see if his mother looks at it. Pointing in order to draw another person's attention to something is characteristically human. Chimpanzees that were raised in an ape environment do not do it, and even among those that were raised in a human environment it is rare. According to Herbert Terrace, a psychologist who investigated the ability of young chimpanzees to communicate in sign language,

> Noticeably absent from an infant ape's reaction to an object is the sheer delight a human infant expresses in contemplating the object and sharing it perceptually with the parent. . . . There is no evidence that suggests that the infant ape seeks to communicate, either to another ape or to its human surrogate parent, the fact that it has simply noticed an object.

Three- or four-year-old human children can use the direction of a person's gaze plus the expression on her face as indicators of what's going on inside her mind. If, for example, the person is looking hungrily at a candy bar, a four-year-old will deduce that she is contemplating eating it. If she has a vacant look on her face and is gazing upward at nothing in particular, a four-year-old will say that she is thinking. We take these mindreading abilities so much for granted that it was only recently that developmentalists noticed them. Still more recently, developmentalists noticed that certain children don't have them. Children with autism don't seem to realize that the eyes are the windows to the soul—in fact, they don't seem to realize that other people *have* souls. In short, autistic children lack a theory of mind. British developmentalist Simon Baron-Cohen calls this deficit "mindblindness." It is what makes autistic people social cripples.

Annette Karmiloff-Smith, another British developmentalist with a hyphenated name, contrasts autism with a rare mental disorder called *Williams syndrome.* Children born with Williams syndrome have a characteristic set of facial features and intellectual deficits. Their upturned noses and fat cheeks give them an appealing elfin look. But their brains are 20 percent smaller than those of normal children of the same age, and their IQs are in the retarded range. These children can't tie their shoes, can't draw pictures, can't do the simplest arithmetic. On the other hand, Karmiloff-Smith and her colleagues reported, they are remarkably verbal and very friendly, and they get along well with others. Although they are retarded, Williams syndrome children do not lack a theory of mind. They are sensitive to the emotions of others and can judge someone's intentions by looking at their face and eyes. Unlike autistic children, Williams syndrome children can tell when a person is joking or being sarcastic.

Williams syndrome children have it and autistic children do not: Karmiloff-Smith calls it a "social module," a department of the brain that specializes in dealing with social stimuli and social behavior. The reason people with autism have so much trouble with language (even if they learn to speak they are poor communicators) is because they don't understand that its purpose is to put thoughts into other people's minds and to get thoughts out of other people's minds.

Life in an Ape Environment

Chimpanzees are not like autistic children—they are more like Williams syndrome children. Gua was very sensitive to the facial expressions of her human surrogate parents and to the direction of their gaze—she would check to see if they were looking at her before doing anything naughty and stop what she was doing if they frowned. Any animal that was adapted by evolution to live with others of its kind needs some sort of social module. Chimpanzees have a social life that is almost as complex as our own.

Watch chimpanzees in their natural habitat, as the admirable Jane Goodall did, and you will see—at least this will be your first impression—a touchy-feely, hail-fellow-well-met bunch of folks. Kids play hilariously with each other, grownups groom each other and schmooze. Small groups come and go, forming and re-forming with frequent exchanges of membership. Two individuals who haven't seen each other for a while will greet each other with hugs and kisses. When nervous, chimpanzees hold hands or give each other reassuring pats. If one manages to kill a bushbuck fawn or a baby baboon, the others flock around the successful hunter with hands outstretched, and each one has a good chance of getting a piece of the spoils.

True, there are struggles for dominance, but they are seldom fatal and usually end with the loser begging forgiveness from the victor and the victor graciously granting it. Even sex generates surprisingly little animosity. Females say yes to just about anyone who asks. Although sometimes a high-ranking male may attempt to restrict access to a particular female, he doesn't always succeed: more often than not, all he can hope for is to be first in line for her favors. Goodall described what happened in the chimpanzee community she was watching when a popular female named Flo went into heat: the males took turns with hardly more pushing and shoving than among commuters on a New York subway platform.

Under the circumstances, no one knows who's anyone's father. Male chimpanzees play no role in rearing their offspring but they generally have a benevolent, albeit detached, attitude toward the young members of their community. Mothers, on the other hand, have very close relationships with their offspring and these relationships may last a lifetime. Female chimpanzees, like female (and male) humans, vary in how maternal they are, but most are indulgent mothers. Sibling relationships also tend to be close and lasting, and if a young chimpanzee loses its mother it may be adopted by an older sibling—even, in some cases, by a male.

There is a limit, however, to the chimpanzees' hail-fellow-well-metness: it extends only to members of their own community. A chimpanzee community is a population of the animals, usually numbering around thirty to fifty, that inhabits a particular territory. Even though the entire community never congregates in one place at one time, they all know each other (many are close relatives) and a stranger is instantly recognized as such.

Chimpanzees don't cotton to strangers. An unaffiliated animal or one from another community that has the bad luck to blunder into their territory is likely to be attacked, unless it is a female in heat. A female carrying a baby and not in heat is almost certain to be attacked, and her baby will be killed and possibly eaten.

Chimpanzees don't cotton to strangeness, either. A polio epidemic struck the chimpanzee community that Goodall was observing and an old male named McGregor was partially paralyzed by the disease. When he returned to the group (after a sojourn alone in the forest) dragging his legs behind him, his former buddies were not glad to see him. At first they were afraid of him. Then fear turned into hostility and one of the healthy males attacked him, pounding on the crippled animal's back while he cowered helplessly. When another male ran toward McGregor flailing a large branch, Goodall couldn't stand it any longer and intervened. Although the other chimpanzees eventually got used to McGregor's strange behavior, they never again accepted him as a member in good standing and he was not welcome at that important chimpanzee social function, the grooming party.

Socially, chimpanzees are a lot like us: they have our faults as well as our virtues. Like humans, they divide the world into "us" and "them." Even a familiar animal may be attacked if it is no longer one of "us" and has become one of "them." The most violent assaults that Goodall witnessed were "perpetrated on individuals who were not completely strange to the aggressors." The victims were members of a new group, the Kahama community, which had split apart from the larger one, the Kasakela community, after many years of close association. For a while the members of the two communities continued to interact from time to time on a friendly basis, but eventually that ceased and they began to avoid each other and, if they met by chance (they occupied adjacent and somewhat overlapping territories), to put on a display of belligerence.

About a year after the members of the two groups stopped being friendly, the first of a series of attacks was launched by the Kasakela community against the upstart Kahamans. It began when a party of eight Kasakela chimpanzees headed southward toward Kahaman territory, moving quickly and silently through the trees. (Chimpanzees are ordinarily very noisy.)

> Suddenly they came upon Godi [a Kahaman], who was feeding in a tree. He leaped down and fled. Humphrey, Jomeo, and Figan [all Kasakelans] were close on his heels, running three abreast; the others followed. Humphrey grabbed Godi's leg, pulled him to the ground, then sat on his head and held his legs with both hands, pinning him to the ground. Humphrey remained in this position while the other males attacked, so that Godi had no chance to escape or defend himself.

After hurling a large rock at the badly wounded chimpanzee, the Kasakelans went home. Godi was never seen again and presumably died of his injuries.

In the same fashion, giving the same impression of malice aforethought, the Kasakela chimpanzees picked off the other Kahamans one by one. Juveniles and adult females were not spared. Only young nubile females were allowed to live, and they were recruited into the Kasakela community. I am reminded of the story of Joshua in the Old Testament. When he and his troops overran the city of Jericho, they killed every man, woman, and child, sparing only Rahab the harlot.

Love and War

"There is no such thing as an instinct to make war," said Ashley Montagu in 1976. The word *war* was in disrepute at the time—people were being exhorted to make love instead, as though the two were incompatible—but the word that Montagu really hated was *instinct.* Now, after a long period of being out of fashion, that word is making a comeback. The psycholinguist Steven Pinker even used it in the title of his excellent book, *The Language Instinct.* Perhaps it is possible to again consider* the hypothesis that humans have an instinct for making war and that we inherited it from our primate ancestors.

Jane Goodall considers that hypothesis very seriously and, though she doesn't put it in exactly those words—she uses "preadaptation" instead of "instinct"—she clearly considers it tenable. She points out that chimpanzees have all the "preadaptations" necessary to permit the emergence of war, including group living, territoriality, hunting skills, and an aversion to strangers. Moreover, she says, male chimpanzees are strongly attracted to scenes of intergroup violence—they appear to be "inherently disposed to find aggression attractive, particularly aggression directed

* Steven Pinker says, in *The Language Instinct,* that it's okay to split infinitives.

against neighbors." Goodall believes that such traits might form a biological basis underlying the more sophisticated forms of warfare practiced by our own species. As Jericho is to Hiroshima, Kahama is to Jericho.

Some theorists get hung up on the apparent contradiction between humans as killer apes and humans as party animals. Charles Darwin, for one, was not bothered by it:

> Every one will admit that man is a social being. We see this in his dislike of solitude, and in his wish for society beyond that of his own family. Solitary confinement is one of the severest punishments which can be inflicted. . . . It is no argument against savage man being a social animal, that the tribes inhabiting adjacent districts are almost always at war with each other; for the social instincts never extend to all the individuals of the same species.

No, never to *all* the individuals of one's species—only to the members of one's own troop, tribe, community, nation, or ethnic group. The commandment "Thou shalt not kill," fresh from Mt. Sinai, did not hinder Joshua in his wholesale slaughter of the inhabitants of Jericho, Ai, Makkedah, Libnah, Lachish, and Eglon. The idea that God might prohibit him from killing *them* never crossed his mind.

History records many such wars, from Jericho and Troy to Bosnia and Rwanda, and archeological evidence proves that waging war and slaughtering our enemies are things we knew how to do long before we learned how to leave written records of our victories. War between groups, says evolutionary biologist Jared Diamond, "has been part of our human and prehuman heritage for millions of years."

Primatologist Richard Wrangham agrees. He believes our species is descended from a primate ancestor that looked and behaved a lot like the modern chimpanzee (which is descended from the same ancestor). From this common ancestor, the chimp and the human inherited their similar lifestyles. Both species live (or used to live) in communities defended by coalitions of males that were born there; the females traditionally transfer to a different community when they reach reproductive age. And in both species the coalitions of males not only defend their territory but also launch offensive attacks on neighboring communities. The pattern of attacking one's neighbors may have begun as a drive for more territory or more females, but once it got going it became self-perpetuating and the original motive became unimportant. Once it got going, there was a new and better motive for killing one's neighbors: let's kill them before they can kill us.

Six million years of evolution divides us from that chimpanzee-like ancestor, and all during that six million years—all but the last little bit of it—we lived in much the same way. We lived in smallish communities composed of our close relatives (in the case of males) or our mate's relatives (in the case of females). We depended on the other members of our group for protection; we weren't designed to live alone. When meat was available—and our appetite for meat soon overtook our appetite for veggies—it was probably shared among the members of the group. And all during those six million years we fought with our neighbors. Successful communities increased in size, split in two, and sooner or later the two halves would go to war against each other. Sometimes one succeeded in wiping out the other. "Of all our human hallmarks," says Jared Diamond, "the one that has been derived most straightforwardly from animal precursors is genocide."

But we are not only killer apes: we are nice guys, too. Darwin pointed out that "A savage will risk his own life to save that of a member of the same community." If the savage risks his life and loses it, he has suddenly become, in Darwinian terms, unfit; therefore an explanation of his behavior is called for. The explanation is that the man who gives up his life to save his group may thereby be preserving the lives of his brothers, sisters, and children—people with whom he shares 50 percent of his genes. If we define fitness in terms of the successfulness of genes in propagating themselves, rather than the successfulness of individuals in living to a ripe old age, altruism toward one's close relatives makes sense.

You may have heard this referred to as the "selfish gene" theory, and perhaps it gave you the idea that the products of evolution are bound to be selfish. Occasionally it has had that unfortunate effect even on its promulgators. "Be warned," declared the biologist Richard Dawkins, "that if you wish, as I do, to build a society in which individuals cooperate generously and unselfishly towards a common good, you can expect little help from biological nature. Let us try to *teach* generosity and altruism, because we are born selfish." But selfish genes do not imply selfish organisms: a gene can be perfectly selfish and yet contain the instructions for building a perfect altruist, if that's what it took to succeed under the conditions the gene evolved in.

Clearly, we are not perfect altruists, any more than we are perfect killer apes. In fact we are a little of each, and that is why writers like Ashley Montagu can see us as flower children while writers like Richard Wrangham see us as born to kill. It all depends on whether you look at our behavior toward the members of our own group or our behavior toward the members of other groups. We are born to be nice to our groupmates because for millions of years our lives and the lives of our children depended on them. And we are born to be hostile toward the members of other groups, because six million years of history taught us to beware of them.

In the thick of battle, our groupmates were our allies, our comrades in arms. Between battles we competed with them for food and for access to desirable mates. But in good times and bad we cooperated with them—call it altruism if you will—because cooperation had long-term survival value. I will help you today if you will help me tomorrow. Such a system tends to give rise to cheaters—those who take and do not give in return. But minds are good for other things than making tools and weapons. Over the millennia, we learned how to watch out for cheaters. Eventually we also learned how to warn our friends against them. Meanwhile the cheaters were getting smarter too. While we were evolving ways to detect cheaters, the cheaters were evolving ways to outwit our cheater-detectors. That led, in turn, to the evolution of ways to detect cheater-detector outwitters. A "cognitive arms race," some have called it.

But cheaters were a minor threat: a graver danger lay on the other side of the hill, where the enemy was assembling its forces. In the words of Jane Goodall,

> The early practice of warfare would have put considerable selective pressure on the development of intelligence and of increasingly sophisticated cooperation among group members. This process would escalate, for the greater the intelligence, cooperation, and courage of one group, the greater the demands placed on its enemies.

When the smoke cleared over Jericho, the cheaters were as dead as the sharers. The cowards were as dead as the fighters. Evolution gives the prize to the winners of such wars. Much as we might deplore their tactics, they are the ones who became our ancestors.

Hominid Evolution

Our ancestors parted company from those of the modern chimpanzee somewhere around six million years ago. That is not very long in evolutionary time; we share 98.4 percent of our DNA with the common chimpanzee, *Pan troglodytes.* The DNA difference between human and chimp is smaller than that between two closely related species of birds, red-eyed vireos and white-eyed vireos.

But it doesn't take many genes to produce a new species; a few changes in the recipe at critical points can produce markedly different results. Our hairlessness, for example, probably resulted from changes in just a few genes and may have occurred over a relatively short period of evolutionary time. Humans have as many hair follicles as apes but most of them produce only vestigial hairs. There is a mutation that causes some of the members of a family in Mexico to grow fur all over their faces, even on their eyelids. It is evidently caused by a single gene.

Walking upright is another human characteristic that may have evolved fairly quickly. *Australopithecus afarensis*—Lucy and her kind—had brains hardly larger than a chimpanzee's, and yet they walked fully erect. That was three and a half million years ago, in Africa.

It was with *Homo habilis,* two and a half million years ago, that things began to get interesting. *Homo habilis* had a much bigger brain than any primate that preceded it. This species was named for its ability to make and use tools, but (as we now know) its members were not the first tool users. The chimpanzee uses rocks as weapons and for cracking nuts, and uses sticks prepared for the purpose to fish for insects in termite mounds.

The next entry was *Homo erectus,* about one and a half million years ago. Some books show *erectus* as a descendant of *habilis,* but the situation is a good deal more complicated than that, because many hominid and pre-hominid species came and went in Africa over the past six million years. It is not easy to figure out, on the basis of a few bones, which species were descended from which other species and which were dead ends. Most, as it turned out, were dead ends.

Homo erectus was not a dead end: it was a highly successful hominid which spread out of Africa into the Middle East, Europe, and Asia. It survived, both north and south of the Sahara, for more than a million years. Eventually it was replaced in Africa by an archaic form of *Homo sapiens,* and then, between 100,000 and 150,000 years ago, by the modern form of *Homo sapiens,* sometimes called *Homo sapiens sapiens.* My guess for when this change took place is 130,000 years ago, during a brief warming spell—the last interglacial period before the one we're enjoying now.

Not long after acquiring the cachet of that extra *sapiens,* the ancestors of modern Europeans and Asians left Africa and moved north into the Middle East. When they got there they found it already occupied by a hominid: the Neanderthals, descended from the northern branch of *Homo erectus* and now spread over much of Europe as well as the Middle East. By this time another Ice Age had started, so we remained in the comparative warmth of the Middle East for a long time, sharing it—not, I am sure, amicably—with the Neanderthals. Then something mysterious happened: Jared Diamond calls it "the great leap forward"; anthropologist Marvin Harris calls it "cultural takeoff." Whatever caused it, its results were soon apparent: with the aid of a greatly expanded technology our species spread out over Europe and Asia, and right about that time the Neanderthals ceased to exist. They had been there for 75,000 years, all through the Ice Age, and then suddenly, just as the weather began to ease up, they disappeared. Hmm.

That left us the victors, the sole hominid to make the cut. Our only surviving close relatives are the gorilla, the chimpanzee, and the bonobo (or "pygmy chimpanzee"), all restricted to small ranges in remote parts of Africa, and the orangutan, found only on the islands of Borneo and Sumatra. All the others are gone. Over a relatively short period of time—around six million years—we went from being apes to being humans, and behind us we left a trail of dust and ashes. We took no prisoners.

Let me tell you how I think it happened. It began when a community of apes got too large and split in two. The two daughter communities (as biologists call them) were now occupying neighboring territories and sooner or later hostility broke out between them. In fact, hostility may have preceded the break and led to its occurrence.

When human groups split up, the chances are good that the daughter groups will become enemies, if they aren't already. As an anthropologist observed, "A village's mortal enemy is the group from which it has recently split." There may be occasional truces for the purposes of trade or matchmaking, but the smallest misunderstanding will set them off and they'll be at each other's throats again. Groups don't need a reason to hate other groups: just the fact that they're *them* and we're *us* is usually enough. And just in case it isn't, there's always territory to fight about. Joshua wiped out all those cities because, he said, God had promised his people the land. But it wasn't merely a land-clearing expedition: there was hatred, too. The king of each conquered city was captured and hanged from a tree, after (at least in some cases) being roughed up.

Joshua was comparatively recent, though—only about 3,500 years ago, well after humans in that part of the world had developed agriculture. For most of the six million years of evolution that separated our line from the chimpanzee's, we lived a catch-as-catch-can existence as hunters and gatherers. Hunter-gatherer societies are reputed to be peaceful and nomadic, with no territory to fight over and no desire to fight. But according to ethologist Irenäus Eibl-Eibesfeldt, that is just another flower-power myth. He reports that the large majority of surviving hunter-gatherer groups are neither peaceful nor lacking in territoriality. It is true that a few groups have given up war (perhaps because they no longer have any territory worth fighting over), but of ninety-nine hunter-gatherer groups that have been studied, "Not a single group was claimed to have never known war."

We hate what we fear because we don't like being afraid. As Eibl-Eibesfeldt points out, human babies in all societies start becoming afraid of strangers when they're about six months old. By then, in a typical hunter-gatherer or small village society, they have usually had a chance to meet all the members of their community, so a stranger is valid cause for concern. What is he here for? Does he want to steal me? Make me a slave? Maybe even eat me? The baby watches its mother for clues; if she seems to think the stranger is okay, the baby is reassured. Eibl-Eibesfeldt calls the baby's reaction to strangers "childhood xenophobia" and considers it the first sign of a built-in predisposition to see the world in terms of *us* versus *them.*

Many people believe that children have to be taught how to hate. Eibl-Eibesfeldt doesn't think so and neither do I. Hating the members of other groups is part of human (and chimpanzee) nature—the most repugnant part. What children have to be taught is how *not* to hate. We are not born selfish, as Dawkins thought, but we are born xenophobic.

Speciation and Pseudospeciation

Evolution, according to biologist Stephen Jay Gould, doesn't proceed in a slow, gradual accumulation of small changes. Species are stable, sometimes for millions of years, and then they disappear and are replaced, rather abruptly (by the standards of evolutionary time), by another species. What leads to speciation is when some little subpopulation of a species splits off and stops interbreeding with the parent species, usually because of geographical isolation. Then this little group evolves different characteristics from the parent species, and if the changes happen to make it more successful than the parent species, it will eventually win the survival-of-the-fittest award and replace them.

It isn't always necessary for the smaller group to be geographically isolated from the larger one, because there are other things that can keep groups from interbreeding. There are two species of grasshopper that coexist in Europe, look alike, and are capable of interbreeding under laboratory conditions. They are considered different species because in the wild they *don't* interbreed. The reason they don't interbreed is that they have different songs. This minor behavioral difference keeps them apart.

When a group of apes or humans splits up, it generally splits along lines of prior association, with individuals tending to go to the side that contains more of their relatives and friends. But inevitably there will be some who have relatives and friends on both sides and could go either way. When Jane Goodall's chimpanzee community split in two, she wondered what made an old male named Goliath throw in his lot with the Kahamans—a decision that cost him his life.

I don't know what Goliath's reasons were, but when human groups split up, individuals tend to choose the side with which they are most compatible: like seeks like. In the case of groups composed of families, such as human communities, most individuals have no choice about which way to go, but those who do will go to the side with which they have the most in common. The result will be, in many cases, a statistical difference between the daughter groups. There might be some minor behavioral difference between the members of the two groups, or some minor difference in appearance. Then again, there might not be.

In humans, hostility between groups leads to the exaggeration of any preexisting differences between the groups, or to the creation of differences if there were none to begin with. You may have thought it was the other way round—that differences lead to hostility—but I believe it is more a case of hostility leading to differences. Each group is motivated to distinguish itself from the other because if you don't like someone you want to be as different from them as possible. So the two groups will develop different customs and different standards of male and female beauty. They will adopt different forms of dress and ornamentation, the better to tell friend from foe in a hurry. They may even develop different languages. Eibl-Eibesfeldt observed,

> Humans show a strong inclination to form such subgroups which eventually distinguish themselves from the others by dialect and other subgroup characteristics and go on to form new cultures. . . . To live in groups which demarcate themselves from others is a basic feature of human nature.

This process is called *pseudospeciation.* If pseudospeciation was also a basic feature of *pre*-human nature, it could have led to a dramatic speeding up of evolution. Groups split, demarcate themselves from each other, and go to war. War puts an end to interbreeding (or drastically cuts down on it) and now we have the preconditions for true speciation. If one of the daughter groups happens to be more successful at waging war, it may wipe out the other. Of course, it may simply outcompete the other group, but that's a lot slower.

New Guinea provides a model for how it could have happened. When European explorers first made their way into the interior of New Guinea, they discovered it was a veritable Tower of Babel. Nearly one thousand different languages, most of them mutually unintelligible, were spoken in an area about the size of Texas. Jared Diamond describes what the island was like before the white man came:

> To venture out of one's territory to meet [other] humans, even if they lived only a few miles away, was equivalent to suicide. . . . Such isolation bred great genetic diversity. Each valley in New Guinea has not only its own language and culture, but also its own genetic abnormalities and local diseases.

Thus, one New Guinea tribe has the world's highest incidence of leprosy, others have high frequencies of deaf mutes or male hermaphrodites or premature aging or delayed puberty. Genetic differences between the tribes, probably due to mutations in one or two genes, underlie these differences. They are small differences, but the groups haven't been separated very long.

Over time, separated groups become more and more dissimilar. In some animals the differences accumulate slowly and at random—genetic drift, biologists call it—but in the genus *Homo* the process may not be random at all and may be speeded up by pseudospeciation. The visible differences between European populations—for instance, the blondness of the Scandinavians versus the darkness of the Italians—evolved so rapidly they are unlikely to be due solely to the health benefits of being blond or dark. Most likely they were helped along by sexual preferences: the first fair-haired people in a population may have occurred by chance, but if they were sought after as mates their descendants would have proliferated. Eventually such traits could serve as markers to distinguish *us* from *them.*

I believe that is how our hairlessness evolved. I think it was a late and relatively rapid evolutionary change: it didn't occur until after the northern branch of *Homo erectus* (the one that gave rise to Neanderthal) stopped interbreeding with the southern branch (our ancestors). Perhaps it didn't occur until around the time when we acquired that extra *sapiens,* a mere 130,000 years ago. The change may have begun with pseudospeciation—a division between a group of less hairy hominids, who got progressively balder as body hair became increasingly unpopular among them, and a group that remained as hairy as the other apes. Hairlessness conveyed no benefits—it simply served to distinguish *us* from *them.* Once we were clear on that distinction, the next step would have been to go to war against the hairy ones and wipe them out.

The Mysterious Disappearance of the Neanderthals

You may be thinking that I was alluding to the disappearance of the Neanderthals, but I was not. I was talking about things that happened (or might have happened) entirely in Africa and that led to the appearance of anatomically modern humans and the disappearance of other closely related groups. What happened in Europe, when *Homo sapiens sapiens* arrived there, was something else again. The two species—modern humans and Neanderthals—had evolved separately, under very different conditions. Neanderthals were adapted to cold weather, humans to warmth. What they had in common was a big brain and a love of meat. But they differed in at least two important respects. Neanderthals probably didn't have our verbal skills (they seem not to have had the right kind of mouth and throat), and they were covered in a heavy coat of fur.

Yes, you heard me right: a heavy coat of fur. Evolutionary biologists and paleontologists like to play this game where they mentally dress up Neanderthal Man in a three-piece suit, set him loose on a London or Manhattan street, and then wait to see if anyone notices. The problem is that they forget to shave him, so of course everyone will notice—he'll be shot with a tranquilizer dart and hauled off to the zoo! The evolutionary biologists and paleontologists, like everyone else, have been too impressed by those artistic pictures that show all of our hominid ancestors lined up in a row, gradually becoming less hairy.

There was no way that Neanderthals could have survived in Ice Age Europe without a heavy coat of fur: they couldn't sew. No three-piece suits, no fur-lined parkas. It has been suggested that they used animal skins to protect themselves against the cold, but did you ever try going out in a snowstorm to hunt game with nothing but a deer skin slung around your shoulders? And they *had* to go out to hunt game nearly every day, because there is no evidence that they stocked up for the future and there was not much in the way of fruits and vegetables in Ice Age Europe. Our own species was certainly no dumber than the Neanderthals, but we didn't manage to make a go of it in Europe until we invented the needle.

We had forgotten our antipathy for hairy hominids by the time we got to the Middle East and spotted Neanderthals. We didn't think they were repulsive-looking people: we thought they were animals. Prey. We didn't think "Yuck," we thought "Yum." And they, no doubt, thought the same of us. Neanderthals are gone, along with most of the other large and tasty mammals that inhabited Europe and the New World before we got there, because we were better predators than they were.

This Is the Brain That Evolution Built

Six million years have gone by since our ancestors parted company with the ancestors of the chimpanzee. We spent most of that time on the ground, not in the trees. We spent it getting along with the members of our own group and fighting with the members of other groups. We spent it honing our ability to detect cheaters and honing our ability to outwit cheater-detectors.

We lived, for most of that time, in small groups of hunters and gatherers. When a group was successful it got bigger, split in two, and then the more successful of the daughter groups outfought or outcompeted the less successful one. That happened over and over again.

What those six million years of evolution bought us was a giant brain —a mixed blessing. It is a prodigious user of energy, makes childbirth risky, and pins down our infants for the better part of a year like a ball and chain. Its fragility and size make it an inviting target whenever push comes to shove.

But consider its advantages. Jane Goodall's chimpanzees had to pick off the members of the neighboring community one by one, but Joshua could slaughter the inhabitants of entire cities in one fell swoop. That was not easy, since most of the cities were walled. The trick with the trumpets worked only once, at Jericho. Joshua had to breach the walls of the other cities without the aid of heavenly intervention. At Ai, he used guile. He sent a small force to attack the city and a much larger force to wait in ambush in a hidden spot. The small force attacked and then retreated, and the people of Ai chased after it, believing that they had defeated their enemies and now had only to administer the coup de grace. They left the city open and unguarded behind them and ran straight into Joshua's ambush.

Guile is one of the things we're good at, and this brings us back to the theory of mind. Joshua was able to guess what the citizens of Ai would do because he could imagine their thought processes. He knew they could be deceived and he was able to think up a complex plan for deceiving them. Another crucial asset was his ability to communicate his plan to his generals.

Of course, the fact that he also commanded a very large army didn't hurt his cause. But that, too, was a kind of cognitive achievement. For the members of a chimpanzee community, *us* includes only individuals that are recognized. An unfamiliar individual is automatically considered one of *them.* By Joshua's time, human groups had gotten so large that not everyone in them knew everyone else—the group had become a concept, an idea. When Joshua met a stranger outside the walls of Jericho, he had to *ask* him, "Art thou for us, or for our adversaries?"—are you one of *us* or one of *them?* The ability to form groups larger than one's adversaries is a cognitive advance with obvious payoffs. One wonders what the outcome would have been if Jericho, Ai, Makkedah, Libnah, Lachish, and Eglon had been able to join together against Joshua. But there was a reason why those cities had walls around them: it was to guard the citizens of each of them against the others.

Although chimpanzees cannot make the cognitive leap involved in considering a stranger one of *us,* many of our other abilities exist, in embryonic form, in that species. Even guile. Jane Goodall witnessed a number of occasions in which chimpanzees used deception to get something they wanted. There was, for instance, the incident of Figan and the banana. During Goodall's first few years in Tanzania, she used to put out boxes of bananas to attract chimpanzees. Usually the high-ranking males would eat most of them. To enable the females and younger males to get their share, she would hide some bananas in the trees. One day a young chimpanzee named Figan spotted a banana hanging in a tree directly above a high-ranking male. If Figan had reached for it, the big male would have taken it away from him. Instead, Figan moved to a spot where he couldn't see the banana and waited. As soon as the big male moved away, Figan retrieved the banana. By sitting in a spot where he couldn't see the object of his desire, he made sure he wouldn't give away the secret with his eyes.

Chimpanzees are not like autistic children; they are aware of the importance of the eyes. After a fight between groupmates, according to primatologist Frans de Waal, the two animals must make eye contact before they can kiss and make up. "It is as if chimpanzees do not trust the other's intentions without a look in the eye."

Do chimpanzees have a theory of mind? That question is not easy to answer, because a theory of mind is not an all-or-none thing. Human children develop it over time, during their first few years of life. The question of whether, or how much, it also develops in chimpanzees is currently a matter of debate. But it is safe to say that chimpanzees are not the equals, in the theory-of-mind department, of human four-year-olds. Whether they are, in this respect, more like human three-year-olds or human two-year-olds is not as important as the fact that there are real differences between the species. These differences are built in, due to nature. Even a chimpanzee reared in a human environment will never be as good a mindreader as a four-year-old human child.

In the six million years of evolution that separates us from the chimpanzee, we didn't get a social module—we already had that when we started out. What we got in those six million years was new and better ways of using our social module. Almost everything we gained was a result of our adaptation to a group lifestyle. Take language, for example. What good is language unless you have someone to talk to? The ability to communicate is so valuable for animals that live in social groups that even bees have evolved a method of conveying information to each other. Perhaps the outcome would have been different for the Kahamans if Godi had managed to drag himself back to his groupmates crying, "The Kasakelans are coming! The Kasakelans are coming!" The message wouldn't have saved Godi but it might have saved his group.

The human brain is a tool, first and foremost, for dealing with the social environment. Dealing with the physical environment is a secondary benefit. Evolutionary psychologist Linnda Caporael points out that we have a default mode for dealing with ambiguous or troublesome things: we try to interact with them socially. We personalize them. We don't treat humans like machines—we treat machines like humans. We say "Start, damn you!" to our cars. We expect our computers to be friendly. And when faced with phenomena we don't understand or can't control, we attribute them to entities called God and Nature, to which we impute human social motives such as vengefulness, jealousy, and compassion.

Parents, Children, and Evolution

One of the purposes that has been attributed to language is the transmission of culture—presumably, according to the nurture assumption, from parent to child. However, as we saw in the previous chapter, in most cultures parents don't teach their children in words. Language is not necessary for the successful rearing of children. The children of deaf couples sometimes do not learn sign language and thus cannot communicate with their parents except in the most rudimentary way, but they turn out just fine. Mammals have managed to rear their children for millions of years without the aid of language.

The nurture assumption implies that children are born with empty brains which their parents are responsible for filling up. Obviously, children do learn things from their parents. But they do not learn *only* from their parents. Although much of what human children need to know is learned after they are born, there are good evolutionary reasons why it wouldn't make sense to allow parents to monopolize that learning. I can think of four reasons why it would not be in an offspring's best long-term interests to allow itself to be overly influenced by its parents.

First, as behavioral geneticist David Rowe has pointed out, a predisposition to learn only from its parents would prevent the offspring from picking up useful innovations introduced by other members of its community. Since young animals, not older ones, are more likely to come up with useful innovations (I will return to this point in Chapter 9), it is to an offspring's advantage to learn from its peers as well as from its elders. What it learns from its peers is also likely to be more timely, better suited to the current conditions.

The second reason has to do with variety. The easiest way to produce young that are exactly like their parents is to clone them, and some species of plants and animals do avail themselves of that method. Cloning is highly efficient. Noah could have filled up the Ark in half the time if he had specialized in species that beget by cloning: he would have needed only one of each kind. But every clone is exactly like its siblings, so anything that killed one of them—a lethal microorganism, for example—could kill them all. Sexual reproduction originated because it introduced variety in the offspring (each combination of egg and sperm produces a unique assortment of genes) and thus enabled larger organisms to keep a step ahead of the smaller ones that plague them. However, variety in the offspring has other advantages as well. In changing times, it increases the chances that one of the offspring will be suited to the new conditions and will survive. In difficult times, it increases the number of ecological niches the members of the family can inhabit. And in good times and bad, variety within a family can provide a wider range of skills and a broader base of knowledge that will be useful to the family as a whole.

Like the other animals that Noah invited onto the Ark, humans inherit many of their behavioral characteristics from their parents. If parents had the power to influence their children by environmental means as well as genetically, the children would be too similar to the parents and too similar to each other. They would be too much like little clones.

The third reason why it wouldn't make evolutionary sense to design children to be programmed by their parents is that children can't count on having parents. We worry about all the children being reared today in single-parent homes and compare it to the halcyon days of fifty years ago, when parents came in Arkable pairs. But having two parents, one of each sex, was not something that children in ancestral times could take for granted. Anthropologist Napoleon Chagnon reports that among the Yanomamö—Amazonian Indians who inhabit the rainforests of Brazil and Venezuela—the likelihood that a child of ten will still be living with both biological parents is only one in three. Although the divorce rate is comparatively low among the Yanomamö—Chagnon estimates that 20 percent of their marriages break up—the death rate is high. In a tribal society, a child's chance of surviving goes down if he or she loses either parent, but it doesn't go to zero. If children required parents in order to learn what they have to learn, losing a parent would have been a death warrant under ancestral conditions.

The final reason has to do with the competing interests of parents and children. As evolutionary biologist Robert Trivers has pointed out, what's best for the parents is not necessarily best for the kids. Take, for example, weaning. A mother may want to wean her child in order to get ready for the next baby, but the child wants to be nursed as long as possible and to hell with the next baby. Trivers uses this conflict of interests to account for the fact that human children often begin to act babyish again after a younger sibling is born. Young apes and monkeys have been observed to do the same thing. Since parental care tends to be given preferentially to the youngest and most vulnerable offspring, the offspring that acts babyish may persuade its parent to give it more than its share. The offspring that can put on the most convincing show of neediness gets fed first.

In other ways, too, the parents' interests may not coincide with the child's. Perhaps the parents would like their daughter to remain with them and take care of them in their old age, or act as nursemaid to her brother's children, or marry the rich old man who would pay them a good bride-price, whereas she has other ideas. Trivers concludes that the offspring's best policy is to watch out for its own interests while trying to remain on good terms with its parents:

> The offspring cannot rely on its parents for disinterested guidance. One expects the offspring to be preprogrammed to resist some parental manipulation while being open to other forms. When the parent imposes an arbitrary system of reinforcement (punishment and reward) in order to manipulate the offspring into acting against its own best interests, [natural] selection will favor offspring that resist such schedules of reinforcement. They may comply initially, but at the same time search for alternative ways of expressing their self-interest.

In many cases, as historian of science Frank Sulloway points out, the conflict between parent and offspring boils down to a conflict between siblings: each child wants more than his share of family resources, while the parent wants to distribute resources where they'll do the most good. Thus, according to Sulloway, brothers are natural rivals, locked in a Darwinian struggle for survival. His model for fraternal relationships is the blue-footed booby, a species in which the largest chick in the nest reduces the competition for parental attention by pecking to death the smallest one.

But we have come a long way from boobyhood. A more informative model is provided by our close relative the chimpanzee. According to Jane Goodall, two male chimpanzees born to the same mother five or six years apart (the usual birth interval in this species) will be playmates in childhood and allies in adulthood. When the younger one is still small, his elder brother will be gentle and protective of him; play gets rougher as the two get older. Eventually there may come a time when the younger one will challenge his elder brother's dominance, but once that issue is resolved, one way or the other, their relationship is likely to become friendly again. Such friendships are of great importance to male chimpanzees because brothers generally support each other in dominance conflicts with other males. "I'll sic my big brother on you" is not an idle threat among primates.

When the Kelloggs set out to rear a chimpanzee in a "civilized environment," they knew they were putting Gua into an environment that evolution hadn't designed her for. It probably never occurred to them, but Donald hadn't been designed for it either. Donald and Gua were both designed for the forests and fields of Africa, not for a house in Indiana with wallpapered walls and indoor plumbing. We are mistaken if we think we are getting a glimpse of primitive human nature when we see our children fighting over the remote control.

Our ancestors spent the past six million years—all but the last little bit of it—as hunter-gatherers, living in small nomadic groups. They survived by triumphing over a hazardous environment, and the greatest hazard in that environment was the enemy group. The lives of hunter-gatherer children depended more on their group's survival than on their parents', because even if their parents died they had a chance of surviving if their group did. Their best hope of success was to become a valuable group member as quickly and convincingly as possible. Once they were past the age of weaning they belonged, not just to their parents, but to the group. Their future prospects depended, not on making their parents love them, but on getting along with the other members of the group—in particular, the members of their own generation, the people with whom they would spend the rest of their lives.

The child's mind—the mind of the modern child—is a product of those six million years of evolutionary history. In the next chapter you will see how it reveals itself in the child's social behavior.

7 US AND THEM

Lord of the Flies, the 1954 novel that won William Golding a Nobel Prize in Literature, is about a couple of dozen British schoolboys who are stranded on a tropical island and left to their own devices. The climate is balmy; food is plentiful; there are no grownups and no homework. And yet they do not have a jolly old time. By the time their hair has grown long enough to tie back in a ponytail, the boys are killing each other.

In view of the bloody picture of human and prehuman history I painted in the last chapter, you might think I approve of Golding's rendition of Life Without Civilization. But I don't. Golding got it all wrong.

In fact, he made a number of errors, not all of them in psychology. He has the boys using eyeglasses to focus sunlight in order to start a fire, but the eyeglasses belong to a boy called Piggy, and Piggy is nearsighted. Only magnifying lenses, used to correct farsightedness, can be used to start a fire. He has the younger boys—the "littluns," he calls 'em—playing with each other all day long and ignoring the older boys, but little boys are fascinated by boys a few years older than themselves and will seek them out even if treated roughly by them. He has Piggy still talking in a lower-class accent—he's the only one with this handicap—after many months on the island. In that amount of time, a real-life boy would have learned to talk like his companions.

But the most important thing Golding got wrong is the way the boys start killing each other. Not the *fact* that they start killing each other, but the way it happens. There are two leaders, Ralph and Jack. Ralph represents, in Golding's heavy-handed symbolism, law and order; Jack represents savagery and chaos. One by one, the boys are won over to Jack's side, except for Ralph, Piggy, and a strange boy named Simon. Simon gets killed, Piggy gets killed, and the mob is hot on the heels of Ralph when a grownup arrives, just in the nick of time.

I am not the first to raise objections to this plot. Ashley Montagu, whose antiwar, anti-instinct views were touched upon in the previous chapter, complained more than twenty years ago that *Lord of the Flies* was unrealistic. He cited an actual case in which six or seven Melanesian children were stranded on an island for several months and got along quite well, thank you. In Montagu's version of the novel, when the grownup appears at the end and renders judgment, it wouldn't be: "I should have thought that a pack of British boys—you're all British, aren't you?—would have been able to put up a better show than that." It would be: "Good show, chaps!"

But Montagu got it wrong, too. The case of the Melanesian children isn't a fair comparison: they had known each other all their lives—they were believed to be part of a single extended family—and there were only six or seven of them. As near as I can figure, there were a couple of dozen schoolboys on Golding's island, and many of them were previously unacquainted.

If you found yourself on an island with some people you'd known for a long time and some others who were strangers, you'd probably gravitate toward the familiar ones. But in Golding's novel, the boys who already knew each other—they were members of a school choir, ruled over (before they got to the island) by Jack—immediately dispersed and some of them became followers of Ralph.

That's not how it would have happened. Jack's choir would have stuck with him and the others would have joined up with Ralph, or the boys from expensive boarding schools would have separated themselves from the ones who attended the local grammar school, and they would have ended up with two groups—the sine qua non of warfare. The boys might have come to blows and even to bloodshed, but it wouldn't have been group against individual: it would have been group against group.

Golding, like the British philosopher Thomas Hobbes, believed that Life Without Civilization would be a dog-eat-dog world: every man for himself and the devil take the hindmost. Montagu, like the French philosopher Jean-Jacques Rousseau, believed that it would resemble a well-run hippie commune: everyone shares the work and the food and there's plenty of time to smell the flowers. I believe all four of them were wrong.

The one who got it right was Darwin. "The tribes inhabiting adjacent districts are almost always at war with each other," he observed, and yet "a savage will risk his own life to save that of a member of the same community." "The social instincts never extend to *all* the individuals of the same species." Whether you see humans as murderous or merciful, selfish or altruistic, depends on whether you are looking at their behavior toward their groupmates or their behavior toward the members of other groups.

The Robbers Cave Experiment

What would *really* happen if you dropped a couple of dozen schoolboys into a wilderness and left them to their own devices? In 1954—the same year *Lord of the Flies* was published—a group of researchers from the University of Oklahoma decided to find out. The experiment was carefully planned in advance: it was to be a study of group relations.

The subjects—twenty-two of them, to be precise—were purposely selected to be as alike as possible. They were all eleven-year-old white Protestant males. Their IQs were all in the average-to-above-average range and so were their school grades. None of them wore glasses. None were fat. None had gotten into any trouble. None were new to the area, so they all spoke with the same Oklahoma accent. And each one came from a different Oklahoma City school, so none of them knew each other before the experiment began.

This homogeneous bunch of twenty-two boys was divided into two smaller groups of eleven. Each group was transported, separately, to a Boy Scout camp in Robbers Cave State Park, a densely wooded, mountainous area in the southeastern part of Oklahoma.

The boys were under the impression that they were being treated to three weeks in summer camp, and so they were. Their experiences at the camp were not noticeably different from the usual camping experiences. Their "counselors" took pains to conceal the fact that they were researchers in disguise, surreptitiously observing and recording the boys' words and deeds.

The two groups, the "Rattlers" and the "Eagles" (they picked those names themselves), didn't know about each other's presence at first. They had arrived in different buses, they ate in the same mess hall but at different times, and their cabins were in different parts of the campgrounds. The researchers' plan was to let each group of boys think they were alone in the camp for about a week. Then they would tell each group about the presence of the other, put them into competition with each other, and observe the results. Competition was expected to lead to hostility. But the boys were way ahead of them. Hostility appeared even before the two groups encountered each other directly. The first time the Rattlers heard the Eagles playing in the distance, they wanted to "run them off." And the boys were so impatient to compete with each other—this was an idea they proposed on their own, the adults didn't have to suggest it—that the researchers had difficulty sticking to their schedule. "Stage 1" was supposed to be the study of *within*-group behavior. Between-group competition wasn't supposed to begin until "Stage 2."

The scheduled events in Stage 2 were normal activities for a boys' summer camp. The two groups played baseball, had tugs-of-war and treasure hunts, and competed for prizes. The counselors acted like real counselors except that they tried to keep a low profile and to step in only when necessary. But push very quickly came to shove. Name-calling was recorded at the first official meeting (a baseball game) between the Rattlers and the Eagles. Before the game the Rattlers had hung their flag on the backstop of the baseball diamond—they thought of the ball field as "ours"—and after the game the Eagles, who had lost, tore down the flag and burned it. The Rattlers were outraged. Soon the counselors were breaking up fistfights.

It got worse. After the Eagles won at tug-of-war, the Rattlers raided their cabin at night. They turned over beds, ripped mosquito netting, and stole—among other things—a pair of blue jeans, which they made into a new flag. The Eagles retaliated with a daring daytime raid and messed up the Rattlers' cabin. They didn't expect to find the Rattlers home at the time but, just in case, they carried sticks and baseball bats. When they got back to their own cabin they prepared a defense against future raids: socks filled with stones and a pailful of additional stones to be used as projectiles. These kids were not just *playing* at war. In a very short time they had gone from name-calling to sticks and stones.

I can imagine the researchers' relief when Stage 2 ended and they could move on to Stage 3, in which the plan was to end the hostility and combine the two warring groups into a single peaceful one. But it is a lot easier to divide people up than to put them back together again. The first thing the researchers tried—bringing the two groups together in noncompetitive situations—did nothing to lessen the antagonism. Having the Rattlers and the Eagles take meals at the same time led to food fights and a mess in the mess hall. It was necessary to create "superordinate goals"—a common enemy too big for either group to fight alone.

The researchers were clever in devising such situations. They pretended that there was a problem with the camp water system and told the boys they suspected that vandals—outsiders—had meddled with it. The entire pipeline had to be inspected and it took all the boys from both groups to do it. A supply truck supposedly broke down and wouldn't start—it was facing uphill and it took the combined pulling power of all the boys to get it moving. The researchers also took the boys away from their familiar camping grounds—grounds the Rattlers and Eagles had fought over—and drove them to a new camping site next to a lake. At the end, a tenuous truce had replaced the open warfare of Stage 2. But if a Rattler had accidentally stepped on an Eagle's toe, or an Eagle had inadvertently knocked over a Rattler's Kool-Aid, I suspect that belligerence might have broken out again.

The Quality of Groupness

Social psychologist Muzafer Sherif, the head of the research team that carried out the Robbers Cave study, never won a Nobel Prize for his work; no Nobel Prizes are given out in psychology or sociology. But his experiment is cited to this day in psychology and sociology textbooks. It was never repeated, partly because it would be too dangerous, partly because it wasn't necessary. Sherif's study had made its point clearly and convincingly. Take a group of boys, allow them time to develop a group identity, and then let them find out there is another group with competing claims to territory they thought of as "ours," and the inevitable outcome is between-group hostility.

But there was still work for later investigators to do. What if the boys hadn't had time to develop a group identity? What if they didn't have territory to fight over? In the wilderness of southeastern Oklahoma, Sherif and his team had had to contend with snakes, mosquitoes, and poison ivy, to say nothing of socks filled with stones. The follow-up work was carried out in the safety and comfort of the laboratory.

The boys who served as subjects in the experiments of social psychologist Henri Tajfel were fourteen- and fifteen-year-olds from a school in Bristol, England. They all were acquainted with each other before they came, in groups of eight, to Tajfel's laboratory. In the laboratory they were given a test of "visual judgment": clusters of dots were flashed on a screen and they were asked to estimate the numbers of dots in each cluster. After completing this task, the boys were told that some people tend to overestimate the number of dots and others tend to underestimate them. Then, after their response sheets were ostentatiously "scored," the boys were taken one by one to another room and each was privately informed which group he belonged to, the overestimators or the underestimators. In fact, group assignment was entirely random: half the boys were told that they were overestimators and the other half that they were underestimators. Their performance on the dot test had nothing to do with it.

The *real* experiment began immediately after this phony information was given. Each boy was seated in an individual cubicle and given a "reward form" to fill out. He was asked to decide how much money should be paid, for participating in the experiment, to several of his classmates. The classmates were identified only by number and by group. For example, a boy who had just been told that he was an overestimator would be asked to check off, from a list of choices, how much money should be given to "member No. 61 of the overestimator group" and how much to "member No. 74 of the underestimator group." Whatever he checked off—this was clearly stated in the instructions—it would not affect his own payment.

The boys didn't know which of their schoolmates were in their own group and which were in the other. They didn't know the identity of the people to whom they were assigning payments. Nonetheless, they gave more money to the members of their own group than to the members of the other. They seemed to be as motivated to *underpay* the members of the other group as to *overpay* the members of their own.

This experiment demonstrated how little it takes to evoke what Tajfel called "groupness." It doesn't require a history of friendship with one's groupmates or of conflict with the members of the other group. It doesn't require territory to fight over. It doesn't require visible differences in appearance or behavior. It doesn't even require knowing who your groupmates are. "Apparently," Tajfel concluded, "the mere fact of division into groups is enough to trigger discriminatory behavior."

People divide up into groups in the blink of an eye, without any help from a researcher. The bus that carried the Rattlers to the Robbers Cave summer camp was a little late getting to one of its pickup points. The four boys who had been waiting there for half an hour had already formed a sense of groupness when the bus arrived. They sat together on the bus and asked whether "us south-siders" could remain together at camp. It took a few days of shared experiences—an encounter with a rattlesnake, the necessity of joining together to erect a tent—to integrate the south-siders with the rest of the group.

In *Lord of the Flies,* the choir makes its first appearance marching in formation, led by Jack. Each is wearing "a square black cap with a silver badge on it." Before the plane crash that landed them on the island, they had been students at what was probably an exclusive boarding school. In those days (the 1950s), British schoolboys who went to exclusive boarding schools were great snobs. They could identify each other by their accents and by their school ties or caps, and they looked down at boys who attended the local grammar school. But the boys on Golding's island did not split along class lines. The ones who had attended the same school did not stick together. All vestiges of their previous lives were dropped: the boys who had been members of the choir never sang another note.

The Rattlers and the Eagles did not leave their previous lives behind them. They had all come from church-going families, and at the Robbers Cave summer camp both groups decided independently to say grace before meals. Despite the animosity between groups, the Rattlers gave "three cheers for the Eagles" after they beat them in baseball—cheering the losers was evidently a tradition in Oklahoma schools. When new groups form, the members look for, and usually preserve, what they have in common.

Novelists are not expected to be social psychologists but they are expected to be good observers of human behavior. Golding got it all wrong. I'm not saying there is no such thing as mob violence: mobs do sometimes attack and kill individuals. But usually the victim is seen as one of *them.* And within groups there may be power struggles and bullying, but these internecine squabbles typically get swept under the carpet when another group—a potential enemy—appears on the horizon. I think what would have happened on Golding's island is that the boys would have split up into two groups. Within each group it would have been more or less like the Melanesian children. Between the groups it would have been more or less like the Rattlers and the Eagles, only with no counselors to step in when push came to shove.

Dividing Up the World

"When we name something," said linguist S. I. Hayakawa, "we are classifying." Naming, classifying, categorizing, pigeonholing, dividing up people or things into groups—whatever you want to call it, we do it all the time. Our brains are built that way. It would be inefficient to have to learn how to deal with each object, each animal, each person individually, so we put them in categories—"cars" and "cows" and "politicians," for example—and then we can apply what we learn about one member of the category to other members of the same category. As a Japanese American who later became a politician, Hayakawa was at pains to point out the dangers of categorization. "Cow$_1$ is not cow$_2$," he reminded his readers. "Politician$_1$ is not politician$_2$."

Hayakawa was a believer in the theory—called the "Whorfian hypothesis"—that the way we cut up the world into categories is entirely arbitrary, and that pinning a name on a category is what causes our brains to pigeonhole things in a particular way. There is some truth in this theory. When Henri Tajfel told a Bristol schoolboy that he was an overestimator, a category was put into the boy's mind that hadn't been there before he walked into Tajfel's lab.

But like many other "laws" of psychology, the Whorfian hypothesis doesn't work for all of the people all of the time, or even for most of the people most of the time. The way we cut up the world into categories is, in general, not arbitrary at all. This is as true for categories that have fuzzy borders as for those that are clearly delineated. *Night* and *day* are as different as, well, night and day, even though it's hard to tell where one ends and the other begins. Children learn quickly and easily to cut up time into *night* and *day* and to use those words appropriately. It takes many years for American children to learn that the twenty-four hours can also be split, unfuzzily, into two halves of twelve hours each, called *a.m.* and *p.m.* The a.m.–p.m. distinction is artificial and unconvincing; the night–day distinction is something we would be aware of even if we didn't have words for it.

The Whorfian hypothesis predicts that babies and animals can't categorize because they don't have words for the categories. This prediction has been soundly disconfirmed. Pigeonholing turns out to be so easy that even a pigeon can do it. Yes, the categorization skills of pigeons have been tested. They earned a passing grade. A pigeon taught to peck at one button when shown a picture of a cow, another when shown a picture of a car, can apply this training to cows and cars it never saw before.*

What makes a category is not a word but a concept. In order to peck at the correct button, the pigeon has to have some sort of concept of what a cow is, so that when it sees a picture it has never seen before, it can match the thing in the picture to its concept of a cow. The pigeon doesn't have to know the word *cow* in order to form the concept of a cow. Babies as young as three months can categorize and hence must be able to form concepts. Jean Piaget, the famous Swiss developmentalist, thought they

* I don't believe pigeons have been tested on pictures of politicians. Just statues of politicians.

couldn't, but he was wrong. In judging the abilities of babies, Piaget was an underestimator.

How do we post-Piagetians know that babies can form concepts? No, we don't make them peck at buttons. Instead, we bore them. Babies are easily bored, so if we show them lots of pictures of cows, pretty soon they stop paying attention to them. If we then throw in a picture of a horse and the baby suddenly looks interested again, we know that she can tell the difference between a cow and a horse.

Using variations of this technique, babies too young to understand words have proved that they can tell the difference between cats and lions, between cars and airplanes, and between men and women. There is also evidence that they can tell the difference between grownups and children: by the second half of the first year they are wary of unfamiliar grownups but unfamiliar children are given the benefit of the doubt. They respond to the facial differences between grownups and children as well as to the difference in size. If you show them a grownup's face on a child-size body, babies are surprised and amused.

Of the three main ways we categorize people, babies know two of them —gender and age—before they are a year old. The third is race, and that takes them a good deal longer. Race is a fuzzy concept, with boundaries arbitrarily drawn. Children can't always tell their classmates' race just by looking at them (adults can't either), and sometimes the only way to know for sure is to ask. But then, the same is true of gender.

Arbitrary or not, categorization has predictable effects, and that was what worried S. I. Hayakawa. Referring to himself in the third person, Hayakawa expressed his dislike of being categorized:

> The writer has spent his entire life, except for short visits abroad, in Canada and the United States. He speaks Japanese haltingly, with a child's vocabulary and an American accent; he does not read or write it. Nevertheless, because classifications seem to have a kind of hypnotic power over some people, he is occasionally credited with (or accused of) having an "Oriental mind."

Contrast and Assimilation

What bothered Hayakawa was not so much the fact that he was classified as an "Oriental" (the term was respectable back then) as the fact that people expected him to have all the characteristics attributed to members of that category. This is one of the consequences of categorization: it causes us to see the items *within* a category as more alike than they really are. At the same time, it causes us to see items in *different* categories as more different than they really are.

The items being classified don't have to be people. If we are considering, for example, the two major categories of pets, *dogs* and *cats*, "dogs" makes us think of the qualities most dogs share and cats don't have, and "cats" makes us think of the qualities most cats share and dogs don't have. We picture the archetypical dog—tongue hanging out, tail wagging, wanting to play ball—and the archetypical cat, tidy and smug. If we were at a dog show, looking at foxhounds, poodles, collies, chihuahuas, and bull terriers, we might be remarking on how much dogs vary in appearance and temperament. But when the categories are *dogs* and *cats*, we see dogs as basically alike and our minds dwell on precisely those characteristics that distinguish them from cats. The tendency to see two juxtaposed categories as more different than they really are is the source of what social psychologists call *group contrast effects*.

All it takes to produce group contrast effects is to divide people into two groups. The groups will inevitably see themselves as different from each other, with the result that any small differences between them will get larger. The interesting case is when the groups start out exactly alike, because if there are no real differences to begin with, the groups themselves will create them. The boys at the Robbers Cave summer camp were purposely chosen to be as alike as possible, so the Rattlers and the Eagles had to find ways of being different. They did this by selectively emphasizing different aspects of characteristics they brought with them to the camp: a shared religious background and the normal tendency for boys to talk dirty among themselves. Here are the Eagles after they won the second baseball game with the Rattlers:

> As the Eagles walked down the road, they discussed the reasons for their victory. Mason attributed it to their prayers. Myers, agreeing heartily, said the Rattlers lost because they used cuss-words all the time. Then he shouted, "Hey, you guys, let's not do any more cussing, and I'm serious, too." All the boys agreed on this line of reasoning.

So the Rattlers became the cussing group and the Eagles stopped using cuss-words and became the praying group. The goody-goodies against the baddy-baddies. And yet none of these boys had been conspicuous for either goodness or badness before the experiment began. The researchers wanted, and had gone to considerable effort to obtain, twenty-two perfectly normal boys.

Categorization causes the differences between human groups to increase but the differences *within* them to get smaller. The tendency for group members to become more alike over time is called *assimilation.* Human groups demand a certain amount of conformity. This is especially true when a contrasting group is in the vicinity, and especially true for the characteristics in which the two groups differ (or believe themselves to differ). At the Robbers Cave summer camp, the Rattlers liked to think of themselves as tough—not a bunch of sissies. An Eagle was permitted (by his fellow Eagles) to cry if he twisted an ankle or bloodied a knee, but a Rattler was expected (by his fellow Rattlers) to bear up stoically. Children's groups use various methods, often quite cruel, to enforce their unspoken rules of behavior. Those who will not or cannot conform to the rules, or who are different in any way, may be excluded, or picked on, or made fun of. "The nail that sticks up will get hammered down," they say in Japan. The nyah-nyah song is heard all over the world. We tend to think of adolescence when we hear the term "peer pressure," but pressure to conform is most intense in childhood. By the teen years it is seldom necessary to punish the nonconformer. Teenagers are not pushed to conform—they are pulled, by their own desire to be part of the group.

A famous series of experiments on group conformity, carried out in the early 1950s by social psychologist Solomon Asch, used college students as subjects. A typical experiment began with eight young men showing up at the laboratory, supposedly to take part in a study of perceptual judgments. Only one of the eight was actually a subject, however—the others were confederates of the researcher, trained to perform a role. Their role was to sit around a big table with the gull—er, the subject—and give incorrect perceptual judgments out loud, with a straight face. They were to show no sign of amusement or surprise when the subject's judgments disagreed with what they had been told to say.

Not all subjects gave in to their desire to conform; in fact, the majority continued to give correct responses even when all seven of the others were aligned against them. The point of these experiments was not to show that people will cave in under the threat of public humiliation: it was to show that a person will question the evidence of his own eyes before he will question the unanimous opinion of his peers. The subject didn't accuse the others of lying or of conspiring against him (though in fact they were). He didn't think there was anything wrong with the other guys

—he thought there was something wrong with *him.* "I began to doubt that my vision was right" was a typical comment.

Within the Group

All this talk of group conformity doesn't mean that human groups are made up of a bunch of clones. I said in the previous chapter that a family of clones would be unlikely to win the survival-of-the-fittest award; the same is true of a *group* of clones. Like families, groups are better off if their members can fill a variety of niches. They must all hang together at times when otherwise they would assuredly hang separately, but when there is no external threat each should be able to contribute to the group in his or her own way. Not everyone in a group can be a leader. In fact, having more than one leader may cause a group to split apart, and that can make them sitting ducks if next door there is a bigger group, led by a single strong leader. Thus, it is the nature of human groups, when not actively engaged in hostilities with other groups, to do some within-group work of a sort called *differentiation.* Differentiation was one of the two processes—the other was assimilation—that the Robbers Cave researchers studied during Stage 1.

One way the members of groups differentiate themselves is through struggles by individual members to achieve dominance or social power. The dominance hierarchy, or "pecking order," is also found in ape and monkey groups; I'll have more to say about it in the next chapter. The other kind of differentiation is peculiarly human. It is encapsulated in this quote from a 1957 developmental psychology textbook:

> The gang is quick to seize on any idiosyncrasy of appearance, manner, skill, or whatever, and thereafter to treat the child in terms of this trait. The stereotype by which the gang identifies the child is often expressed in his nickname: "Skinny," "Fatso," "Four-eyes," "Dopey," "Professor," "Limpy."

There were no fatsos, four-eyes, or limpies among the Robbers Cave boys, but in the week before the two groups came into contact the boys had already begun to carve out specialty niches for themselves. One niche that is always available in a boys' group, and is usually filled, is the role of group clown. The Rattlers had a group clown named Mills:

> Baseball work-out followed with members accepting decisions of the rest of the group on plays, excepting Mills who changed a decision in his own favor. During the rest period, Mills started tossing pine cones and ended

up in a tree being pelted by all others and shouting "Where's my fellow men?" A boy replied, "Look at our leader!" (The "clown" role often kept him in the center of attention.)

Another Rattler, Myers, got tagged as the group exhibitionist when he became the first one to swim in the nude—an audacious act that earned him the nickname "Nudie."

What Is a Group?

You may have noticed that I've said quite a lot about groups but still haven't said what a group is. That's because the definition depends on one's theoretical outlook. I will ally myself with a particular theoretical outlook by defining *group* as a social category, a pigeonhole with people in it. Often, a social category will bear a label—Japanese American, Rattler, female, child, Democrat, college graduate, physician—but it doesn't have to, because a category is defined by a concept and a concept can exist without a label. This definition can also apply to animal groups. If a pigeon can have a concept of a cow, it can have a concept of its group.

Groups can be big or little, but they generally have more than two people. Two people are not usually called a group; the technical term for two people is *dyad,* as in "dyadic relationship." To put it in less technical terms, two's company, three's a crowd.

Human groups can come into existence in many ways. A researcher can tell a boy that he is an overestimator, and immediately he will identify with an anonymous group of people called "overestimators." Five people get stuck in a stalled elevator; if they are rescued in a few minutes they are still just five people, but if it takes half an hour they have become a group. Shared fate—the sense of "We're all in this together"—is one of the things that create groupness. Note that the elevator group doesn't have a name—social categories depend on concepts, not labels—and note also that the people in the elevator don't all behave alike. Stuck elevators, too, have their group clowns.

One of the basic and enduring types of groups is the family. In tribal societies, when villages split up and the two sides go to war against each other, families almost always stick together, and people who have relatives on both sides feel torn and are reluctant to fight. One of the ways that small groups such as villages can coalesce into larger groups is by forming family alliances. If the headman of one village gives his daughter in marriage to the headman of the other village, then her children will have grandparents on both sides. It's sometimes enough to avert a war. Just think: if Romeo and Juliet had lived and had had a child, the Montagues and the Capulets might have come together in peace at the christening. Then again, they might not have.

When groups fall apart, they often fall apart into families. In November 1846, a wagon train led by a farmer named George Donner got stranded in a snowy mountain pass in California. The Donner Party, as it came to be called, soon ran out of food. Of the eighty-seven people they started out with, forty died that winter or were killed, and some of the bodies were eaten by other members of the party. The death rate among the women was only about half that of the men, but it was not chivalry that saved them: there was no "women and children first" rule at the Donner Pass. What saved the women was the fact that all of them belonged to family groups, whereas many of the men were single. Of the sixteen unattached men in the Donner Party—most of them healthy and in the prime of life—only three survived. According to evolutionary biologist Jared Diamond, "The Donner Party records make it vividly clear that family members stuck together and helped one another at the expense of the others." Some of them survived by resorting to cannibalism, but they did not eat the flesh of their sisters or brothers, their children or parents, their husbands or wives.*

It's All in Your Head

The basic phenomena of group relations that I have touched on in this chapter—preference for one's own group, hostility toward other groups, between-group contrast effects, and within-group assimilation and differentiation—are so robust, so easy to demonstrate in the laboratory or observe in natural settings, that social psychologists soon found themselves with little left to do but clean up the crumbs. It was the success of social psychology, not its failure, that led to the decline of the field in the wake of the brilliant research carried out in the 1950s.

Okay, so that wasn't the *only* reason for the decline of social psychology. The other reason was the popularity of Skinnerian behaviorism. In the psychology department where I was a graduate student before they kicked me out in 1961 (see the Preface), B. F. Skinner was the most prominent

* If the Donner Pass sounds to you like the world according to Thomas Hobbes, consider what a *real* Hobbesian world might be like. Here's Homer Simpson, of the TV show *The Simpsons,* being abducted by aliens: "Don't eat me! I have a wife and three kids! Eat *them!*"

professor and most of the other grad students were his disciples. There was no social psychology at all: social psychology was off in another department, called "Social Relations." We in the *real* psychology department were wont to sneer at the softheads in Soc Rel.

It took me thirty-three years to see this, but my fellow grad students and I were wrong to sneer. Skinner's idea was that he could explain behavior by looking at the reinforcement history—the rewards received or not received—of the individual organism. He called them "organisms" because he didn't see any important differences between species: they all dance to the same tune. The problem (I should say *one* problem) with this approach is that you can't explain the behavior of individuals by looking at them in isolation, if they happen to belong to a species that was designed by evolution to live in groups. Skinner's students studied how pigeons behave if you put them in a box, give them a button to peck at, and occasionally feed them a few grains of corn when they peck at the button. But pigeons weren't designed to live alone in little boxes: they were designed to live with other pigeons.

Some ornithologists in Arizona made the same mistake. They hand-reared eighty-eight thick-billed parrots, members of an endangered species, and released them in the pine forests where they once had thrived. The birds all died or disappeared. In the wild, these parrots normally flock together, but the hand-reared ones showed no interest in seeking the company of birds of their feather. A solitary bird quickly falls prey to hawks, and that is apparently what happened to the thick-billed parrots raised in captivity.

Today, Skinnerians are dying out like thick-billed parrots, while social psychologists are proliferating like pigeons. But social psychology has changed: it has less to do with behavior and more to do with what goes on inside people's heads. The important data have already been collected; what is needed now is a theoretical framework to put them in. Constructing theories of group relations and arguing their merits keeps many present-day social psychologists occupied.

Here are some of the questions that their theories are designed to answer. What makes people favor their own group and feel hostile, at least some of the time, toward other groups? What motivates them to be similar to their groupmates, even if there is no pressure to conform, and different from the members of other groups? What motivates them to differentiate themselves from their groupmates—to carve out their own niches, strive for individual success and recognition? What determines which of these two contradictory processes, assimilation or differentiation, will prevail? And how do people decide which group they belong to when there is more than one choice? What made Mary Breen, one of the survivors of the winter at Donner Pass, think of herself as a member of the Breen family rather than a member of the Donner Party?

Human group behavior is very complex. People in our society identify themselves—*self-categorization,* Australian social psychologist John Turner calls it—with many different groups. Mary Breen's great-great-great-granddaughter might categorize herself, depending on the circumstances, as "a woman," "a Californian," "an American," "a Democrat," "a student at Berkeley," "a member of the class of 2002," or "a member of the Breen family." The other members of these groups don't have to look familiar to her; she doesn't even have to know who they are. She can switch her allegiance from one group to another, inside her head, without moving an inch—she doesn't have to move to Kahama to become a Kahaman. All these things make human group behavior look very different from the group behavior of nonhuman animals. No one, as far as I know, has tried it, but it seems unlikely that we could evoke groupness in a chimpanzee by whispering in its ear, "You are an overestimator."

Nevertheless, human group behavior is clearly something we inherited from our primate ancestors. Like thick-billed parrots, we weren't designed to live alone.

The theories of group relations constructed by social psychologists are theories of what goes on inside the human mind. Skinner erred by assuming that human behavior can be explained with the same simple mechanism that he used for explaining the behavior of rats and pigeons. I believe that modern social psychologists make the opposite error: they construct theories of group behavior that cannot be applied to animals, even though many of the same behaviors can be observed in animal groups. John Turner's theory, for instance, says that the reason we prefer our own group and denigrate other groups is that we are motivated to increase our self-esteem. Thinking that our own group is better increases our self-esteem. Even if you're willing to grant a desire for self-esteem to a chimpanzee, it seems too puny a motive to account for the immense power of group behavior. People kill, people die for their groups! I don't believe that the fierce emotions and warlike behavior of the eleven-year-old boys at the Robbers Cave summer camp were driven by a desire for self-esteem. As a motivator, it's not even strong enough to make an eleven-year-old boy do his homework.

The strong motivators are things that have to do with survival or reproduction. For many millions of years (long before our own species stepped onto the stage) primates have been living in groups. For all that time—all but the last little bit of it—the individual's survival depended on the survival of the group, and the members of the group were close relatives. A willingness to die for others who carry your genes makes sense in evolutionary terms. Many animals do things that appear to be self-sacrificing—the bird squawks to alert its fellows, though its squawk may make it the predator's target—because even if they die, their sisters and brothers, their children and parents, might be saved. The individual may be lost but the genes it shares with its kin are preserved and passed on.

In a human hunter-gatherer group, everyone was related to everyone else, either by blood or by marriage. Human groups no longer consist solely of people who are related to each other, but the motivator that powers group behavior doesn't seem to know that. Underneath the embellishments provided by our recently acquired cognitive abilities are deep evolutionary roots. The emotional power of groupness comes from a long evolutionary history in which the group was our only hope of survival and the members of the group were our sisters and brothers, our children and parents, our husbands and wives.

Recognizing Your Relatives

Many kinds of animals are capable of what biologists call *kin recognition.* It tells them which members of their species to be nice to and which to be nasty to. A paper wasp, for instance, decides whether another paper wasp, seeking admission to the nest, is one of *us* or one of *them* by smelling it. If the newcomer smells like *us,* it is allowed to enter. Tiger salamanders can recognize their own siblings, also on the basis of smell. If you rear them among nonsiblings they often become cannibals. They don't mind eating other salamanders, but they'd rather not eat their own sisters and brothers. Kin recognition by means of odors is based on a biochemical mechanism similar to the one that permits your immune system to distinguish between "self" and "nonself."

Humans recognize kin, not on the basis of odors, but on the basis of familiarity. A sister or brother is someone you grew up with. People don't marry their sisters or brothers, not because it's against the law, but because they don't want to. Israelis who grew up in a kibbutz where boys and girls were reared together, treated like brothers and sisters, don't marry each other.

But people are nonetheless attracted to others who are similar to themselves. Husbands and wives are, on average, much more alike than they would be if Cupid fired off his arrows at random. The ways in which married couples tend to resemble each other include race, religion, socioeconomic class, IQ, education, attitudes, personality characteristics, height, breadth of nose, and distance between the eyes. Married couples don't come to look alike as they grow older: they look alike to begin with.

Similarity also serves as a basis for friendship. Even in nursery school, a child is attracted to others "like me." In grade school, children who are good friends are likely to be of the same age, sex, and race, and to have similar interests and values.

I believe the tendency to be drawn to people who are similar to yourself has its remote origins in kin recognition. If you were a hunter-gatherer, someone who looked like you and spoke your language was more likely to be a member of your group, possibly a relative, than someone who didn't look like you and spoke a language you couldn't understand. If you are an educated North American, you find yourself wanting to trust someone who looks like you, talks like you, and thinks like you.

The stranger is instinctively distrusted, by paper wasps and human babies, because he may be up to no good. If he is a cannibal—cannibalism is found in many species, including our own—he might eat you, because you are not his kin. The first reaction to the stranger, or to the one who is behaving strangely, is fear. Fear turns into hostility because being afraid is unpleasant. Remember the polio-stricken chimpanzee who dragged himself, crippled, back to his group? His groupmates reacted first with fear and then with anger—they attacked him. Damn you for giving us such a fright!

We don't need a fancy cognitive explanation for hostility toward other groups—evolution provides a good one, and it works for animals as well as people. Group contrast effects, which exaggerate the differences between groups or create differences if there are none to begin with, are not (as far as I know) found in animals, but they are a direct consequence of the human and animal tendency to feel hostile toward other groups. If you fear and dislike someone, you are motivated to be as different from them as possible. Humans—adaptable creatures that they are—are ingenious at finding ways of being different from the members of other groups.

How and Why We Categorize Ourselves

In the modern world, group affiliations still involve the "they are like me, I am like them" response—the perception that you are similar in some way to the other people in the group, that you and they have something in common. What you have in common can be almost anything: you live in the same state, you voted the same way in the last election, you are the same age or sex, you came to camp on the same bus, you are stuck in the same elevator.

Social categories nest within each other like layers of an onion or overlap like a plateful of fried onion rings. The number of choices a person has in a complex modern society is mindboggling. Earlier I said that Mary Breen's great-great-great-granddaughter can categorize herself as "a Californian," "an American," "a Democrat," "a woman," "a student at Berkeley," "a member of the class of 2002," or "a member of the Breen family." Yet another alternative open to her is to categorize herself as none of the above, but as "*me,* a unique individual." Of the many self-categorizations available to Mary VI, which will she choose? Which will direct her thoughts, feelings, and actions? Now, I'm afraid, we do need to turn to the social psychologists and their fancy cognitive theories.

The approach that has most influenced my own thinking is that of Australian social psychologist John Turner, whom I mentioned before in this chapter. Turner studied under Henri Tajfel, the inventor of overestimators and underestimators, and his theory is based on earlier theoretical work by Tajfel.

What I like about Turner's theory is the part that has to do with self-categorization. Turner says that we can categorize ourselves in a variety of ways and on a variety of levels, ranging in inclusiveness from "me, a unique individual," to very big categories such as "an American" or even "a human being." Self-categorization can vary from moment to moment: it is highly dependent on social context—on where we are and who is with us. What causes us to adopt one self-categorization rather than another is the relative salience, at a given moment in time, of the various social categories.

Salience means prominence, conspicuousness, the quality things have when they demand your attention. But it's a slippery concept, hard to define without lapsing into circular reasoning—an ever-present danger for academic psychologists. Why did you adopt that particular self-categorization? Because it was salient. How do we know it was salient? Because that's the self-categorization you adopted.

Turner gets out of this endless loop by specifying a condition that makes a social category salient: when a comparable or contrasting category is simultaneously present. Thus, the social category *adult* is not salient when you're in a roomful of adults, but as soon as some children enter the room it becomes salient. The category *Rattler* acquired instant salience when the Rattlers found out that there was another group of eleven-year-old boys sharing the camp grounds with them. If they had discovered, instead, a group of eleven-year-old *girls* on the other side of the camp, the salient social category would have been *boys.*

When a particular social category is salient and you categorize yourself as a member of it—that is when the group will have the most influence on you. That is when the similarities among the members of a group are most likely to increase and the differences between groups to widen.

John Turner calls it the *psychological group;* an older term is *reference group.* It is the group with which, at a given moment in time, you identify yourself. Here is how Turner defines it:

> A psychological group is defined as one that is psychologically significant for the members, to which they relate themselves subjectively for social comparison and the acquisition of norms and values . . . from which they take their rules, standards and beliefs about appropriate conduct and attitudes . . . and which influences their attitudes and behavior.

Acquisition of norms and values. Rules, standards, and beliefs about appropriate conduct. Influences their attitudes and behavior. But that is what families are supposed to do to children! That is a description of socialization!

Occasionally families do socialize their children. But usually they don't and I will show you why.

Families and Other Groups

Within a group of monkeys or apes there are frequent quarrels, usually quickly resolved, as individual animals attempt to improve or defend their position in the dominance hierarchy. The members of the group, observes primatologist Frans de Waal, "are simultaneously friends and rivals, squabbling over food and mates, yet dependent on one another."

These within-group squabbles cease abruptly when the group is threatened by a predator or by another group of monkeys or apes. To put it in human terms, the outside threat has increased the salience of the group. The consequence—just as in human groups—is that differentiation (in this case the struggle for dominance) is put on the back burner and the group comes together to confront the common enemy.

Even monkeys and apes are clever enough to use the threat of the common enemy as a way of reducing within-group tensions. Frans de Waal has seen wild baboons resolve a dispute by jointly threatening the members of another baboon troop, and chimpanzees in a zoo making aggressive "wraaa" calls in the direction of the cheetah enclosure, though no cheetah was visible. "The need for a common enemy can be so great that a substitute is fabricated," says de Waal. "I have seen long-tailed macaques run to their swimming pool to threaten their own images in the water; a dozen tense monkeys unified against the 'other' group in the pool."

In the absence of a common enemy, or of a common goal that can be achieved only if everyone pulls together, groups tend to fall apart into a collection of individuals or smaller groups. The people in the stuck elevator each behave differently, vying for leadership and adopting roles such as group pessimist and group clown.

Aside from the Donner Party, there were no other people at the Donner Pass that winter. Had they met up with another group of pioneers or a hostile tribe of Native Americans, it would have united them. The social category "Donner Party" had low salience because categorization requires more than one category: it takes a *them* to make an *us.* So the group split apart into families. If the weather hadn't been so harsh and everyone so hungry, the Donner Party might have split up in a different way: into adults and children.

There was no children's play group at the Donner Pass, but that was because the circumstances were exceptional. Normally, whenever groups of families come together, the children seek each other out. Sometimes the families split apart again—this happens in hunter-gatherer societies when group tensions mount or when spread-out resources make it difficult for larger groups to find food—and it's hard on the kids. The grownups are the ones who make the decision to split up, not the children. Ethologist Irenäus Eibl-Eibesfeldt describes a pair of !Kung San brothers getting in each other's hair and explains that "the !Kung group had dissolved into individual families at the time," so "the older brother could find no outlet in the children's play group, where he normally would have been."

American pioneers did not always cross the country in large groups. The family of Laura Ingalls Wilder, the author of the *Little House* books, set out alone: just Ma, Pa, and their three daughters, Mary, Laura, and Carrie. Did that make "the Wilder family" a salient category for Laura? No, because it was the only family around. For Laura the salient categories were *children* and *parents.* She was socialized, perforce, by her family, but "the Wilder family" didn't become a salient category until they settled down in a place where there were other families.

Within her family, Laura didn't learn to behave like her parents. She learned from her parents how to do many things, but she also learned that she was not expected to behave like them—she was expected to behave like a child. The rules for children's behavior, by the way, were quite different in those days, as were the rules of parenting. The *Little House* books, which are nothing like the TV show, provide vivid evidence of how styles of parenting change over time, and how different styles of parenting can produce equally satisfactory results.

The world that Laura Ingalls Wilder grew up in—the one depicted in the books, not the TV show—was different in many ways from our own world. But the houses we live in today have one thing in common with Laura's isolated little house on the prairie: they're private. In the privacy of the modern home, the family is not a salient social category, because *it's the only family around.*

When people categorize themselves, they put themselves into a pigeonhole with other people like themselves—that is, with people they perceive as being like themselves. Children don't perceive adults as people like themselves, not if there are any other children around to make the distinction clear. To a child, an adult might as well be a member of another species. Grownups know everything and can do whatever they want. Their bodies are enormously big and strong and hairy, and they bulge out in odd places. Though grownups can run, they are usually seen sitting or standing. Though they can cry, they hardly ever do. Different creatures entirely.

Modern children are provided, by our laws of universal schooling, with a ready-made group of people "like me": their classmates. They interact with their families only when they're at home, and when they're at home the family isn't salient because it's the only family around. When they're at home, big families fall apart into the children and the grownups, and small families fall apart into individuals, each looking for recognition and a personal niche to fill.

Like the child in the hunter-gatherer or village play group, children in developed societies are socialized in a group of children. This is the group they see as being "psychologically significant" for them, the one to which they "relate themselves subjectively," the one from which "they take their rules, standards and beliefs about appropriate conduct and attitudes," as Turner put it.

I call my theory, for want of a better name, "group socialization theory." It is not just about socialization, however; it is also about the way children's personalities are shaped and changed by the experiences they have while they are growing up. It is what I am offering as a replacement for the nurture assumption. I'll tell you about it in the next chapter.

Einstein once said that a chief motivation for constructing new theories is a "striving toward unification and simplification." There are simple, unified theories in psychology: Skinner's theory is a prime example. My theory, I'm afraid, is not like that. The mind of the child is too complex; it cannot be forced into the procrustean bed of a simple theory. I hope you will judge my theory, not on the basis of its simplicity or lack thereof, but on its ability to explain things that the nurture assumption cannot explain.

8 IN THE COMPANY OF CHILDREN

I was, by all accounts, a rambunctious child in my early years. Today such a child might be labeled "hyperactive"—unusual, but not unheard of, in girls. I was fearless, adventurous, outgoing, and loud. I was one of those kids who, if there was a hole to fall into, fell into it. I was persona non grata in restaurants because I couldn't sit still.

It drove my parents crazy. A "little lady" was what girls were supposed to be in those days, and I was not a little lady. My mother bought me frilly dresses and I got them dirty and torn; there was always an untied bow straggling down the back of my bare legs and bandaids on the front of them. Jeans would have been more sensible, but they hadn't started making jeans for little girls and it never occurred to my mother to dress me in boys' clothes. Or perhaps she continued to hope that the frilly dresses would eventually work their magic and somehow turn me into a little lady.

They didn't. Nothing worked. My parents despaired of me. Kindergarten came and went; first grade, second grade, and third. We moved around a lot, those early years, and several times I was taken out of a classroom in the middle of the school year and put into another one, but I had no trouble making new friends. My high spirits and outgoing nature made me popular with my peers, both boys and girls.

Then we moved once more—as usual, after the school year had begun—and everything changed. I found myself the youngest and smallest child, and one of the few who wore glasses, in a fourth-grade classroom in a snooty suburb in the Northeast. The other girls were sophisticated little ladies, interested in hairstyles, proud of their pretty clothes. I wasn't like them, and they didn't like me.

My family remained in that place for four years, and they were the worst four years of my life. I went to school each day with children from my neighborhood, but not one of them would play with me or talk to me. If I dared to say anything to them, it was ignored. Pretty soon I gave up trying. Within a year or two I went from being active and outgoing to being inhibited and shy. My parents knew nothing of this—they saw no major changes in my behavior at home. The only thing that changed, as far as they were concerned, was that I was spending a lot of time reading. Too much time, in their opinion.

Then, a couple of months after the beginning of eighth grade, my family moved once more, and my days as an outcast were over. We moved back to Arizona, where I had spent my early years. The kids there were not snooty or sophisticated; I had friends again, though just a few. And the years of solitude, of seeking solace in books, were beginning to pay off: my classmates were referring to me as a "brain" and I started making good grades—this was something new for me—and seeking other brainy kids to pal around with. But I remained inhibited and insecure. The kids in the snooty suburb had accomplished what my parents could not: they had changed my personality.

Children are born with certain characteristics. Their genes predispose them to develop a certain kind of personality. But the environment can change them. Not "nurture"—not the environment their parents provide —but the outside-the-home environment, the environment they share with their peers. In this chapter I will show you how it happens.

Letting Go of Mommy

The other day I went to the post office and found myself at the end of a long line. It was during school hours so no school-age children were present, but two of the women ahead of me had their toddlers with them: a girl and a boy, each about two years old. They were standing next to their mothers, like squirrels next to their trees, and an arm's length below the level of adult eyes they were gazing at each other. Finally the little boy let go of his mother's hand, toddled over to the little girl, and stood in front of her. "You're the most interesting person in this place" was no doubt beyond his verbal capacity, so he didn't say anything—just stood there, looking at her expectantly. But at that point the line moved and his mother grabbed him and pulled him away.

Young humans gravitate toward others of their kind, "their kind" being defined first of all by age. The same is true of other young primates. An infant monkey, as soon as it is able to move around on its own, will leave its mother for sessions of rollicking play with its peers. A young chimpanzee, hearing the sounds of other young chimpanzees playing in the distance, will try to persuade its mother to travel in that direction and will whine and pester until she gives in. The strong desire of young primates to find other young primates to play with can override the divisions between groups and even those between species. A young baboon or rhesus monkey may temporarily transfer to another troop if no playmates are available in its own. Jane Goodall saw young baboons playing with young chimpanzees in Tanzania, and we saw an infant chimpanzee playing with an infant human in Chapter 6. Playfulness is a prime primate trait, and, though it is not completely lost in adulthood, it always seems to be more fun for a young creature to play with another young creature than to be entertained by an adult of its species.

Developmentalists Carol Eckerman and Sharon Didow have described what happens if you put a pair of unacquainted human babies, along with their mothers, into a room strewn with toys. Year-old babies—at an age when they're wary of strange *adults*—smile at each other and babble. One baby might offer a toy or accept a proffered toy from the other. They sit near each other on the floor; sometimes one gently touches the other. Sometimes the touches are not so gentle or there is a struggle over a toy, but most of the interactions are friendly—at least they're *meant* to be friendly. These early gestures of friendship are often inept: one baby might, for example, offer a toy to the back of the other one. And their interest in each other seems to wax and wane, not always in synchrony, perhaps because contact with another baby is so stimulating it has to be taken in small doses. Nevertheless, of all the things in the room—the toys, the mothers, the researcher with her clipboard—what each baby typically looks at most is the other baby.

They look at their mothers too, of course, but mainly to make sure that she's still there. Infant primates, including humans, like to have their mothers nearby when they are playing; developmentalists say that the mother provides "a secure base from which to explore." Among monkeys and chimpanzees, the mother can step in if play with peers gets too rough, as it often does. Since these groups usually have a spread of ages, and sometimes the older ones are bullies, it helps to have a mother on one's side. Primate infants scream when they are hurt and that brings Mommy on the double.

The relationship between a primate baby and its mother is a close one; for humans and chimpanzees it often lasts a lifetime. Jane Goodall described a full-grown chimpanzee that remained with its mortally injured mother for five days, keeping off the flies, until the mother (a Kahaman) died of her injuries; and an adolescent chimpanzee that went into a deep depression when its mother died of old age. Goodall also described female monkeys risking their own lives in a desperate but futile attempt to get their babies back from chimpanzees that had stolen them: "One of these mothers even tried to reach her infant (who was being eaten) while she herself was being killed." Life in the jungle may be bloody but it is not devoid of love and loyalty.

Ethologist Irenäus Eibl-Eibesfeldt believes that the mother–infant relationship forms the evolutionary basis for all friendly dyadic relationships (relationships between two individuals). Fish and reptiles may gather in groups but no bonds of love or friendship exist between members of these groups. It was only after warm-blooded creatures began to care for their young, says Eibl-Eibesfeldt, that long-lasting affectionate relationships between individuals became possible. The evolution of maternal care led to the ability of animals to recognize and remember individual members of their species, as well as the motivation to be nice to them.

The ability of a bird or mammal mother to recognize its offspring works in different ways in different species. Recognition may be built in or learned, quick or slow, based on vision, smell, or hearing. The offspring's ability to recognize its mother also relies on different mechanisms in different species. Ducklings and goslings are renowned for their eagerness to become "imprinted" on any moving thing they set eyes on after they are hatched. This works out well if the moving thing happens to be their mother, noticeably less well if it happens to be the guy who mows the lawn, still less well if it happens to be the lawnmower.

Imprinting is a crude and chancy device; primates have a more sophisticated one, known as *attachment.* The infant primate takes some time to get to know its mother: weeks (in monkeys) or months (in chimpanzees and humans). By the time a baby monkey can move through the trees on its own, or a human baby can crawl, it is attached to its mother. When a baby monkey is frightened or hurt, it leaps toward its mother and clings to her. When a baby human is frightened or hurt, it clings to its mother in much the same way. The jungle is a dangerous place for small tasty creatures, so evolution provided a device—a sort of psychological leash—to keep them from straying too far.

The leash gets longer as the small creature gets larger, and eventually it breaks. For young chimpanzees the break comes relatively late: they are eight or nine years old—almost adolescents—before they are willing to be out of earshot of their mothers for any length of time. Human children achieve this level of independence considerably earlier: typically by the age of three. Most three-year-old humans will part with their mothers with little or no protest after a brief period of introduction to a nursery school or day-care center. My older daughter, whose unpropitious entry into nursery school was described at the end of Chapter 5, was okay after that first day, though for several years she remained somewhat timid around other children—especially active, noisy ones. (She is not the least bit timid as an adult, by the way.)

Notice that I started out as a bold child and my biological daughter started out as a somewhat shy one. The fact that children inherit their parents' genes does not mean that they necessarily inherit all their parents' characteristics. We tend to think of heredity as producing similarities between biological relatives, but heredity can also be responsible for the differences between them. One sibling can have blue eyes, the other brown, and this difference between them is genetic. My daughter and I were not at all alike as three-year-olds, due at least partly to genetic differences in our temperaments.

Genetic differences in temperament can help to explain why some children find it easier to let go of Mommy at the door to the nursery school classroom, and why some are more interested than others in socializing with their peers. But genes are unlikely to be the whole story—the children's experiences, too, surely play a role. The question is, which experiences? According to the nurture assumption, the answer must be: experiences with parents. Socialization researchers have labored long and hard to find evidence that children's relationships with other children depend upon their earlier relationships with Mommy and Daddy. A popular strategy for this kind of research is based on the work of developmentalist Mary Ainsworth.

Ainsworth's goal was to find some way young children vary in their attachments to their mothers, so that these variations could be linked to —that is, correlated with—how well these children do in other areas of their lives. The problem is that you can't just look to see whether a child is or is not attached to his mother, because *all* normal children are attached to their mothers (assuming they have a mother to be attached to). Even children whose mothers have neglected or abused them are attached to their mothers. It is a sad and paradoxical fact that abuse may actually increase a child's clinginess, because attachment is most evident when a child is frightened or in pain. The abused child may go for comfort to the very person who abused her.*

Since testing for the presence or absence of attachment was found to be useless, some other measure was needed. Mary Ainsworth's contribution was to devise a way of testing what she labeled the *security* of the child's attachment. The test is generally given to toddlers between the ages of twelve and eighteen months—the period when attachment reaches its peak. Here's how it works. The toddler and his mother are brought into a laboratory room strewn with toys—no second baby this time—and after a few minutes the mother leaves. In fact, she leaves twice: the first time there is another woman (a researcher) in the room; the second time the baby is briefly left alone. Most babies cry when the mother leaves, but the moment of truth comes when she returns. How does the baby react to her reappearance? How glad is he to see her? Some babies—the ones judged to be "securely attached"—crawl or toddle to the mother and are comforted by her presence. Others—the "insecurely attached" ones—ignore her, or continue to scream inconsolably, or alternate between clinging to her and angrily pushing her away.

I am in agreement with the attachment researchers in believing that these differences in the child's behavior really do indicate something important about the mother–child relationship. What they indicate is how helpful the mother has been in the past when the child was upset. If the child has, in the past, found his mother to be a source of comfort when he was frightened or unhappy, he will expect her to continue to be a source of comfort. At this point, though, the attachment researchers and I part company: they believe that these expectations color the child's subsequent relationships and I do not. Yes, the child has learned to expect certain things from his mother, but he would be foolish indeed to generalize these expectations to other people he might meet up with in the future. Cinderella would never have gotten to the ball if she thought that everyone was going to treat her the way her stepmother did.

It was the British psychiatrist John Bowlby who proposed that the mother–child relationship forms a sort of template for all later relationships. Fueled by the nurture assumption, the idea took off. The baby, said Bowlby, develops an "internal working model" (a kind of concept) of its

* The same is true in other species. A researcher who studied imprinting in ducklings noticed that if he accidentally stepped on the feet of a duckling that was imprinted on him, the duckling followed him more closely than ever.

relationship with its mother, and then expects other relationships—with father, siblings, peers, babysitters, and so on—to follow the same pattern. An appealing theory but, not to put too fine a point on it, wrong. There may well be a working model of the Mommy–Baby relationship in the baby's mind, but if so, it is trotted out only when Mommy is around. The model is useless for predicting how other people will behave and whether or not it is safe to trust them. Knowing what to expect from Mommy is of no use at all in dealing with a jealous big sister, or an indifferent babysitter, or a playful peer. It definitely comes in handy, though, for dealing with Mommy.

In the twenty years since Mary Ainsworth devised the test for measuring security of attachment, thousands of toddlers have been subjected to the Where's-Mommy?-Oh-here-she-is! procedure and hundreds of articles have been published reporting the results. The purpose has been to show links between security of attachment and something—anything—else. Not surprisingly, most of the published articles have reported significant correlations. Some have reported that securely attached preschoolers have an easier time relating to their peers and an easier time with a variety of other developmental tasks—problem-solving, for example. But other researchers have reported contrary results. Developmentalists Michael Lamb and Alison Nash took a clear, cold look at the security-of-attachment data and concluded:

> Despite repeated assertions that the quality of social competence with peers is determined by the prior quality of infant–mother attachment relationships, there is actually little empirical support for this hypothesis.

The one convincing result that has come out of the attachment research is that children's relationships are, to a large extent, independent of each other. Toddlers who are securely attached to their mothers are not necessarily securely attached to their fathers, and vice versa. Children who are securely attached to their caregivers at the day-care center are not necessarily securely attached to their mothers, and vice versa. Security of attachment does not reside in the child, it resides in the child's relationships. The child's mind holds not just one working model but many of them—one for each relationship.

Though these relationships are largely independent, they are not entirely so, because the child contributes something to each of them. Characteristics the child is born with—including how sociable, how amiable, and how good-looking he is—are going to affect his relationship with his mother, with his father, with his other caregivers, and with his peers.

It's the same child, with the same genes, who participates in all these relationships, so it's not surprising that the attachment researchers occasionally find correlations between them.

The child lets go of Mommy's hand in order to join his peers, but he takes his genome with him.

Motherlessness versus Peerlessness

Don't get me wrong; I am not underestimating the importance of the mother–child relationship. I think these early relationships are essential, not just for normal social development, but even for normal brain development. As big as the human brain is when it makes its hazardous escape from the womb, it is still only a quarter of its eventual size. In order to complete its development, the brain requires certain inputs from the environment. The visual system, for example, requires patterned stimuli to both eyes during the first few months of postnatal life; if it doesn't get it, the child (or monkey or kitten) will later be found to lack three-dimensional vision. The problem is not in the eyes; it's in the brain. You might say that the developing brain "expects" certain stimuli to be present in the world outside the womb and relies on them in producing the finished product. Since this expectation is usually met, the visual system usually develops normally.

In the same way, I believe that the developing brain "expects" the baby to be taken care of by one person, or a small number of people, who provide food and comfort and are around a lot. If this expectation is not met, the department of the brain that specializes in constructing working models of relationships might not develop properly. Primatologists Harry and Margaret Harlow reared infant rhesus monkeys in cages by themselves, with only a terrycloth-covered doll and a bottle of formula to keep them company. As adults, these motherless monkeys were highly abnormal in their social behavior—extremely fearful and either indifferent or aggressive toward others of their kind.

But we primates are adaptable creatures. Rhesus monkeys that are reared from birth without a mother but kept in cages with three or four other infant monkeys turn into reasonably normal adults. They are miserable as babies—at least they *look* miserable, clinging to each other desperately—but by the time they are a year old they are behaving normally. There is no law of nature that says misery has to have sequelae. The things that make babies (or grownups) miserable do not necessarily have long-term consequences.

Nor does contentment today arm us against the morrow. Monkeys reared *with* mothers but *without* peers are happy enough in infancy but have serious problems later on, when they are caged with other monkeys. The peerless ones, reported Harlow and Harlow, show "no disposition to play together" and are abnormal in their social behavior—in fact, only the monkeys raised in total isolation are more abnormal.

Although a mother cannot act as a substitute for peers, peers can sometimes act as a substitute for a mother. This was demonstrated in our own species fifty years ago, in a poignant story reported by Anna Freud (Sigmund's daughter). It involved a group of six young children who had survived a Nazi concentration camp. The children—three boys and three girls, all between three and four years of age—were rescued at the end of the war and brought to a nursery in England, where Anna had a chance to study them. They had lost their parents soon after they were born and had been cared for in the concentration camp by a series of adults, none of whom survived. But they had remained together—the only scrap of stability in the chaos of their young lives.

When Anna Freud met them, they were like little savages:

> During the first days after arrival they destroyed all the toys and damaged much of the furniture. Toward the staff they behaved either with cold indifference or with active hostility. . . . In anger, they would hit the adults, bite or spit. . . . They would shout, scream, and use bad language.

But that was how they behaved toward *adults.* Toward each other they behaved quite differently:

> It was evident that they cared greatly for each other and not at all for anybody or anything else. They had no other wish than to be together and became upset when they were separated from each other, even for short moments. . . . The children's unusual emotional dependence on each other was borne out further by the almost complete absence of jealousy, rivalry, and competition. . . . There was no occasion to urge the children to "take turns"; they did it spontaneously since they were eager that everybody should have his share. . . . They did not tell on each other and they stood up for each other automatically whenever they felt that a member of the group was unjustly treated or otherwise threatened by an outsider. They were extremely considerate of each other's feelings. They did not grudge each other their possessions, on the contrary lending them to each other with pleasure. . . . On walks they were concerned for each other's safety in traffic, looked after children who lagged behind, helped each other over

> ditches, turned aside branches for each other to clear the passage in the woods, and carried each other's coats. . . . At mealtimes handing food to the neighbor was of greater importance than eating oneself.

That last sentence is the one that always tears me up. Unbelievable that little children could come out of a *concentration camp* more concerned about feeding their companions than feeding themselves! But, you see, each of these children was responding to the neediness he or she perceived in the others. It was like a game of House that never ended—each child playing the role of Mommy or Daddy to the others, while simultaneously maintaining a real-life identity as Baby.

In 1982, when the six would have been around forty years old, an American developmentalist wrote to Sophie Dann, Anna Freud's collaborator, and asked her what had happened to the concentration camp children. Evidently they had turned out all right; she replied that they were "leading effective lives."

They turned out all right because they had managed, against all odds, to form lasting attachments before they reached the age of four. Children who spend the first four years of their lives in old-style orphanages do not, as a rule, turn out all right. This is a puzzle, because after all there are plenty of other children in an orphanage to become attached to. But evidently the policies of old-style orphanages discourage children from becoming too attached to each other, perhaps out of a misguided attempt at kindness: children keep leaving as adoptive homes are found for them, so maybe we'd better not let them become too fond of each other. American researchers recently visited a Romanian orphanage that held five groups of children, each with their own room and their own caregivers. But, the researchers reported, individual children were frequently switched around from group to group, which meant that any attachments they managed to make would soon be broken.

Children who spend their early years in an orphanage do not lack social skills; if anything, they are overly friendly. What they lack is the ability to form close relationships. They seem to be unable to care deeply about anyone. The department in their brain where working models are constructed either has never learned to construct them or has given up the job as futile. "Use it or lose it" is a saying most appropriately applied to the developing brain, not the aging one.

Children who enter an orphanage *after* the age of four seem to be okay as adults, even if they spend the remainder of their childhood years in the institution. In the war-torn African nation of Eritrea, many children have lost their parents and are being cared for in institutions; others have suffered various disruptions but have managed to remain with their parents. Some American researchers recently compared a group of institutionalized Eritrean orphans with a matched group of children who were living with their parents and found "relatively few clinically significant differences" between them. The only important difference was that the orphans were unhappier.

No question about it, children without parents are unhappier. An Australian researcher named David Maunders interviewed a number of adults who had spent most or all of their childhood years—but not the first four—in Australian, American, or Canadian orphanages. What he found out about life in an orphanage reminds me of the early chapters of *Jane Eyre.*

> Entering the institution was confusing and traumatic, and little was done to ease the transition. Life was characterized by discipline and corporal punishment, though this was tempered in recent times. Household chores dominated daily routines. There was little possibility of love and affection.

These children had started out with parents, so they knew what they were missing. One of Maunders' informants, who had been placed in an institution at the age of five, told him,

> I can remember going to bed every night and thinking "when I wake up this dream will be over." And I would wake up and it wasn't. But I would do that every single night that I lived there.

The remarkable thing about these orphanage-reared people is that as adults they are leading what Sophie Dann called "effective lives." They have husbands and wives. They have children and careers. They didn't have parents during most of their childhood but they did become socialized.

It is harder to find reports of people who did have caring adults in their lives but who lacked the normal opportunities to be with other children. Those raised on isolated farms, for example, usually have siblings to keep them company. Nonetheless, these people do show some subtle signs of social impairment. Consider, too, the abnormal childhood experiences of the little princes and princesses of bygone European kingdoms and ask yourself whether those people turned into normal adults. Another unfortunate group consists of those who were kept at home during childhood due to chronic physical disorders. As young adults these people are, as one report puts it, "at high risk for psychological symptoms."

Finally, there are the prodigies. Prodigies are often depicted as peculiar and perhaps their reputation is not undeserved. I am not talking about garden-variety gifted children—those children do fine. But the ones who are off the charts, who have nothing in common with other children of their own age, have a high rate of social and emotional problems.

Take, for example, the sad case of William James Sidis. His parents (who named him after the famous psychologist) thought their only child was so special that they devoted their lives to educating him. William was born in 1898, a time when enthusiasm for education was unbounded and authorities were saying that any child could be made into a genius if he or she received the proper training. William learned to read at eighteen months; by the age of six he could read in several languages. At that point he was required by Massachusetts law to go to school. In six months he had moved through all seven grades of the local public school, so his parents took him out of school and William spent the next couple of years at home. Then he spent three months in high school and a couple more years at home.

At the age of eleven, William James Sidis entered Harvard University. A few months later he delivered a speech on "Four-Dimensional Bodies" to the Harvard Mathematical Club. Those who attended it were astounded by the boy's brilliance.

That was the high point of William's life—from there on it was downhill all the way. Although he received a bachelor's degree from Harvard at the age of 16, he never put it to use. He spent a year in graduate school and then went to law school, but never received a degree from either place. He obtained a position teaching math at a university but that didn't work out either. Reporters kept following him around in search of stories on the theme of "Early ripe, early rot." The paparazzi were a nuisance but they cannot be blamed for the quirks in his personality.

As an adult, William turned against his parents—he even refused to attend his father's funeral—and against the academic world in general. He spent the rest of his life working at mindless, low-paying clerical jobs, moving from one to another. He never married. His hobby was collecting streetcar transfers; he wrote a book on the topic, described by someone who read it as "arguably the most boring book ever written." People who met him in his later years gave various descriptions of his personality: one said he "was possessed of the chronic bitterness common to lonely roomers," another that "he had a certain childlike charm underlying his intense, erratic manner." William James Sidis died of a stroke at the age of forty-six, alone, obscure, penniless, and terminally maladjusted.

William's situation was similar to that of the monkeys reared with a mother but without peers. As adults, those monkeys were more abnormal than the ones reared with peers but without a mother. The worst-off monkeys, of course, were those that had neither. Thankfully, such cases are extremely rare in humans. Two that spring to mind are Victor, the Wild Boy of Aveyron, and Genie, the California child who spent her first thirteen years alone in a small room, tied to a potty chair.

Victor and Genie turned into highly abnormal adults. We will never know, however, whether their abnormalities were due to a lack of parental love or to a lack of other children to play with; a third possibility is that there was something wrong with them to begin with. But a case study from Czechoslovakia provides a clue. A pair of twin boys lost their mother at birth and were placed in an orphanage. When they were about a year old their father remarried and brought the boys home—to a stepmother who made Cinderella's look like Mister Rogers. For the next six years the boys were kept in a small unheated closet, undernourished and periodically beaten. When they were discovered at the age of seven they could barely walk and had less language than an average two-year-old. But they turned out all right. They were adopted into a normal family and by the age of fourteen they were attending public school and had caught up with their classmates. They had "no pathological symptoms or eccentricities," according to the researcher who studied them. In their first seven years they hadn't had a mother's love—nor, it would appear, a father's—but they had had each other.

Playmates

Twins are in an unusual situation: they have an agemate to play with from Day One. They do not, however, play with each other from Day One. Playing with an agemate is a skill that takes time to develop. The two unacquainted babies in the laboratory room, described earlier in this chapter, were interested in each other but their attempts at friendliness were clumsy and sometimes counterproductive. Poking one's finger in the eye of a new acquaintance is not the best way to begin a relationship.

It is easier for a baby to play with a parent or sibling: the older person structures the game and, through repetition, teaches her to respond appropriately. By the first birthday, the average American baby can play patty-cake or peek-a-boo with her parents. An agemate, a peer, is not nearly so helpful or understanding. Even with the best of intentions, a year-old baby cannot play games with another baby her age.

But a two-year-old can. Carol Eckerman and her colleagues have studied the development of play between agemates, using the same two-babies-in-a-laboratory-room procedure with babies of various ages. What they saw was a steady increase in the use of imitation as a way of getting along with others. Two babies coordinate their activities, and signal their interest in each other, by imitating each other's actions. Imitation is a human specialty; no species is as good at it as we are. That was what went wrong with Dr. Kellogg's experiment (described in Chapter 6) and with Dr. Kellogg's son: the child imitated the chimpanzee more than the chimpanzee imitated the child.

For the two unacquainted human babies in the laboratory room, imitation begins around the time they learn to walk. At first it is a simple matter of playing side by side doing the same thing. One baby picks up a ball, then the other baby picks up a ball. If there is only one ball, the second baby may try to take it away from the first one.

By the age of two, imitation has turned into something more elaborate and a great deal more fun. One child runs around the room, or bangs two toys together, or does something silly like falling over or licking the table, and the other does the same. Then the first player either repeats the action or thinks up a new one, in which case it becomes a game of Follow the Leader. These imitation games go on for only a few turns, but while they last they are enjoyed tremendously by both parties.

At two and a half the children can use words as well as actions to coordinate their play, and at three they are capable of playing games like House, which require coordinated imagination as well as coordinated action. At this point the children are no longer just imitating each other: each party in these shared fantasies plays a different role.

The other thing that happens in the period between one and three is that children start forming true friendships—they have constructed working models of their relationships with a number of their peers and have decided that they like some of them better than others. In a nursery school or day-care center you see children playing day after day with the same companions. In settings where there is a range of ages, these little cliques tend to consist of children of roughly the same age, because older children prefer not to play with younger ones if they have any choice in the matter. The cliques also tend to consist of children of the same sex. By age five they are almost entirely single sex.

What I am describing here is the development of play with peers among children living in industrialized, urbanized societies like our own. In such societies, parents take it for granted that their children should have opportunities to play with other children and they go out of their way to provide them. Parents who don't send their children to nursery school or a day-care center arrange play groups for them or make friends with people who have offspring of the same age. Whether they are college graduates or high school dropouts, behavioral geneticists or socialization researchers, few parents doubt that experiences with peers are important to their child's development.

Unlike the belief in the nurture assumption, the belief in the importance of playmates is held around the world. But before societies became industrialized and urbanized it was rare for a young child to have others of the same age to play with, and this is still true in some parts of the world. In tribal and small village societies, the young child graduates from her mother's lap into a play group of children with a range of ages, and she starts off as the youngest one in the group. The range of ages may be from two and a half to six, or from two and a half to twelve—it depends on population density. If there are enough children in the vicinity, the older ones go off and form their own groups.

I have already described, in an earlier chapter, the mixed-age play group of traditional societies. In such societies extended families tend to cluster together, so the play group is generally composed of children who are related to each other. Children play with their siblings, their cousins, and their younger aunts and uncles. The older children in the group are responsible for the younger ones—it is they, to a large extent, who teach the younger ones how to behave and how to play the local games. Their instruction is not gentle—teasing and ridicule are prevalent, as is the use of force—and it is not based on reasoning. The five-year-old does not tell his little sister that she shouldn't throw sand at Bisi because "How would you like it if Bisi did that to *you?"* And yet fights and serious aggression are uncommon. Even in Western societies, children tend to be less aggressive when they're playing together by themselves than when they're being watched over by parents or teachers. Perhaps they fight more when adults are present because they know they can count on the adults to stop them before things go too far.

Children in traditional societies also learn their language in the play group—at two and a half, they're just beginning to talk. They do not learn it from their parents because their parents do not talk to them much. Their conversational partners are other children. Older children simplify their speech a bit when they're talking to younger ones, but they do not provide the kind of language instruction that parents give their toddlers in our society—the question-asking, the patient rephrasing of the learner's poorly phrased statement, the smile or pat when something is said exceptionally well. So the children in traditional societies learn language at a slower rate. But learn it they do. They all become competent users of the language that is spoken in their community. And they all become socialized.

Even after they graduate from their mothers' laps into the play group, children in most traditional societies remain emotionally attached to their parents, just as they are in our own society. They turn to their parents for food, for protection, for comfort, and for advice. The bond between parent and child—the love that children have for their parents and parents have for their children—typically lasts a lifetime. In most traditional societies, a young man remains in his native village and builds a house near his parents and brothers. A young woman generally leaves her native village when she marries, but she is likely to come home for visits or welcome her parents when they come to visit her.

Nevertheless, when the children in traditional societies graduate from their mothers' laps into the play group, in some sense they stop being their parents' children and become the community's. Any adult in these societies can admonish a child if he or she sees the child doing something wrong. It takes a *village* to raise a child.

But the reason it takes a village is not because it requires a quorum of adults to nudge erring youngsters back onto the paths of righteousness. It takes a village because in a village there are always enough kids to form a play group. "It is in such play groups that children are truly raised," observes Irenäus Eibl-Eibesfeldt. "The child's socialization occurs mainly within the play group." Eibl-Eibesfeldt is talking about the traditional societies he specializes in—inhabitants of places like sub-Saharan Africa and the highlands of New Guinea. But I believe similar statements can be made for children in complex, urbanized societies like our own.

In our society we place great stress on the bond between parent and child. We talk about spending "quality time" with our children; the children of divorced parents shuttle back and forth between two households so they can put in some quality time with each parent. But if spending time with their parents is so all-fired important to children, why is it so hard to get them to come home? Why do we need curfews?

In Chapter 5, I described a young Okinawan boy who came home during the day only to stuff his face; then he was off again—his friends were waiting for him, he told his mother. Among the Chewong, who eke out a living in the rainforests of the Malay Peninsula, children voluntarily detach themselves from their parents well before they reach their teens.

"At about the age of seven," reports an anthropologist who studied these people, "children can be seen to shift gradually away from parents in order to join a peer group consisting of older children of the same sex." Once the transition is complete—the anthropologist doesn't say how long this takes but I gather it is no more than a year or two—the adults in the community "do not seek actively to teach anything at all" to their offspring. "A child is left to perform various tasks whenever it chooses to, and to approach an adult if and when it requires specific guidance."

As British ethologist John Archer has observed, "Many features found in young animals are not precursors of adult ones, but serve to aid survival at that point in development." The fact that a close attachment to a parent (or parent substitute) is a necessity for babies and toddlers does not mean that it is a necessity for an older child.

Socialization by Proxy

In nonhuman primates, a lot of social behavior is built in. A chimpanzee that grows up in the Mahale Mountains of Tanzania behaves much the same—though, interestingly enough, not exactly the same—as one that grows up in the Gombe Stream National Park. But in humans, group contrast effects (described in the previous chapter) can produce noticeable differences in social behavior even between groups that live right next door to each other. An anthropologist studied two Zapotec villages not far apart in southern Mexico. Their inhabitants speak the same language, plant the same crops. But in La Paz, aggression is rare and disapproved of; in San Andres it is pervasive and accepted as a fact of life. The homicide rate in San Andres is more than five times that of La Paz. The anthropologist saw two San Andres brothers throwing rocks at each other. Their mother, he reported with ill-concealed disapproval, "did nothing to stop this rather dangerous activity and simply remarked that her boys always fought."

We know that social behavior in humans isn't built in, because it varies so much from one group to another. It has to be learned. We know that children learn it, because most of them end up behaving more or less like the other people in the society they grew up in. Not necessarily the one they were *born* in—the one they grew up in.

How do they do it? Back in the days when Freudian theory was an influential force in psychology, it was easy: the child learned to behave by identifying with his father or her mother. Identification led to the formation of the superego, and the superego kept him or her on the straight and narrow.

Even after Freudian theory went out of fashion, many psychologists continued to believe that children tailor their behavior to that of the same-sex parent. Pictures of fathers shaving, and little boys pretending to shave, festoon developmental psych textbooks—including, I admit, my own.

Sure, children imitate their parents. We humans are the champion imitators of the animal kingdom: we have to be, since so much of our social behavior has to be learned. And American parents think it's cute when little boys pretend to shave. We don't think it's so cute, though, when they light matches, or chop down the cherry tree, or smack their younger brother, or use the F word, even though those behaviors, too, are imitative. We want our kids to behave like good little children, and good little children don't behave like grownups.

As a way of getting socialized, imitation of the parents doesn't work any better in other parts of the world. If you think American children have a hard row to hoe, consider what it takes to learn proper social behavior in, say, village societies in the Polynesian islands. Polynesian children are expected to behave in a restrained and self-effacing manner with adults: the adult is supposed to initiate and control all interactions, the child is supposed to be compliant and undemanding. With their peers they are allowed to behave in a more assertive fashion. I pointed out in Chapter 1 that the children cannot learn these rules by observing their parents. Polynesian parents do not behave in a restrained and self-effacing manner, either with other adults or with children. Kids who imitated their parents' social behavior would be way out of line.

Children can also get into trouble behaving like their parents if their parents happen not to be normal members of their society. They may be eccentrics or alcoholics or criminals. Or they may simply be immigrants, untrained in the local rules of social behavior. We think of the immigrant parent as a modern phenomenon but in all probability it is an ancient one. Consider a little girl born into a tribal society that is always warring with its neighbors—a traditional lifestyle, older than our species. This hypothetical child is the daughter of a woman who was neither born into the tribe nor reared in it, but who was abducted during a raid on an enemy village. She, the captive, is now the trophy wife—or one of the trophy wives—of a successful warrior. But she is ignorant of many of the customs of her new tribe and she speaks a different dialect. The daughter would be ill advised to copy the social behavior and speech of her mother.

When children do imitate their parents, they don't do it blindly: they are careful about it. They do it only when they think the parent is behaving normally or typically, the way other people in their society behave. They become conscious of such things at a surprisingly early age. A German-born colleague told me that his four-year-old daughter refuses to speak German to him when they are in the United States but she is quite willing to do so during visits to Germany. Children also decide, at an early age, that women and men are "supposed to" do different things. One of my daughters, at around age five, announced that fathers are not supposed to do the cooking.

"And mothers are not supposed to do the hammering and sawing?" I asked her.

"That's right," she said, though she had the grace to look embarrassed. In her home, the father did about half the cooking, the mother did all the hammering and sawing.

Kids probably get these ideas partly from television and story books. But they check them for accuracy in the fantasy games they play with their friends in the nursery school or day-care center. When children play games like House or Fireman, they are not pretending to be their parents (even if Daddy happens to be a fireman): the roles are stereotypes, painted with a broad brush and agreed upon by a committee of children. Such games are less common among children in traditional societies where there is no privacy and everyone knows what everyone else is doing. In places where all the women do pretty much the same things and so do all the men, there is no need for children to hold committee meetings to discuss the job specifications.

Adaptable creatures, kids are. An only child who lives with her parents in a place where there are no other children would perforce model her behavior on that of her parents. If she were raised by apes, like Tarzan,* or by wolves, like a pair of young girls reportedly found in a wolves' den in India, she would behave, to the best of her ability, like an ape or a wolf. But usually there is a choice. Usually children have a number of potential models and they don't all behave alike, so whose behavior do they imitate?

Donald Kellogg, whose infancy I described in Chapter 6, wasn't raised by apes—he was raised, for the better part of a year, *with* one. Gua went back to the zoo when Donald's parents realized that the ape was influenc-

* If Tarzan had *really* been raised by apes and had not been discovered until he was full grown, he would probably be something like Genie or Victor. His English would never get beyond the "Me Tarzan, you Jane" stage, and he would not be housebroken. Living in the trees it doesn't matter, except maybe to the one just below you in the tree.

ing Donald more than they were influencing the ape. At nineteen months, Donald could speak only three English words but he had gotten pretty good at communicating in Chimpanzee. Why did Donald imitate the chimp's language in preference to the language of his parents?

I think the answer is that Donald already had a rudimentary sense of social categories. He perceived—correctly—that he and Gua were in the same social category, one based on age. Babies can categorize, as I said in the previous chapter. They categorize people by age and by sex before they are a year old. Perhaps they already have some inkling of which category they themselves are in. If monkeys and apes can do it, why shouldn't a human one-year-old?

Donald and Gua were like siblings. The Kelloggs treated them alike—dressed them in the same clothes, fed them the same foods, disciplined them the same way. When there is a choice, young children preferentially imitate certain models, and older siblings are among their favorite models. Gua was actually a couple of months younger than Donald, but chimpanzees mature more rapidly. To Donald, Gua was like an older sibling.

Consider the Polynesian children who have to learn two different sets of social rules. How do they learn the rules for interacting with adults? Not by hearing lectures from their parents on Polynesian etiquette. In traditional cultures, parents do very little lecturing and provide very few explicit guidelines. Mostly, children just get reprimanded or smacked if they do something wrong. They are expected to learn by observation—and they do. B. F. Skinner said that organisms have to be rewarded in order to learn, but children can learn without being rewarded—and, for that matter, without being punished. They can learn by observing others like themselves and seeing what happens to them. A child does not have to burn her own fingers on the hot stove to learn not to touch it: all she has to do is to witness what happens when her brother touches it. A Polynesian child can learn the rules for behaving with adults by watching children a little older than herself. And those children, in turn, watch the children a little older than *them.*

The other day my sister-in-law was cutting up a sweet red pepper and offered a piece to my nephew. He put it in his mouth. His little sister said "I want some too!" Then my nephew decided he didn't like it and asked permission to spit it out. My niece instantly changed her mind. She decided, without having tasted it, that she didn't like red pepper either.

Her parents like red pepper. But that didn't matter to my little niece—all that mattered was whether her brother liked it. A developmentalist named Leann Birch noticed that children of preschool age—an age notorious for its picky eaters—cannot be cajoled by their parents into eating foods they dislike, or think they dislike. Parental propaganda and persuasion don't work; the child remains intransigent. There is only one way to get a preschooler to learn to like a disliked food: seat her at a table with a group of children who do like it and serve it to all of them.

Preschoolers' preferred models are other children. By the age of three or four they have begun to tailor their own behavior to that of their nursery school playmates and, what's more, they have begun to bring that behavior home. The easiest way to see this is to hear it: they are starting to pick up the accents of their peers. The daughter of a British psycholinguist was "speaking black English* like a native" after four months of attending a nursery school in Oakland, California. Not all the kids in the nursery school were black, but the ones she played with were. Although this child probably spent more time with her British mother than with her African-American playmates, it was their accent, not her mother's, that was influencing her speech.

"Us" versus "Me and Thee"

In the previous chapter I described the experiment by social psychologist Henri Tajfel in which boys were told that they were either underestimators or overestimators. That was all it took to make a boy favor his own group over the other one. Tajfel coined the word *groupness* to refer to this feeling of affiliation with one's groupmates.

John Turner, Tajfel's student, went on to specify some of the characteristics of groupness. People don't have to like all the members of their group. For that matter, they don't have to *know* all the members of their group. For that matter, they don't have to know *any* of the members of their group. All it takes is the knowledge that you and they are in the same social category. It's a matter of self-categorization:

I am an X.
I am not a Y.

From these simple premises, our evolutionary history has predisposed us to draw a simple corollary: that we prefer Xs to Ys. We also conclude, as a result of the categorization process itself, that we are similar to other Xs and different from Ys. These mental activities go on at a level not ordinarily accessible to the conscious mind, but they have some visible conse-

* Sometimes called "Ebonics."

quences. We become more similar to the other members of our group through the process of assimilation. The differences between our group and the other one become exaggerated due to group contrast effects. And, under some conditions, hostility toward the other group emerges—the "us versus them" effect.

What I am describing here is not at all like relationships between individuals. The capacity to form dyadic relationships is present at birth. Groupness takes longer to develop. Dyadic relationships are based on things like dependency, love and hatred, and enjoyment of the other's company. Groupness is based on recognition of basic similarities—we are alike in some way—or of shared fate—we are all in the same boat. Dyadic relationships involve two people; three's a crowd. Groupness almost always involves more than two people; there's no upper limit on the number. If this description makes groupness sound like a purely intellectual sort of thing, don't be misled: deep and powerful emotions are involved. Over the course of our species' history, many more people have died for their group than have died for their personal relationships.

In Chapter 6, I spoke of a "social module"—it is the part of the brain that doesn't work properly in autistic children. In the same way, one could speak of the "visual system"—the system that doesn't work properly in blind children. But the visual system has a number of separate components, and something can go wrong with one of them and not the others. There are brain-injured people who can see where things are but not *what* they are, and there are those who have the opposite problem. There are people who can visually identify objects but not faces, and people who can see perfectly with either eye but cannot combine the two images into a three-dimensional picture. What we call the visual system is actually composed of a number of subsystems that are more or less independent, require different kinds of input and produce different kinds of output, and are assembled at different times and in different ways during early development.

I believe the same is true of the social module. It is composed of at least two subsystems: one that specializes in dyadic relationships—this one is ready to go at birth—and one that specializes in group things—this one takes a little longer to assemble.

Groupness and personal relationships not only work independently: they can work in opposition to each other. I used to wonder why it is an insult when someone says, "Some of my best friends are Jews." The reason is that the speaker is making a distinction between friendship—a personal relationship—and his feelings about a group. He can like his friends without liking the group they belong to, and one gets the impression that this is the case.

Groupness and personal relationships sometimes make conflicting demands. In wartime, for example, people sometimes have to choose between remaining with their loved ones or leaving them in order to defend their group. Different people resolve such dilemmas in different ways.

According to my theory, it is the groupness department of the mind that enables children to be socialized and their personalities to be modified by the environment. Groupness is involved whenever there are long-term changes in children's behavior. The department that deals with personal relationships may give rise to some very powerful emotions, but it produces only temporary changes in behavior.

Group Socialization Theory

The central question of this book is: How do children get socialized—how do they learn to behave like normal, acceptable members of their society? What shapes the raw material of the infant's temperament into the finished product of the adult's personality? These may sound like two separate, almost unrelated questions—indeed, they are the subject matter of separate, almost unrelated schools of psychology—but from my point of view they are two sides of the same coin. For children, socialization consists largely of learning how to behave when they're in the presence of other people. And an adult's personality consists largely of how he or she behaves in the presence of other people. In a social species like our own, most behavior *is* social behavior. I am sitting here all by myself, but nonetheless I am engaging in social behavior. If you weren't ever going to read what I'm typing into my computer, what would be the point?

Children have to learn to behave in a way that is appropriate for the society they live in. The problem is that people in their society don't all behave the same way. In every society, people behave differently according to whether they are children or adults, males or females, single or married, princes or peons. What children have to do first is to figure out what sort of people they are—which social category they belong in. Then they have to learn to behave like the other members of their social category.

Figuring out which social category they belong in is the easy part. Even a three-year-old can tell you, in case you were misled by her unisex clothing and unisex name, "I'm not a boy, I'm a girl!" She also knows she is a child—she will be amused if you pretend to mistake her for a grownup, deeply offended if you call her a baby. Age and gender are the only categories that matter at this point. Race doesn't matter to a three-year-old. The daughter of the British psycholinguist didn't notice, or didn't care, that her favorite playmates at the day-care center had darker skin than she did.

The psycholinguist's daughter ended up talking like her African-American playmates because, from an early age, children tailor their behavior to that of others in their group, others they perceive as being "like me." If that is the case, you may be wondering, then how do the *others* learn how to behave? The answer is that children's groups operate by the majority-rules rule: whoever comes to the group with behavior that is different from the majority is the one who has to change. The African-American children learned their language at home or in their neighborhood, and when they came to the day-care center they found many others who spoke the same way. The daughter of the British psycholinguist found that she was a party of one: no one talked like her. So she changed and her playmates did not. And then she brought her new language home. This, she was saying, is how people like me are supposed to talk. Of course, she didn't actually *say* that. For children, socialization is largely an unconscious process.

My theory of how children become socialized and how personality gets modified during development is called "group socialization theory." At least, that is what I called it in my article in the journal *Psychological Review.* I'm not entirely happy with that name for two reasons. First, my theory has to do with personality development, not just with socialization. Second, the word "socialization" is misleading, because it suggests something that is done *to* children. What I'm talking about is something that children, to a large extent, do to themselves.

Children get their ideas of how to behave by identifying with a group and taking on its attitudes, behaviors, speech, and styles of dress and adornment. Most of them do this automatically and willingly: they *want* to be like their peers. But just in case they have any funny ideas, their peers are quick to remind them of the penalties of being different. School-age children, in particular, are merciless in their persecution of the one who is different: the nail that sticks up gets hammered down. The hammering sometimes makes the child aware of what he's doing wrong and almost always motivates him to change it. Psycholinguist Peter Reich still cringes when he recalls a childhood experience at a Boy Scout Jamboree. He grew up in Chicago, where the word *Washington* is pronounced *Warshington.* Boy Scouts from other parts of the country would come up to him, ask him to say the name of the capital of the United States, and would "double up with laughter" when he did so. "I can still remember," says Reich, "practicing hard to change the pronunciation of this and other words that marked my dialect."

Laughter is the group's favorite weapon: it is used around the world to keep nonconformers in line. Those for whom laughter alone does not do the job—those who don't know what they're doing wrong or who will not or cannot conform—suffer a worse fate, expulsion from the group. That was my fate for four years.

Perhaps you are wondering how I could be expelled from a group when girls don't usually gather in groups. School-age girls usually have friends, not groups—they split up into twosomes and threesomes. I have confused the issue by using the word *group* to mean both a play group—a group of real children playing together—and a social category. It is the social-category meaning that is relevant here—what John Turner called the "psychological group" and earlier theorists called the "reference group." Although as a fifth grader I didn't interact at all with the other fifth-grade girls in my class, I still identified with them. They were my psychological group and they rejected me, so in that sense I was expelled from the group.

My absence from their group meant that I had no opportunity to influence them. Yet they were still able to influence me. You don't have to actually interact with the members of your psychological group in order to have them influence you. I was a fifth-grade girl too, and even though the others wouldn't talk to me, I watched them very closely. It wasn't as good as being a participating member, but it was better than nothing.

The peer group may not accept the child, but that does not prevent the child from identifying with the peer group. When he was six years old, an American boy named Daja Meston was dumped in a Tibetan monastery by hippie parents who had spent the previous six years wandering around Europe and Asia. The boy remained in the monastery until he was fifteen—he was one of a group of boys in training to become Buddhist monks. The others were all Tibetans. Daja stood out like a sore thumb: too tall, too white. He had no close friends; he was teased by his peers for being different. But they were his psychological group and he did get socialized. Now Daja lives in the United States; he is married to a Tibetan woman he met in this country. His appearance is misleading, he tells an interviewer: "a white body that houses a Tibetan."

Daja identified with his peers at the monastery because he had no other choice. It was clear to him, even if it wasn't clear to them, that they were all in the same social category. So he became a Tibetan like them—he learned to behave and talk and think like a Tibetan. If he had been accepted by his peers he would probably have turned into a different sort of Tibetan (a point I'll return to later), but, accepted or rejected, a Tibetan he was bound to become.

I don't believe that Daja would have turned into a different sort of Tibetan if he had had a close friend at the monastery. His sojourn there would have been considerably happier, but friendship (or lack thereof) leaves no permanent marks on the personality. Identification with a group, and acceptance or rejection by the group, do leave permanent marks on the personality. Researchers have studied the long-term effects of grade-school friendships (or lack thereof), and the long-term effects of peer acceptance or rejection. They found that peer acceptance or rejection was associated with "overall life status adjustment" in adulthood; having or not having a friend in grade school was not.

Friendship is a dyadic relationship. One can have a talent for friendship even if one has no talent for commanding the attention or respect of one's groupmates. Children who have low status in the peer group, or no status at all, often have successful friendships. During my sojourn in the snooty suburb I did have one friend. She was three years behind me in school and two years younger, and she lived next door. As far as I know, our unequal friendship had no long-term effects on either of us. Children accommodate their behavior to their friends in the same way they accommodate their behavior to the standards of their peer group, but for the friends the accommodations are short-lived and specific to the relationship, managed by the part of the mind that specializes in working models—the relationship department, not the groupness department. Sometimes friendships *appear* to have long-term effects, but that is because most friendships are between children who are members of the same psychological group.

Girls versus Boys

The most important psychological groups during childhood are the gender categories. Even three-year-olds identify themselves as girls or boys, and even three-year-olds generally prefer playing with others of their own sex. By age five, they are playing in little groups that are almost completely segregated by sex. They are able to divide up like this because urbanized societies like our own provide children with so many same-age peers: it allows them to be choosy. At home or in neighborhood play groups, where there are fewer kids, they will play with anyone they can get. Even a chimpanzee.

One reason girls and boys prefer playmates of their own sex is that from nursery school on, they tend to have somewhat different styles of play. They naturally gravitate toward others who share their play interests. But I don't think it's just a question of different play interests: I think it's also a result of self-categorization—of seeing themselves as members of a particular group. Because they're in it, they like their group best.

And because they're in it, they want to be like the other members of their group and not like those of the other group. Little girls want to be like other girls (and not like boys); little boys want to be like other boys (and not like girls). The four-year-old daughter of a colleague refuses to wear what used to be her favorite sneakers because one of her friends told her they were "boy shoes." Another father overheard his little girl telling her toy stegosaurus that only boys can play with guns—a notion, he said, she picked up in the day-care center. Being philosophically opposed both to guns and to sexism, the father was in something of a bind.

> I tried explaining to my daughter that (a) boys or girls could play with guns; (b) I didn't like them regardless of who was playing with them; and (c) really, even though she was a girl, she could have a gun except that I didn't want her playing with any guns.

Good try, Dad. But relax: it isn't your opinion that matters to your little girl. My colleague's four-year-old doesn't care if her *parents* think it's okay to wear the disputed sneakers. Her opinions on such issues are not based on what she hears her parents say; they have never, for example, announced that "Boys are yucky" or that "He can't play with us, he's a boy." And sex-typed behavior such as playing with guns is not something children catch, like a virus, from their same-sex parent. Even in America, the fathers of most little boys do not play with guns. Nor do the mothers of most little girls play hopscotch or jump-rope.

For older children, the most stringently enforced rules of behavior pertain to how they are expected to act toward members of the opposite sex. An eleven-year-old girl explained to some researchers what would happen if she should violate her group's taboos by sitting down next to a boy in school. "People would not be my friends," she said. "They would scorn me." It would be like "peeing in your pants," she told the researchers. "You would be teased for *months* about this. But if you wore your shoes backwards you'd only be teased for a few days."

During middle childhood other things—such as whether one's skin is brown or white—become increasingly important, but they are never as important as the gender distinction. A sociologist who spent some time observing sixth graders in a racially integrated school noted that it was rare for a kid to sit down at a lunch table next to someone of a different race, but virtually unheard of for a kid to sit down next to someone of a different sex. Students, she reported, will risk the wrath of their teachers rather than join a group of the "wrong" sex.

> Mr. Little instructs the students to form groups of three for a science experiment. None of the groups formed are sexually integrated. Mr. Little notices a group of four boys and instructs one of its members, Juan (black), "Go over and work with Diane" (Diane's group has two black girls in it). Shaking his head, Juan says, "No, I don't want to!" Mr. Little says quietly but with an obvious edge in his voice, "Then take off your lab apron and go back to the regular class." Juan stands absolutely still and doesn't reply. After a long heavy silence, Mr. Little says, "Okay, I'll do it for you." He unties Juan's apron and sends him out of the room.

Perhaps Mr. Little would have had more sympathy for Juan if he had known that, for kids of this age, sitting down next to someone of the opposite sex is as bad as peeing in your pants.

Because girls and boys form separate gender groups during middle childhood, socialization is specific to gender. A child doesn't get socialized to behave like an American—he gets socialized to behave like an American *boy,* or she gets socialized to behave like an American *girl.* The norms of behavior differ for the two groups. Timidity and shyness, for instance, are acceptable in girls' groups but unacceptable in boys' groups. On the other hand, loudness and excess exuberance are frowned upon by children of both sexes: the ideal in Western societies is to act "cool."

Some researchers in Sweden tracked a group of children from the age of eighteen months to sixteen years. A few of these children started out timid and shy; a few were the opposite—boisterous and uninhibited. These characteristics did not change much between eighteen months and six years, but from six to sixteen two things happened: the boisterous children of both sexes calmed down and became more moderate in their behavior, and the boys who had started out shy and timid were no longer distinguishable from the other boys. The shy, timid girls didn't change much, but the shy, timid boys changed a lot. Timidity is acceptable among girls but unacceptable among boys, and a boy who acts that way

—remember Mark, in Chapter 2?—will be teased and bullied by his peers until he learns to master his distress.

I saw it happen in the family I grew up in. My brother was a child like Mark and I was a child like Audrey. We were biological siblings with the same parents, but we were nothing alike. As a toddler, my brother was afraid of everything, especially strangers and loud noises. Thunderstorms terrified him (I loved them). My mother pampered him, my father was annoyed with him, and neither had any more effect on him than they had had on me. When my brother started first grade he was still a timid child. But by the time he was twelve, this boy who had been afraid of thunderstorms was doing experiments with gunpowder in the company of his friends. Came damn near to killing himself. As a grownup my brother is brave, calm, and low-keyed. A typical Arizonan.

My peers taught me the opposite lesson. My brother became bolder and I became more inhibited. After passing through the refining fires of childhood, we are much more alike, my brother and I, than we were as kids.

"Us" versus "Them"

The most troublesome effect of self-categorization is the tendency to dislike the category the self isn't in. Intergroup hostility is not an inevitable result of categorization into two contrasting groups, but it is a common one.

A boy plays with the girl next door when there's no one else to play with, but he nails a "No Girls Allowed!" sign on the clubhouse he builds with his male peers. At times and in places where the salient social categories are *girls* and *boys,* hostility toward members of the opposite sex is detectable in nursery school and increases during the elementary school years. Over five years of coeducation, from kindergarten to fourth grade, girls' ratings of how much they like their male peers, and boys' ratings of how much they like their female peers, go steadily downhill. A researcher asked some grade-school boys to name (privately) the girls in their classroom whom they disliked. Several of the boys refused to answer the question, reported the researcher. "They disliked *all* of the girls in their classroom," they insisted.

Most boys don't really dislike *all* girls, and most girls don't really dislike *all* boys. At the same time that these group-against-group animosities are erupting in mock battles on the school playground and the crisis in Juan's science lab, children of both sexes are forming crushes on individual members of the opposite sex. Some of the boys even have girlfriends! Ah, but those are relationships. A different thing entirely. Juan and Diane may be friends somewhere else, but not in the classroom. Gender categories are too prominent in the sixth-grade classroom.

But gender categories aren't the only ones that are prominent during childhood. There are also the age categories: kids versus grownups. Unless you've led a very sheltered life you are no doubt aware of the animosity between adults and *teenagers,* but I'm not talking about teenagers here—I'm talking about children. Even little children.

Children are dependent on adults. They love many of the adults in their lives; sometimes they even love their teachers. But these are relationships. When they are in a social context that evokes their groupness, and the salient categories are *grownups* and *kids,* you can see, if you know where to look, signs of us-versus-them effects even at the tender age of four. Here is sociologist William Corsaro's description of children in an Italian *scuola materna,* a government-sponsored nursery school:

> In the process of resisting adult rules, the children develop a sense of community and a group identity.

(I would have put it the other way around.)

> The children's resistance to adult rules can be seen as a routine because it is a daily occurrence in the nursery school and is produced in a style that is easily recognizable to members of the peer culture. Such activity is often highly exaggerated (for instance, making faces behind the teacher's back or running around) or is prefaced by "calls for the attention" of other children (such as, "look what I got" in reference to possession of a forbidden object, or "look what I'm doing" to call attention to a restricted activity).

I detect in this description not only us-versus-them effects but group contrast effects as well. Children see adults as serious and sedentary, so when the salient social categories are *kids* and *grownups*—as they might be, for instance, when the teacher is being particularly bossy—they become sillier and more active. They demonstrate their fealty to their own age group by making faces and running around.

As children get older, demonstrating their fealty to their own age group becomes more and more important. It always amuses me to see preadolescents walking with their families in a shopping mall. They walk ten paces in front of their parents or ten paces behind them. In case any of their peers should see them, they want to make the situation perfectly clear: I am not with these people. I am not one of *them.* This has nothing to do with whether or not they love their parents. Some of their best friends are grownups.

Follow the Leader

Though signs of groupness are visible in the nursery school, and though even a four-year-old can switch back and forth between thinking of herself as a "kid" and thinking of herself as a "girl" (depending on whether age or gender categories are more salient), the fancier aspects of human groupness do not come into play until middle childhood. Middle childhood—the elementary school years—is when I think the most important things happen. It is when children get socialized for keeps and when permanent changes are made in their personalities. And yet it is also the period most likely to be ignored by psychologists. Sigmund Freud called it the "latency period"—the period in which nothing much happens. Shows you how much *he* knew.

The social and intellectual advances that occur around the age of seven are recognized all over the world. Parents in many societies believe that this is the age when kids "get sense." The Chewong children are not the only ones who bid their parents goodbye at around this age. In Europe during the Middle Ages, children were often sent away from home when they were seven or eight years old. The offspring of the rich served as pages in the homes of noblemen, those of the poor worked as apprentices or domestic servants. This tradition didn't die out completely: even today it is common for the sons of upper-class British parents to be sent away to boarding schools at the age of eight.

During middle childhood, children become more alike, more similar to their peers of the same sex. They learn how to behave in public—to not hit (if they're girls) or not cry (if they're boys), to act polite to grownups (if they're girls) but not *too* polite (if they're boys). Some of the rough edges get smoothed off their personalities as social behaviors unacceptable to their same-sex peers give way to more acceptable behaviors. The new behaviors become habitual—internalized, if you will—and eventually become part of the public personality. The public personality is the one that a child adopts when he or she is not at home. It is the one that will develop into the adult personality.

But assimilation—taking on the group's norms—is only part of the story. The other part is differentiation. At the same time children are becoming more like their peers in some ways, they are becoming *less* like them in others. Some of the characteristics they have when they enter middle childhood get exaggerated, rather than toned down, as a result of their experiences in the peer group.

How can these two contradictory processes go on over the same period of time? For an answer I turn once again to the theory of John Turner. Turner writes about adults, not children, but I think by the age of eight most humans are capable of the sort of mental gymnastics he describes.

According to Turner, people sometimes categorize themselves as "us" and sometimes as "me," depending on the social context. When groupness is salient, they see themselves as members of whichever group is in the spotlight at that moment. When groupness is not salient, they see themselves as unique individuals, sui generis. But most of the time they are at neither of these extremes—most of the time they are hovering (mentally) somewhere in the gray area between "us" and "me." So most of the time they are susceptible both to the urge to conform and the urge to be different. The usual solution is to conform in most ways and to find a few ways to be different.

Of course, the best way to be different is to be better. But "better" has different meanings in different groups. In boys' groups in most parts of the world, it means bigger, tougher, able to make others do what you want. In girls' groups in most parts of the world, it means prettier, nicer, able to make others like you.

Up to now I have spoken as though each child in the group had equal power to influence the others—the majority-rules rule implies one member, one vote. But within a group, some are more equal than others. One of the things that interested the researchers in the Robbers Cave study (described in the previous chapter) was how groups—boys' groups, they meant—choose their leaders. Among the Rattlers a boy named Brown was the biggest and strongest, and during the first few days in camp the others regarded him as their leader. Leadership in a boys' group, as in a chimpanzee troop, often boils down to a question of who can dominate whom. But boys are not, after all, chimpanzees. Brown lost status because he was too aggressive with his groupmates and too bossy. "We're tired of just doing the things he leaves over," complained one of the smaller boys. So Brown was demoted and replaced by Mills, who proved to be capable of leading with a bit more finesse.

Iron muscles do not a leader make, not even in a boys' group. Force of personality, imaginativeness, intelligence, athletic ability, sense of humor, and a pleasing appearance can also win votes. Aggressive children tend to be unpopular with their peers and may even be rejected by them. Not all aggressive boys are unpopular, however, and there are some who are widely liked. I suspect that boys can get away with aggression if they apply it discerningly. It is the ones who do not play by the rules—who flare up in anger unpredictably and lash out at inappropriate targets—who become the rejects.

The Robbers Cave researchers talked about "dominance hierarchies"—the infamous "pecking order"—but that term is less often used nowadays, partly because things are seldom as clear-cut as the word *hierarchy* would suggest, partly because the word *dominance* implies something one-way, something that the higher-up one is doing to the lower-down one. Even the Robbers Cave researchers recognized that leadership among humans is more a matter of being chosen than of feeling the call. They judged leadership by watching which boy the others addressed when they were making suggestions.

A newer and better term is "attention structure." Which children do the other members of the group pay attention to? Which ones do they look at when they're not sure what to do? Someone who is high in the attention structure has privileges only dreamed of by the lower-downs. He or she can be an innovator, not just a follower. The penalties for being different are mainly imposed on those in the middle and lower ranks of the attention structure. Those on top don't have to imitate anyone: they are the imitatees.

Unlike dominance hierarchies, attention structures are as visible in girls' groups as in boys'—maybe more so, because what gets imitated is not just behavior but also things like clothing and hairstyles. The higher-ups among the girls get to decide, for instance, when to switch from winter to summer clothes. If the girls lower down in the attention structure—the less popular girls—show up in school still wearing sweaters when the higher-ups have already switched to short sleeves, they have committed an embarrassing faux pas. Switching *before* the higher-ups would be equally embarrassing. The only solution is to switch on exactly the same day. Getting it right requires spending a fair amount of time on the phone, I would imagine.

Where groups are composed of children of the same age, as they generally are in our society, those who have the highest status tend to be those who are the most mature. This hearkens back to the mixed-age groups of our hunter-gatherer ancestors, where the older children were in charge of the younger ones and the younger ones learned how to behave by watching the older ones. For boys, it may hearken back even further, to our primate ancestors. Young male chimpanzees can't learn the rules of proper chimpanzee behavior by watching their fathers, because they don't, so far as they know, have fathers. And they can't learn the rules for proper *male* chimpanzee behavior by watching their mothers. Perhaps for these reasons, young male chimpanzees are strongly attracted to older male chimpanzees and seek them out even though they are likely to be buffeted around by the older males. The same is true of young male humans. Little boys seek out the company of older boys even if the older boys are quite rough with them.

Older children have higher status than younger ones, and that is why children who are mature for their age tend to have higher status among their agemates than those who are slow to mature. Children who are close friends are likely to be fairly even in status, so those who have high status among their agemates often have friends older than themselves, while those who have low status often have younger friends. During the years I was rejected by my classmates, my only friend was two years my junior. I was rejected by my classmates partly because I was young for my class and small for my age. I looked—and no doubt acted—like a younger child and therefore I had no status at all among my peers. Maturity for children is like money for adults: it can raise or lower popularity independently of anything else. The rich ugly guy wins as desirable a wife as the poor gorgeous one.

I believe high or low status in the peer group has permanent effects on the personality. Children who are unpopular with their peers tend to have low self-esteem, and I think the feelings of insecurity never go away entirely—they last a lifetime. You have been tried by a jury of your peers and you have been found wanting. You never get over that. At least, I didn't.

It is not easy to prove, however, that adults' insecurities (or other psychological problems) date from experiences in their childhood peer groups. Inevitably there are cause-or-effect uncertainties. Let's say a kid named Ralphie is unpopular with his peers and later he turns into a psychologically messed-up grownup. Are his adult problems the result of his having been rejected as a child, or was there something wrong with Ralphie to begin with? Maybe he was unpopular with his peers because they noticed something odd about him, something wrong with his personality. Maybe his parents also noticed it, and maybe *they* weren't very nice to him either. If Ralphie is a mess as an adult, is it because his peers rejected him, because his parents rejected him, or because whatever was wrong with him as a kid hasn't gotten any better?

I have found some evidence that it is indeed the experiences in the peer group that are responsible for the later problems: it involves children who are small for their age, either because they are slow maturers or because they are destined to be short adults. Short children, especially if they are boys, tend to have low status among their peers. There is no reason, other than their size, why these children should be rejected by their peers, and no reason at all to expect them to be rejected by their parents—if anything, parents are more protective of smaller children. And yet short children are considerably more likely than tall ones to suffer from low self-esteem and a host of other psychological problems.

Though they may outgrow their smallness, their other problems are not so easy to leave behind. A researcher followed two groups of boys—slow and fast maturers—into adulthood. The slow maturers were small for their age all through childhood and adolescence, but they eventually caught up: as adults they were only half an inch shorter, on the average, than the men who had matured more rapidly. But the differences in personality persisted. The early maturers tended to be poised and self-confident; several became successful executives. The late maturers were less sure of themselves, more prone to "touchiness" and "attention-seeking."

In parts of the world where mixed-age play groups still exist, issues of size and status aren't so important. A child starts out being the youngest and smallest in his play group and gradually moves up in the ranks. He has the experience of being pushed around by everyone and, later, the experience of having younger and smaller children look up to him. Children in urbanized societies don't get to run this gamut of experience. At home they remain the oldest or the youngest among their siblings. In school they are likely to remain for years, if they are lucky, at the top of the totem pole or, if they are unlucky, at the bottom.

Know Thyself

Somewhere around the age of seven or eight, children start comparing themselves to their peers in a way they hadn't done before. Ask a bunch of little boys in a nursery school "Who is the toughest boy in this room?" and they'll all jump up and shout "Me! Me!" At eight they are wiser. They'll point to the biggest boy in the room, or the most aggressive, and say "Him."

What these eight-year-olds have done is forever beyond the capacity of a chimpanzee: they have formed an internal working model, not just of the significant others in their lives, but of themselves. They can compare this model—this self-image—to something quite abstract: the group as a whole. A chimpanzee knows which members of its troop it can beat up and which it had better defer to, and so does a child in nursery school. But I don't think even the alpha chimp knows it is the alpha chimp. All it knows is that you'd better get out of its way if you know what's good for you.

It is during middle childhood that children learn about themselves. How tough they are. How good-looking. How fast. How smart. The way they do this is by comparing themselves to the others with whom they share a social category—the others in the group of people "like me."

"Social comparison" is the technical term for finding out about yourself by comparing yourself to others. "Oh wad some power the giftie gie us, To see oursels as ithers see us!" said the poet Robert Burns.* But what if the ithers see us as a nerd, a weirdo, a schlemiel? I don't want to look a giftie horse in the mouth but it's not always a treat, seeing oursels as ithers see us.

Fortunately, it has its saving grace: *we* get to choose which group to compare ourselves to. A fourth grader can consider himself tough if he's tougher than most of the other fourth graders. He doesn't have to compare himself to the fifth and sixth graders.

If he discovers he is not the toughest boy in the room, there are plenty of other niches in the fourth-grade classroom he can try out for. Class clown, for example. Middle childhood is when children get typecast into roles that might last them the rest of their lives. They choose these roles themselves or get nominated for them—or forced into them—by others. When it happens, the characteristics a child starts out with tend to become exaggerated. The funny child gets funnier, the brainy child gets brainier. Humor and intellect have become their specialties.

This is all very well and good for those who are different on purpose or different in ways that the group finds acceptable. But what about the unfortunate children who are different and can't help it? The girl with a hearing aid? The boy who was too tall and too white? When a chimpanzee was stricken with polio and returned to his troop as a cripple, the members of the troop attacked him. Dislike of strangers translates very easily into dislike of strangeness. If you are different you are not one of *us*.

As children get older they become more aware of all the ways that people differ from each other. More and more things become the basis for splitting up into separate, smaller groups. Friendships between children of different races or different socioeconomic classes become steadily less

* "To a Louse," 1786.

common over the elementary school years. Academic achievers pal around with other academic achievers, trouble-makers with other trouble-makers. By fifth grade, children are associating with each other mostly in little cliques of three to nine members, and these cliques are themselves differentiating themselves from other cliques. Within them, meanwhile, the members are becoming more similar to each other.

Developmentalist Thomas Kindermann studied some cliques in a fifth-grade classroom and found that children who belonged to the same cliques had similar attitudes toward schoolwork. Well, that's not too surprising: the kids probably belonged to the same cliques *because* they had similar attitudes. But in fifth grade, cliques haven't solidified yet—children can still move into them or out of them. This gave Kindermann the opportunity to study what happens when a kid moves into or out of a clique of academic achievers. What he found was that children's attitudes toward schoolwork change if they switch from one group to another over the course of a school year. If a child moves into a clique of academic achievers, her attitude toward schoolwork is likely to improve; if she moves out of it, her attitude gets worse. Kindermann's findings demonstrate that children's attitudes toward achievement are influenced by their group affiliations. The changes he measured could not have been due to changes in the children's intelligence or in their parents' attitudes, since neither is likely to reverse direction over the course of a single school year.

As children get older, they have more and more freedom to choose the company they keep. This is yet another way that the characteristics they start out with can become exaggerated. A bright child is more apt to join a clique of academic achievers, a not-so-bright child a different kind of clique. The influence of his companions motivates the bright child to do well in school and as a consequence he may become still brighter. It is a vicious cycle which in this instance is not vicious at all. Loops of this sort turn up over and over again in development. Psychologists have a name for them: Matthew effects, they're called, after the passage in the New Testament that says, "For to him who has will more be given, and he will have abundance." Whoever said that life is fair?

Sometimes it is, though. During four years of childhood I was rejected by my peers. For those four painful years I have been paid back in abundance. If those "little ladies" in the snooty suburb had accepted me, I probably would have turned out just like them.

9 THE TRANSMISSION OF CULTURE

What is a culture? Margaret Mead defined it as "the systematic body of learned behavior which is transmitted from parents to children." In this definition, "learned behavior" covers a lot of ground. It includes social behavior such as acting assertive or self-effacing, emotional or cool, aggressive or affectionate. It includes skills such as the ability to chip an arrow point out of a stone or to operate a microwave oven. It includes knowing how to speak the local language and which words to use on which occasions. And—now we're really stretching the word "behavior" but surely Mead didn't mean to exclude things of this sort—it includes beliefs such as how your remote ancestors came to exist and who or what was responsible for their existence.

Mead assumed that learned behavior was "transmitted from parents to children" because she could see that children in different societies acquire different learned behaviors—in one they learn to speak Italian, in another, Japanese; in one they learn how to make arrows, in another, how to operate microwave ovens—and that these behaviors are, to a first approximation, similar to those of their parents. How else could a culture be transmitted from one generation to the next? How else could a culture be preserved, sometimes for hundreds of years, other than by being "transmitted from parents to children"?

Margaret Mead was an anthropologist, not a psychologist, but that didn't make her immune to the nurture assumption. Her assumption that culture is something parents teach their children is just that—an assumption. In this chapter I will present an alternative way of looking at how cultures are passed down from one generation to the next.

Take This Culture and Pass It On

In the previous chapter I mentioned two Mexican villages not far from each other in distance but very far apart in social climate. The inhabitants of the villages an anthropologist dubbed "La Paz" and "San Andres" speak the same language (Zapotec) and plant the same crops, but they behave differently. In La Paz the people are peaceful and cooperative; in San Andres they are aggressive and prone to violence.

Margaret Mead described a pair of similarly contrasting cultures in one of her early books, published in 1935. She studied two tribes located within a hundred miles of each other in New Guinea: the mountain-dwelling Arapesh and the river-dwelling Mundugumor. The Arapesh, she reported, are gentle and peace-loving; the Mundugumor are hostile and warlike. I would like to say that Mead wondered what made these two groups behave differently and studied their cultures in order to find out, but I suspect she had made up her mind before she ever set foot on the island of New Guinea.* Freudian psychology was just coming into its own and Mead was prepared in advance to look at child-care practices such as weaning and toilet training. Here is Mead, asking rhetorical questions about the Arapesh and promptly answering them:

> How is the Arapesh baby moulded and shaped into the easy, gentle, receptive personality that is the Arapesh adult? What are the determinative factors in the early training of the child which assures that it will be placid and contented, unaggressive and non-initiatory, non-competitive and responsive, warm, docile, and trusting? It is true that in any simple and homogeneous society the children will show the same general personality-traits that their parents have shown before them. But this is not a matter of simple imitation. A more delicate and precise relationship obtains between the way in which the child is fed, put to sleep, disciplined, taught self-control, petted, punished, and encouraged, and the final adult adjustment. Furthermore, the way in which men and women treat their children is one of the most significant things about the adult personality of any people.

The Arapesh, said Mead, are kind and indulgent with their babies. Weaning is gentle and so is toilet training. In contrast, the Mundugumor —"a group of cannibals and head-hunters," as she described them—use a recipe for infant care right out of Alice in Wonderland: "Speak roughly

* She evidently did the same thing in Samoa. See Freeman, 1983.

to your little boy and beat him when he sneezes." The angelic Arapesh and the malevolent Mundugumor. I think I've seen this movie.

Though it makes a good story, it doesn't hold up to closer scrutiny. In fact, the Arapesh engage in warfare too, and most warlike peoples—even those who are downright nasty to everyone else—are very nice to their babies. Anthropologist Napoleon Chagnon lived for several years among the Yanomamö, a "fierce people"—their own description of themselves—who dwell in the Amazon rainforest of Venezuela and Brazil. These people are almost constantly at war with their neighbors. The men beat their wives with sticks if they're a bit slow to fetch dinner and shoot arrows into their nonessential parts for more serious transgressions. But the babies are breast-fed on demand and treated with indulgence by both parents.

And the babies turn into fierce children and then into fierce adults, like their parents. As Mead pointed out, children tend to show "the same general personality-traits" as their parents. Taking that as our starting point, let us examine, with an open mind, some possible explanations for it.

The first and simplest is that these personality traits are inherited: like father, like son. Within our own society, measures of aggressiveness show about the same degree of heritability as other personality traits—that is, roughly half of the variation in aggressiveness can be blamed on the genes. Although these results do not permit us to draw conclusions about differences between groups, they at least suggest the possibility that genes can play a role in aggressive behavior.

Consider this: Chagnon found that Yanomamö men who have killed in battle have about twice as many wives, and twice as many children, as men of the same age who have never killed. These people pride themselves on their fierceness, and men who live up to the Yanomamö ideal have higher status in the tribe. Like most tribal peoples, the Yanomamö permit polygyny: high-status men win extra wives. Consequently, they have more children. For who-knows-how-many generations, the Yanomamö have been systematically breeding warriors. The men who go gladly into battle have many children; the men who come down with stomachaches on the big day—yes, such men do exist among the Yanomamö—have fewer or none (none because where some men have extra wives, others must remain wifeless). It is not implausible that such a system would produce a race of people who are outstanding for their fierceness.

Not implausible but, to me, not very interesting. Although heredity may be an arguable explanation for differences in aggressiveness, it cannot account for most of the other differences between cultures. It cannot explain why some children (like their parents) grow up speaking Italian while others grow up speaking Japanese, and some learn how to make arrows while others learn how to operate microwave ovens. It cannot explain why the Yanomamö boys (like their fathers) tie their foreskins to a string worn around the waist—a fashion Chagnon assures us is deucedly uncomfortable*—or why the parents in this society (like the grandparents) attribute infant deaths to witchcraft perpetrated by their enemies.

Although personality is partly inherited, culture is not. The attitudes, beliefs, knowledge, and skills that are part of a culture are not passed down from one generation to the next by way of the genes. I accept the part of Margaret Mead's definition that says culture is learned. But how is it learned? Who are the teachers?

In the Mexican village of San Andres, and among the Yanomamö of the Amazon rainforest, adults behave aggressively and so do children, and the children grow up to become aggressive adults. Aside from heredity, I can think of four possible explanations—four environmental mechanisms—that might be responsible for the similarities between the children's behavior and the adults'.

The first is that parents encourage aggressive behavior, or at least fail to punish it. Among the Yanomamö, children who come complaining to their parents that one of their playmates hit them with a stick are provided with a stick of their own and told to return the favor: go hit 'em back. In contrast, in a peaceful society like the Mexican village of La Paz, children are discouraged even from play-fighting.

Acquiring behavior approved by the culture "is not a matter of simple imitation," said Margaret Mead, but maybe she was wrong about that, too. The second alternative is that children may imitate their parents' behavior. The third—this is the explanation favored by Douglas Fry, the anthropologist who studied the inhabitants of La Paz and San Andres—is that children may imitate *all* the adults in their society. The final alternative is the one I proposed in the previous chapter: children may imitate other children, preferably those who are a little ahead of them in age or social status. In this case the influence of the adult society would have to be an indirect one.

How can we decide among these four alternatives? My answer may surprise you: in most cases, we can't. Under ordinary conditions there is

* Whatever they pay anthropologists, it isn't enough.

no way to distinguish among them. Any one, two, or three of these mechanisms, or all four of them together, may be producing the observed effects on the children's behavior. In the kinds of societies anthropologists study, all the parents use pretty much the same child-rearing methods: child-rearing methods are a part of the culture. And the parents behave pretty much alike in other ways as well (they all behave in the manner approved by their culture), so how could we tell if children are imitating their own parents or all the adults? True, there are small variations in behavior within a culture—not all Yanomamö men are equally enthusiastic about going to war—but these could be due to genetic differences within the population. If the son of a reluctant warrior also turns out to be timid by Yanomamö standards, it can't be used as evidence for alternative 2, that children imitate their parents. It could just be heredity. Thus, the small variations in behavior within a culture can't help us in our efforts to distinguish among the four environmental alternatives.

The trouble is that under ordinary conditions all the aspects of a child's environment are correlated—they all vary together—so it is impossible to tell which aspect of the environment is having the effect on the child. We cannot tell whether the San Andres children are more aggressive than the La Paz children due to their parents' child-rearing methods, or to imitation of their parents, or to imitation of other adults, or to imitation of other children—or, for that matter, to genetic differences between the inhabitants of these two communities—because all the possible influences work in the same direction: toward greater aggressiveness in San Andres, toward greater peacefulness in La Paz.

The same confounding of influences occurs within our own multicultural society. Imagine a hypothetical couple: he is a lawyer, she is a computer scientist; they met at the same Ivy League college that their fathers attended. They have two designer children. They live in a suburb where all the homes are expensive, all the parents are well-educated, and all the children are above average. The kids get trips to the museum and the zoo and the library. Their home is full of books, and when they were small their parents were always willing to read to them. The parents also spend a lot of time reading books and magazines of their own. The other kids in the neighborhood come from similar homes, and so do most of the kids in the school they attend.

If the designer children turn out to be excellent students and gain admission to the same exclusive Ivy League school their parents and grandfathers attended, to what should we attribute their academic success? Their genes? The fact that their parents read to them and encouraged intellectual activities? The fact that the parents themselves engaged in intellectual activities? The fact that other adults in their community also engaged in intellectual activities? Or the fact that the other kids in their neighborhood and their school were similarly inclined?

When all these factors vary together, as they do here, it is like trying to tell why poodles and foxhounds behave differently while continuing to rear all the poodles in apartments and all the foxhounds in kennels. The only way we can tell what is really going on is to look at cases in which the various influences work in opposition to each other. We did that in Chapter 2 for heredity versus environment: we raised poodles in kennels and foxhounds in apartments. We looked at adopted children, whose genes came from one set of parents and whose environment was provided by a different set.

Now I am saying that separating genetic influences from environmental influences is not enough: we also have to separate the various environmental influences from each other. Just as heredity and environment tend to vary together, environment and environment tend to vary together. Children who are reared in a culture where aggressive behavior is the norm may be rewarded for aggressive behavior with attention or approval. They see their parents behaving aggressively, they see other adults in their society behaving aggressively, and they see other children behaving aggressively. As long as all these forces are pulling in tandem, there is no way of telling which is moving the wagon. We have to look at cases in which the forces are pulling in different directions.

Psychologists and anthropologists haven't done this. They haven't realized it is necessary. They make pronouncements about which environmental factor is important on the basis of intuition—that is, on the basis of which version of the nurture assumption is currently in vogue. The evidence they use to support their position is useless, because it cannot distinguish among the various alternatives.

The only way we can tell which environmental factors are having an effect is to look at cases in which they do not work together, and that is why I keep coming back to the immigrant family. When the parents belong to one culture and the rest of the community belongs to a different culture, we can at least distinguish the effects of the parents from the effects of outside-the-family influences.

Environment versus Environment

Tim Parks is a British writer who has lived for a number of years in Italy and is rearing his three children there. His book *An Italian Education* is about his experiences as an immigrant father. He wrote it, he says, in the hopes that

> by the time we got to the last page of such a book, both the reader and, far more important, I myself would have begun to understand how it happens that an Italian becomes Italian, how it turns out (as years later now it has turned out) that my own children are foreigners.

As far as I could tell, Parks never does figure out how it happens that an Italian becomes Italian. But he is very good at describing the feelings of a father who watches his children becoming card-carrying members of a different culture.

> Then Michele comes in and says to me, in English, "Oh, don't be so fiscal, Daddy. Don't be so fiscal." He's complaining about my sending them to bed on time, and what he means is *fiscale. Non essere fiscale, Papà.*

The Italian word *fiscale,* Tim Parks explains, is a pejorative term meaning "too severe," or "perversely exacting." Don't be so uptight, Daddy. Don't be such a fussbudget.

> "Don't be fiscal," Michele says, knowing I like him to speak English. "We'll be good if you let us stay up." What he means is, These rules (which he doesn't know are typically English) don't need to be applied to the letter (a flexibility typically Italian).

With a mixture of pride and regret, Parks sees his son becoming a full-fledged member of a society in which he will forever remain an outsider. He must have figured on Michele becoming an Italian—otherwise why would he have given him an Italian name? And yet he is sad to see it happen. He is losing his child, even more than most parents lose their children.

I think all immigrant parents experience this mixture of pride and regret as they watch their children become members of a different culture, but in some the pride is the stronger emotion, in others it is the regret. I know a Japanese woman, married to a European American and living in the United States, who never spoke Japanese to her children because she was afraid it would interfere with their learning of English (it wouldn't have). On the other hand, I know a Jewish woman whose Orthodox grandparents immigrated to the United States from Poland and then took their children back to Poland when they saw them turning into godless Americans. The grandparents and all but one of their children perished in the Holocaust.

It is possible for Orthodox parents to rear children in the United States without having them turn into godless Americans. In Brooklyn, New York, there are Hasidic Jews who have preserved the religion, customs, and even styles of dress and adornment that came from Eastern Europe several generations ago. The way they do it is to educate the children themselves. The children go to religious schools called yeshivas; they do not mingle with children from other cultures either in school (where all the children are offspring of Hasidic Jews) or in the neighborhood (most of their neighbors are also Hasidic Jews).

Another group that has managed to keep its children from being assimilated into the majority culture are the Hutterites of Canada. These people live communally, practice adult baptism, dress in old-fashioned clothing, and have strict rules of comportment. Each colony has its own school where children are taught "the fear of God, self-discipline, diligence, and the fear of the strap," as a British journalist put it. The journalist, who spent some time in a Hutterite colony, explains:

> At stake in the question of Hutterite education is nothing less than the continued existence of the Hutterites as a separate social entity in Canada. The continuity of Hutterite communal life depends not on God or religious belief but on their retaining control of the children's education. "We could never hold them if they went to school out there," an elder confessed.

But most children whose parents are not members of the majority culture do go to school "out there." What happens, at least for a time, is that the children become bicultural. In effect, they become citizens of two different countries, that of their parents and that of the Out-Theres. Bicultural children may blend their two cultures, or they may switch back and forth between them. Switching back and forth is called code-switching; I described it in Chapter 4.

Why do some children code-switch and others blend? Why does it sometimes take only one generation for the immigrants' culture to be lost and sometimes three generations? With all that has been written on the topic of the "melting pot," sociologists and psychologists have still not paid much attention to the things that make a difference. That is why the evidence I must use to support my position is mainly anecdotal.

When immigrants come to the United States from another country, they often move to areas where there are others of the same national background. There are Chinatowns and Koreatowns; there are neighborhoods in which most of the adults came from Puerto Rico or Mexico. In the past there were neighborhoods that were predominantly Italian or Irish or Jewish, and parts of the Midwest that were predominantly Swedish or Norwegian or German. The children of immigrants who grow up in such areas are surrounded by peers who come from similar homes—homes in which English might not be spoken, in which chopsticks might be used instead of spoons and forks.

In such areas, children blend their two cultures. They acquire American ways with a foreign flavor. They learn English but they may speak it with an accent. In a Princeton University student newspaper a few years ago, a freshman complained that her classmates kept asking her what country she came from. She was a Mexican American, born and reared in Texas, and was offended by the question. She didn't realize that the reason they asked was because she spoke English with a Spanish accent. In the Arizona high school I attended, there were many Mexican-American kids. Most of them belonged to Mexican-American peer groups and spoke English with a Spanish accent.

Immigrant cultures are generally lost after one, two, or at most three generations. Sociologists regard this as a gradual process, but it only appears to be gradual. It is gradual for the group as a whole but not for individual families. The old culture is lost in a single generation as soon as a family moves away from the Chinatown or the Mexican-American neighborhood to an area where they are no longer surrounded by people of the same national background. What makes it look gradual is that families don't all move away at the same time. Some go as soon as they can afford to, others wait a generation or two.

When the immigrants' child joins a peer group of ordinary, non-ethnic Americans, the parents' culture is lost very quickly.* A Chinese father who came to California from Hong Kong laments the loss of his daughter's Chinese identity:

> "All her friends at school were Caucasian girls," he says of his youngest child. "That's fine while you're growing up. But Caucasian girls marry Caucasian husbands and observe western customs. Then you start to feel the differences between you, but it is too late. When you pay too much attention and spend too much time with your Caucasian friends, you tend to ignore your own group."

* The last aspects of the old culture to disappear are the things that are done only at home. Styles of cooking, for example, may survive for several generations. Children do not ordinarily learn to cook in the presence of their peers.

Because her friends were European Americans and not Chinese Americans, the daughter of the immigrant from Hong Kong would have been a code-switcher, rather than a blender of cultures. At home she might have spoken Chinese and used chopsticks; with her friends it would have been English and a knife and fork. The code-switching child toggles between her two cultures as she passes through the door of her home. Click. Click.

But the two cultures of a code-switcher, though separate, are not equal. The children of immigrants bring the culture of their peers home to their parents; they do not, as a rule, bring the culture of their parents to their peers. The daughter of the British psycholinguist (mentioned in the previous chapter) brought Black English home—she didn't teach her friends at the day-care center to speak with a British accent. A psychologist reared in Canada by Portuguese immigrant parents reported that for the greater part of her childhood she refused to speak Portuguese: when her parents addressed her in that language she would reply in English. She became interested in relearning Portuguese only when her family spent a summer in her parents' native land.

Tim Parks doesn't realize how lucky he is that his Italian-born son is still willing to speak to him in English. Michele is a typical code-switcher: he doesn't mix together his two languages. He doesn't tell his father, "Don't be *fiscale,* Daddy." Because he lacks an English word to serve his purpose, he uses an Italian word, but he translates it into the closest English equivalent he can find—one that, unfortunately, doesn't have the right connotation. Though Michele makes a valiant effort to stick to English, his English vocabulary is not keeping up with his Italian, and that, too, is typical of code-switchers. Children who speak one language at home, another language outside the home, continue to improve in their outside-the-home language while their home language gets stuck at a level that is just barely adequate for conversing with their parents. Linguist S. I. Hayakawa, reared in Canada by Japanese-born parents, confessed that he "speaks Japanese haltingly, with a child's vocabulary."

When the code-switching toggle clicks each time the child goes through the door of his home, it is an unstable situation that is eventually resolved in favor of the outside-the-home code. But there is another kind of code-switching that may have more staying power: it involves two different outside-the-home codes. An anthropologist who studied Mesquakie Indian boys from a Native American community in Iowa reported that they behaved differently when they were in the nearby Anglo-American town than when they were in the Mesquakie community. Peer groups—gangs, the anthropologist called them—of Mesquakie Indian boys switched between Anglo-American norms of behavior when they were in town and Indian norms in their own community. The difference between these boys and classic code-switchers like Michele is that the Mesquakie boys had peers with whom they shared *both* cultures.

When in Rome, do as the Romans do. For children it's more than that: when in Rome, they become Romans. Even if their parents happen to be British or Chinese or Mesquakie. When the culture outside the home differs from the culture inside it, the outside culture wins.

I conclude that neither the parents' child-rearing methods nor imitation of the parents by the child can account for the way cultures are transmitted from one generation to the next. That still leaves two possibilities: that children imitate all the adults in the community or that they imitate other children. In order to distinguish between these alternatives, it is necessary to find cases in which children have a culture that differs from that of the adults in their community. Such cases exist.

The Culture of the Deaf

"Language, I realized, is a membership card for belonging to a certain tribe." The realization is that of Susan Schaller, a teacher and interpreter of American Sign Language. ASL is the language used by the Deaf* in the United States—the membership card of their culture. It took Schaller a while to catch on to the groupness, the "us versus them" aspect, of the Deaf culture.

> For someone who identifies with the Deaf culture, it is foreign and ludicrous to desire hearing. When I first met Deaf people, I would have never understood this. My ignorance of Deaf culture prevented me from understanding almost every signed joke I saw. Translation from ASL to English didn't help, because I still thought of Deaf people as people who couldn't hear, and the punch lines always related to cultural differences. Finally I began to catch on when someone joked about a mixed marriage between a hearing woman and a Deaf man.

There is nothing unusual about this attitude; it is characteristic of all minority groups—of all groups, in fact, when groupness is salient. What makes the Deaf culture unique is that it *can't* be passed down from parents to children. A large majority of deaf children are born to hearing parents

* *Deaf* is capitalized when it refers to a culture or to membership in a group. It is not capitalized when it refers simply to the inability to hear.

who know nothing about the world of the Deaf. And a large majority of the children born to deaf parents can hear, and these children become members of the hearing world.

And yet the Deaf have a robust culture, as durable as the hearing culture but different from it in a number of ways. They have their own rules of behavior, their own beliefs and attitudes.

The profoundly deaf children of hearing parents get their behaviors and beliefs in the same place they get their language: in the schools for deaf children. Where else could they get them? Not from their homes, since typically—at least in the past—there was little communication between deaf children and their hearing families. What communication there was consisted of primitive gestures, pantomimic in nature, called "homesigns." These signs bear little resemblance to the abstract, flowing, grammatically complex language called ASL.

Researchers who study bilingual hearing children and observe that the home language eventually gets dropped in favor of the one used Out There often blame it on the relative prestige of the two languages. They say, for example, that the reason Hispanic children in the United States eventually stop speaking Spanish is that Spanish has no prestige—it is not valued by the Out-Theres. "Under these circumstances," alleged one set of researchers, "the language of the economically and culturally more prestigious group tends to replace the minority language."

For many years in this country, misguided educators from the hearing culture did their damnedest to provide deaf children with a language that has high economic and cultural prestige: spoken English. And yet, for some reason, the little rascals were not grateful. They persisted in learning sign language, even though in some schools they were beaten for using it. In those schools they used it surreptitiously, in the dormitories and playgrounds. Despite the earnest efforts of their teachers to teach them to speak aloud and to read lips, sign language became their native language —the language they thought in and dreamed in. It was the language they used as adults to communicate with their friends in the Deaf community. It was the language most of them used to communicate with their hearing children.

How did they learn sign language if their teachers would not teach it to them? In most cases, they learned it from the few deaf children in the school who came from Deaf families. Such children have high status among the Deaf, because their early introduction to a language gives them an edge they never lose. They are the eloquent ones, the skilled communicators of the Deaf community. Though they are a small minority—usually about 10 percent—of the students in a deaf school, the language they bring with them to the school has higher prestige among their classmates than the language of the Out-Theres, the language their teachers try so hard to give them.

Even when a school contains no children who came in already knowing sign language, the children still manage to acquire it. Susan Schaller tells a story of a school for the deaf on the island of Jamaica. Signs and gestures were prohibited in this school, yet the children had nonetheless learned sign language. How, Schaller asked a colleague who had visited the school and interviewed some of its graduates, had they managed to learn it?

> "The laundry woman," he answered. Generations of deaf students passed through that school, and a few of each generation were employed as janitors, cooks, and assistants. The children picked up signs and grammar from these adult signers, adding their own vocabulary and idioms every generation. For the group he had met, the laundry woman was the head sign teacher.

"The language of the economically and culturally more prestigious group tends to replace the minority language," alleged the researchers. But for the children in the Jamaican school, it was the language of the laundry woman. They didn't learn it so they could communicate with *her* —they learned it so they could communicate with each other. True, sign language came a lot easier to them than the arduous business of reading lips and trying to produce sounds they could not hear. But if they had really wanted to behave like the majority of the adults in their community, they would have eschewed signing and concentrated on learning spoken English.

In some places there is no one—not even a laundry woman—to teach deaf children to sign. There are places where, until quite recently, no sign language existed, because there were no schools for deaf children. Such children remained isolated in their families, unable to communicate with anyone except in the most rudimentary way. Other children wouldn't play with them. Some of them ended up in institutions for the mentally retarded.

When children who lack a common language come together for the first time, what happens then is something like a miracle. Psycholinguist Ann Senghas and her colleagues are studying the birth of a language in the Central American nation of Nicaragua, where the education of deaf children dates back only to the early 1980s. Here, in Senghas's words, is how it happened:

Only sixteen years ago, public schools for special education were first established in Nicaragua. These schools advocated an oral approach to deaf education; that is, they focused on teaching spoken Spanish and lip-reading. Nevertheless, the establishment of these schools led directly to the formation of a new signed language. Children who previously had had no contact were suddenly brought together to form a community, and they immediately began signing with each other. The first children to arrive at these schools ranged in age from four to fourteen. They all entered with separate means of communicating that they had used with their families. Some had a lot of miming and gesture skills, some had homesign systems that were slightly more elaborate, but none entered with a developed sign language.

The children rapidly developed an interlanguage among themselves, a kind of signed pidgin, that is not exactly a full language, but which has many shared conventions, and could serve their communication needs pretty well. Since that time, the children have been creating their own indigenous sign language. The language is not a simple code or gesture system; it has already evolved into a full, natural language. It is independent from Spanish, the spoken language of the region, and is unrelated to American Sign Language (ASL), the sign language used in most of North America.

Something similar happened many years ago in Hawaii, but the product was a spoken language rather than a sign language, and no psycholinguists were lucky enough to be around while it was happening. Derek Bickerton, the psycholinguist who studied the creation of this language by children in Hawaii, had to reconstruct its history from evidence he collected many years after the fact. By then, the creators of the language were elderly adults.

They had been the children of people who came to Hawaii in the late 1800s to work on the sugar plantations. The immigrant generation came from many different countries, including China, Japan, the Philippines, Portugal, and Puerto Rico. They had no language in common.*

In the biblical story of the Tower of Babel, the workers threw down their tools and walked off the job because each spoke a different language and they couldn't understand each other. But people who need to communicate with each other find a way of doing it. What normally happens

* That was probably the idea. Indentured laborers who worked long and hard for very little pay might have gotten together and organized a strike if they had been able to express their opinions to one another.

under such conditions—this is what happened in Hawaii—is that a pidgin language springs into existence, created over a relatively short period of time by its sundry speakers. Pidgins are makeshift languages that lack prepositions, articles, verb forms, and standardized word order. Each speaker of a pidgin speaks it a little differently; the native language of each is still detectable, peeking out from behind the skimpy list of vocabulary words that the speakers have in common.

The immigrant generation to Hawaii spoke either pidgin or the languages they had brought with them to the island. But their children spoke something else—something that linguists call a *creole.* A creole is the offspring of a pidgin, but it is a genuine language, with standardized word order and all the other things a pidgin lacks. Unlike a pidgin, it is capable of expressing complex, abstract ideas.

The creole-speaking children had not learned their language at home. They had not learned it from their parents—their parents couldn't speak it. According to Bickerton, the children had created the language themselves. He was able to trace its development back to the two decades between 1900 and 1920 by interviewing (during the '70s) elderly adults who had been born during that period. Those who had immigrated to Hawaii as adults still spoke pidgin; those who were raised there spoke creole. It was a language that didn't exist until about 1905. The children who created it had carried it with them to adulthood. They had, says Bickerton, "adopted the common language of their peers as a native language in spite of considerable efforts by their parents to maintain the ancestral tongue."

Derek Bickerton studied only their language, but the children of the Hawaiian immigrants would have had to create a common culture as well. In Nicaragua, Richard Senghas (the brother of psycholinguist Ann Senghas) is recording the development of a Deaf culture among the first generation of Nicaraguan sign language users. These people can communicate with each other now; they keep in touch after leaving school and there is a growing sense of groupness. Even though their culture is derived from that of hearing Nicaraguans, contrast effects are starting to appear. Deaf Nicaraguans pride themselves on their punctuality, whereas hearing Nicaraguans (like most Central and South Americans) have a casual attitude toward time. It is exactly the opposite in the United States, where the hearing are generally punctual and the Deaf have a more relaxed attitude toward time.

At the beginning of the chapter I said there were four ways, aside from heredity, that cultural behaviors could be passed from the older generation to the younger one. I have now eliminated three of those alternatives. Cultures are not passed on from parents to children; the children of immigrant parents adopt the culture of their peers. That eliminates the first two alternatives, the parents' child-rearing methods and imitation of the parents by the child. The third alternative was that children imitate all the adults in their society, but that explanation doesn't work in cases where the children's culture differs from the adults'. I conclude—this is one of the tenets of group socialization theory—that culture is transmitted by way of the children's peer group.

My theory unites three different realms of academic research: socialization, personality development, and cultural transmission. All these things happen in the same way and in the same place: the peer group. The world that children share with their peers is what shapes their behavior and modifies the characteristics they were born with, and hence determines the sort of people they will be when they grow up.

Children's Cultures

The evidence is there but psychologists and anthropologists have long ignored it. The reason, I believe, is that they have misconstrued the goal of childhood. A child's goal is not to become a successful adult, any more than a prisoner's goal is to become a successful guard. A child's goal is to be a successful child.

At the risk of pushing the analogy too far, I would like to look more closely at the parallels between childhood and imprisonment. Within a prison there are two different social categories, prisoners and guards. The guards have the power. They can, for example, arbitrarily and abruptly transfer a prisoner to another prison, as I was moved from one part of the country to another as a child, against my will. Because the guards have power over them, prisoners try to keep on reasonably good terms with their guards. But what really matters to most of them is how they are regarded by their fellow prisoners.

Prisoners are aware that, sooner or later, they will probably become free people like the guards. But that is in the hazy future. Right now they are involved with the day-to-day job of getting along as a prisoner. Regardless of what they were in the past and what they might be in the future, right now they are categorized—by themselves and by others—as members of the group *prisoners.*

Like other groups, prisoners have their own culture—a culture that persists over time even though individuals keep leaving and new ones come in. They have their own slang terms and their own standards of morality. They have great scorn for those who suck up to the guards or who rat on their fellow prisoners. They have to obey guards' orders or suffer the consequences, but at the same time they don't want to knuckle under completely—they want to preserve some modicum of autonomy. So they delight in outwitting the guards, in beating the rules in little ways they can get away with. This attitude is part of the prisoners' culture, and those who succeed in outwitting the guards take pleasure in revealing their little triumphs to their fellow prisoners.

How do prisoners learn to be prisoners? How do they acquire the culture and learn the rules of prison behavior, which no doubt vary from one prison to another? One way is by making mistakes: the guards will punish them if they break any of the guards' rules, and the other prisoners will mock or shun or attack them if they break any of the prisoners'. But for those who are observant and keep on their toes, it is possible to become a "successful" prisoner without ever getting negative feedback: they can learn by watching the others. Although prisoners keep leaving the prison and others keep coming in, the new ones always have the ones who came before them to serve as their models. They cannot learn how to behave by imitating the guards, because prisoners are not allowed to behave like guards, but they can learn by imitating the other prisoners.

That said, let me hasten to add that childhood differs from imprisonment in important ways. Most children—though alas, not all—lead pleasanter and happier lives than prisoners. And children love many of the people who watch over them, and their feelings are reciprocated, as feelings usually are. A final difference is that prisoners may be back on the streets in a year or two and then they can cast away—if they so choose—the behaviors and attitudes they learned in prison. Children are in for the duration and what they learn they learn for keeps.

Though childhood is a time of learning, it is a mistake to think of children as empty vessels, passively accepting whatever the adults in their lives decide to fill them up with. It is almost as far off the mark to think of them as apprentices, struggling privately and individually to become full-fledged members of the adults' society. Children are not incompetent members of the adults' society: they are competent members of their own society, which has its own standards and its own culture. Like the prisoners' culture and the Deaf culture, a children's culture is loosely based on the majority adult culture within which it exists. But it adapts the majority adult culture to its own purposes and it includes elements that are lacking in the adult culture. And—like all cultures—it is a joint production, the creation of a committee. Children cannot develop their own cultures, any more than they can develop their own languages, except in the company of other children.

The committee meetings start early—in the play groups of the younger children in traditional societies, in nursery schools and day-care centers in our own. Sociologist William Corsaro, who has made the study of children's cultures his specialty, has spent years observing three- to five-year-olds in nursery schools in Italy and the United States. He describes how children in this age range delight in outwitting the teachers by breaking rules in little ways that the teachers don't notice, or choose not to notice. For example, there is a rule in most nursery schools against bringing toys or treats from home.

> In both the American and Italian schools the children attempted to evade this rule by bringing small, personal objects that they could conceal in their pockets. Particular favorites were small toy animals, matchbox cars, candies, and chewing gum. While playing, a child often would show his or her "stashed loot" to a playmate and carefully share the forbidden object without catching the teachers' attention. The teachers, of course, often knew what was going on but simply ignored minor transgressions.

Showing the hidden object to another kid turned an act of personal defiance into an expression of groupness—us kids against the grownups—and made it much more fun. The strategies by which children "mock and evade adult authority" are a highly valued part of the nursery school culture, according to Corsaro.

Mocking and evading adult authority seem to be universal in children's groups. Each new generation of kids discovers the strategies on its own—they don't have to learn them from older kids. But some traditions are passed along from older kids to younger ones and in that way become part of the children's culture. In an Italian nursery school where William Corsaro spent many months as an observer, the children ranged in age from three to five and some of the five-year-olds had been attending since they were three. This overlap of generations—of "cohorts," as psychologists call them—made it possible for traditions to form and to be passed along to the younger ones. Corsaro discovered that the children in this school had a tradition the teachers didn't know about: when they heard the garbage truck picking up the trash outside the wall of the play yard, the kids would climb up on the jungle gym, look over the wall, and wave to the man driving the truck. He would wave back. They thought this was great fun.

Languages can be passed along in the same way. Nyansongo children in Africa have a private language of dirty words for describing certain parts of the body. These words are not used by adults and are forbidden in their presence. Little children learn them from older ones and pass them along, when their turn comes, to still younger ones. The words are part of the children's culture. They are not part of the adults' culture.

Then, of course, there are children's games. British researchers Iona and Peter Opie spent their lives documenting the games that children play when they are out of doors and out of the purview of parents and teachers. "If the present-day schoolchild was wafted back to any previous century," said the Opies, "he would probably find himself more at home with the games being played than with any other social custom." They found English, Scottish, and Welsh schoolchildren still playing games that date back to Roman times.

> When children play in the street . . . they engage in some of the oldest and most interesting of games, for they are games tested and confirmed by centuries of children, who have played them and passed them on, as children continue to do, without reference to print, parliament, or adult propriety.

These games are not taught to children by adults; they're not even taught to children by teenagers. When a child becomes a teenager, said Iona and Peter Opie,

> a curious but genuine disability may overtake him. He may, as part of the process of growing up, actually lose his recollection of the sports that used to mean so much to him. . . . Older children can thus be remarkably poor informants about the games. . . . Fourteen-year-olds, re-met in the street, from whom we wanted further information about a game they had showed us proudly a year before, have listened to our queries with blank incomprehension.

I don't for a moment believe that a fourteen-year-old has so short a memory. Embarrassment, not forgetfulness, was what made the ex-informant clam up. It is as embarrassing for a teenager to be identified with a children's group as it is for a nursery school child to be called a baby. "I'm not one of *them,*" the fourteen-year-old was telling the Opies. "You can't expect me to know what *they're* up to." Because self-categorizations operate in the here-and-now, it is as hard for a teenager to admit that he once was a child as it is for a child to believe that he will ever be a grownup.

Games, words, strategies for outwitting adults, mini-traditions—a children's culture is a mixed bag. They can throw anything they like into the bag—anything that is approved by the majority of the children in the group. They can pick and choose from the adult culture and each group will come up with different choices. In the Robbers Cave study the Rattlers specialized in being tough and manly, the Eagles in being holier-than-thou: two different aspects of the culture that all the boys had in common. Within the space of fourteen days they created two contrasting cultures and adapted their behavior to the requirements of those cultures.

For children who share more than one culture, the range of options is even broader, because they can pick and choose from each of their cultures. During the long summer evenings in Alaska, girls in a Yup'ik Eskimo village play a traditional Eskimo game called "storyknifing": telling stories while illustrating them with pictures drawn in the mud with a dull knife. As the story progresses, pictures are smoothed over with the knife and new ones drawn. The stories used to be told in the Yup'ik language—the language of the girls' grandparents—but the children in this village are bilingual and English is the language they use among themselves. So now, as they scratch pictures in the mud, the Yup'ik girls tell stories in English, and some of the stories they tell are based on characters and plots they saw on television.

The Child Is Father to the Man

Cultures can be changed, or formed from scratch, in a single generation. Young creatures are more likely than older ones to be innovators and to be receptive to new ideas. It was a four-year-old monkey named Imo, a member of a troop of Japanese macaques on the island of Koshima, who invented a new way of separating grains of wheat from grains of sand. Imo threw the wheat into the ocean: it floated, the sand sank. Imo's playmates copied her and soon the whole troop—all but its oldest members—had learned to cast their wheat upon the water.

Another cultural innovation followed, begun by a two-year-old female named Ego. Ego introduced swimming to her peer group, and before long all the young monkeys were splashing in the surf and diving underwater for seaweed. Most of the adults in the troop didn't cotton to this new sport, but little by little they died off and the younger ones grew up and replaced them, and swimming in the ocean became part of the culture of the Japanese macaques of Koshima Islet.

In the fullness of time, the younger generation becomes the older one. It may be different from the one that came before it or it may be much the same. From the beginning of the nineteenth century through the middle of the twentieth, generations of upper-class British men closely resembled—in behavior, attitudes, and accent—their fathers. And yet their fathers had practically nothing to do with their upbringing. This is one of the puzzles I mentioned in the first chapter of this book.

Sir Anthony Glyn, whose father was a baronet, had a traditional upper-class British upbringing. He was born in 1922 and spent the first eight years of his life being tended by nannies and governesses. In those days it was fashionable for upper-class Brits of both sexes to proclaim that they couldn't stand kids. The rule that children should be seen but not heard wasn't enough for them: "The true Britishman," observed Sir Anthony, "feels that children should not be seen either. A lecture each holiday on fortitude, fitness, and trying hard at games is almost all the parental contact required."

At the age of eight, little Anthony was sent away to an exclusive boarding school—a preparatory school—and from there he went on to Eton. Until he graduated from Eton at the age of eighteen, he came home only during the breaks in the school year. His contact with his father, I gather, consisted mainly of those semiannual lectures on fortitude, fitness, and trying hard at games.

"The school's the thing," said Anthony Glyn, "particularly if it has a long tradition and is known to produce a good type of boy." His tone is sarcastic; I don't think he was happy at school. But he cannot deny that Eton does produce a good type of boy. The Duke of Wellington, in explaining his victory over Napoleon at Waterloo, said that the battle was won "on the playing fields of Eton." That was where the character of the British officer was formed: on the playing fields of Eton. Not in the classrooms but on the playing fields—the places where boys play together with a minimum of supervision from their teachers. It wasn't their education the Duke was extolling. It was their culture.

"The object of a public school education," Glyn reported, "is not to learn anything useful or indeed to learn anything at all. It is to have the character and mind trained, to have the right social image, and to make the right friends." And to acquire the right accent. Glyn described the long, slow decline of the younger sons of aristocratic families in Great Britain, and the sons of those sons. Because of the rule of primogeniture, the younger sons became, in adulthood, "poor relations." They couldn't afford to send their own sons to the schools that they had attended and as a result the sons drifted downward in social class: "Their language, their accents, became less noticeably aristocratic."

"Language," said Susan Schaller, the teacher of American Sign Language, "is a membership card for belonging to a certain tribe." For the British, it is accent. The proper accent is a membership card for belonging to the upper class. In *Lord of the Flies,* the character named Piggy was handicapped in three ways (trust Golding never to know when enough is enough). Piggy was fat, he wore glasses, and he had the wrong accent. It was Jack, the villain of the piece, who came from the fancy boarding school. A poke in the nose for the Duke of Wellington.

The boys who went to those fancy boarding schools didn't get their aristocratic accents from their nannies, who tended to be of lower-middle-class origins, or from their governesses, who might have been Scottish or French. They didn't get their accents from their brief and impersonal interactions with their parents. They didn't get them from their teachers, who were unlikely to be of the manor born. They got them from each other. The accents were passed down from the older boys to the younger ones, generation after generation, at places like Eton, Harrow, and Rugby. Other aspects of upper-class British culture—the stiff upper lip, the stern sense of moral rectitude, the refined aesthetic tastes—were passed along in the same way. These boys didn't get their culture from their father's little lectures on fortitude and fitness. They got it in the same place their father got his.

At the prep schools and "public" schools to which British aristocrats send their sons, there is a children's culture that is passed down, in the same manner as the Opies' games, from the older kids to the younger ones. Before the invention of television the kids at those schools had little contact with the adult culture; what went on in the world outside the school had little impact on them. They had limited access to radios or newspapers and there was no source of novelty other than what they themselves could dream up. Each new cohort of kids was much like the last one; the culture continued almost unchanged while generations of kids passed through it. The reason the sons were so much like the fathers was that they had both been socialized in the same way and in the same place. The sons took their culture along with them as they grew up, just as their fathers had before them. It happened to be more or less the same culture.

We think of the younger generation getting its culture from the older one, but in this case it was the other way around. The children had very little contact with the adults' culture, but all the adults were exposed to the children's culture. Every one of them was a former child.

The Parents' Peer Group

Deaf children, the offspring of immigrants, the sons of British baronets—all right, I admit it, these are exceptional cases. These are cases in which the children cannot, for one reason or another, get their culture from their parents. But what about ordinary children? Most children do, after all, live with their parents and communicate freely with them in the same language used by their neighbors.

And most parents communicate freely with their neighbors. One of the things they talk about is children: how they're turning out, how to rear them, what you're doing right or wrong. These are topics on which almost everyone has an opinion and, though almost no one realizes it, the opinions are very much a product of the culture. Upper-class British in Anthony Glyn's day would say—out loud, right in front of their kids—that they couldn't stand children. The Yanomamö worry that their enemies will cast an evil spell on their children and cause them to sicken and die, but they don't worry about letting them fight with each other with little bows and arrows. Each group has its own concerns, its own attitudes and beliefs about children.

These attitudes and concerns are passed from parent to parent through what I call the parents' peer group. It is not only children who have peer groups. Adults have them too, and—though the penalties for being a nonconformer aren't quite so devastating to their recipients—there still are penalties. But adults, like children, seldom need to be pushed to conform to their group's standards. They do it voluntarily and automatically, usually without even realizing what is going on.

Within a group—among the participants in a culture or subculture—child-rearing methods and attitudes about children tend to be quite uniform. A foreigner can see that more easily than a native. In Italy, observes the fiscal father Tim Parks, parents worry about whether their children are getting enough to eat and force-feeding is not uncommon, but the notion "that there comes a moment when parents actually force their little children to go to bed" is "unthinkable." When Michele said "Don't be fiscal" about bedtime rules, what he meant, according to his father, was:

These rules (which he doesn't know are typically English) don't need to be applied to the letter (a flexibility typically Italian).

Michele may not know that strict bedtimes are typically English, but he certainly knows they are not typically Italian. Tim Parks doesn't feel required to abide by Italian rules of child-rearing because he is not an Italian, but his son's protests nonetheless make him uneasy. Parents don't like to be different from their friends and neighbors in the way they rear their kids. They worry about it. And the kids sense this vulnerability and are quick to put it to their advantage. "None of the other kids have to phone home." "All the other guys are getting new Nikes." Though parents scoff at these transparent ploys, they are not completely immune to them.

In Chapter 5, I mentioned the nineteenth-century German girl who was treated with leeches and made to hang each day from a horizontal bar because her mother was afraid she was getting crooked. Here is her description of how fear of crookedness spread like an epidemic through her mother's group of friends and relatives:

> Suddenly, instigated by the newspapers, or God only knows what publications, epidemic fear for the deformity of their children began to spread among our mothers. The fact that our posture was straight, and there was nothing noticeably wrong with us, did not reassure our mothers at all, and was of no help to us. In all families domiciliary visits were made to search for inchoate crookedness; a true misfortune had befallen us, and before we knew what was happening, we were one and all of infirm health, and our ranks were decimated for the purpose of the cures that were to be undertaken with us. Three of my cousins, daughters of the same house, were sent to the newly established Königsberg orthopedic institute; a couple of girls from the Oppenheim family were taken to Blömer in Berlin; this one and that one of my girl friends were given fabulous machines to wear in their families, and at night were strapped into orthopedic beds at home.

The German girls with their fabulous machines got off easy. They didn't know the horrible things parents can do to their children just because all the other parents in their neighborhood or village or tribe are also doing them. I have here in my hands an article entitled "Female Genital Mutilation," published in 1995 in the *Journal of the American Medical Association.* It describes the procedures, known euphemistically as "female circumcision," that are performed on girls in Africa and parts of the Middle East and in Muslim populations elsewhere. The surgery is performed without anesthesia; the girl—typically around seven years old

—is likely to be told that screaming will bring shame to her family. Girls sometimes bleed to death afterwards, or die more slowly from tetanus or septicemia. Long-term complications can lead, in adulthood, to sterility or difficult childbirths. Intercourse may be painful and of course is unlikely to be pleasurable—that's the point of the operation.

The reason the parents are doing this terrible thing to their daughter—jeopardizing her life and health, her ability to have children—is that everyone else is doing it. Their friends and their neighbors, their siblings and cousins, are doing the same to *their* daughters. They risk the scorn of all these people if they don't go along with the practice. They risk being stuck with a daughter whom no one will marry because, according to their culture, nice girls don't have clitorises.

Although female circumcision is traditional in the parts of the world where it is performed, such practices are not necessarily passed down from parents to children. The German women who worried about their children's crookedness got their fears from the newspapers and from each other; it was not something their mothers worried about. People rear their children the way their friends and neighbors are doing it, not the way their parents did it, and this is true not only in media-ridden societies like our own. When anthropologists Robert and Barbara LeVine studied the Gusii people of Africa in the 1950s, the custom was to force-feed millet gruel to an infant by holding its nose so that it would have to suck in the stuff in order to take a breath. When anthropologists Robert and Sarah LeVine (his second wife) revisited the tribe in the 1970s, this "risky and wasteful way of feeding" was no longer being used—all the mothers had switched to feeding the millet gruel out of plastic bottles with rubber nipples.

Bottle-feeding has gone over big in the Third World and the change is not always a benign one. On the Yucatán Peninsula of Mexico, Mayan women who as infants had been fed in the traditional way—on milk from their mothers' breasts—are now feeding their own infants on formula from a bottle. The grandmothers of these babies do not approve: they believe that breast-fed babies are healthier and plumper. As it happens, the grandmothers are right.* A researcher found that the bottle-fed babies were prone to gastrointestinal infections and as a result tended to be scrawny. "Why," asked the researcher, "have Yucatecan parents given up an old, adaptive breast-feeding practice in favor of a new, maladaptive bottle-feeding one?" Because that's what their friends and neighbors are

* Grandmothers, 1; mothers, 0.

doing. So what if that's not how Mama used to do it? So what if she disapproves?

Within a multicultural society like that of the United States, parenting practices vary across subcultural groups. Breast-feeding is currently most common among educated, white, financially well-off women. In some African-American communities it has been so long since anyone nursed a baby that members of the younger generation are sometimes unaware that it is possible to feed a baby that way. The director of a New Jersey program designed to encourage breast-feeding among economically disadvantaged mothers reported, "I've had women say to me, 'Oh, you mean you can actually get milk out of there?' "

The fads in infant feeding, the fears of crookedness, the beliefs in the dangers of evil spells or the efficacy of hugs, are passed from one woman to another through what psychologists call "maternal support networks." Fathers have their networks too. Some adult male peer groups have an antidomestication ethos: they discourage their members from staying home and helping their wives with child-rearing chores. Bye, honey, I'm going out with the guys.

Researchers have reported that middle-class American parents who do not belong to support networks are more likely to violate cultural norms by abusing their kids. But not all parents' peer groups frown on the use of harsh physical punishment—this is something that varies from one cultural or subcultural group to another. The residents of La Paz and San Andres, the two Mexican villages I mentioned earlier, had very different views on discipline. "In San Andres," observed anthropologist Douglas Fry, "parents advocated and employed markedly more severe types of physical punishment than did parents in La Paz." Fry saw San Andres parents beating their children with sticks; he never saw that done in La Paz. It is to Fry's credit that he doesn't blame the aggressiveness of the San Andres villagers on the beatings they received as children. He sees the beatings as a symptom, rather than a cause, of the prevailing atmosphere of the village, and so do I.

Within our own society, attitudes toward the use of physical punishment differ from one neighborhood to another and from one subcultural group to another. Physical punishment is used more often in economically disadvantaged neighborhoods than in wealthy ones, and it is used more often by parents who belong to ethnic minorities than by European-American parents. These cultural differences in child-care practices are spread by way of parents' peer groups.

From the Parents' Peer Group to the Children's

My husband and I reared our daughters in a small, pleasant town in New Jersey. We lived there for almost twenty years, from the mid-sixties to the mid-eighties. In our neighborhood of middle-class homes, there were many people who had children the same age as ours. Most of us were of European ancestry and we were fairly evenly matched in income and in lifestyle. None of us mothers had jobs while our children were small; even when they were old enough to attend the nice elementary school a couple of blocks away, we worked only part-time.

We saw each other often, the other mothers and I. We had something in common: our kids. That's what we mainly talked about, our kids. We were Catholic, Protestant, and Jewish; we had high school diplomas or graduate degrees; it didn't seem to matter. Though I didn't realize it at the time, we all had similar views on how to raise kids. None of us worried about crookedness or evil spells cast by our enemies; we worried about how our kids were doing in school. None of us went in for force-feeding. None of us believed in letting little kids share their parents' beds. We all believed in bedtimes, though we varied in how fiscal we were in enforcing them. We all believed that an occasional smack, administered at the right time and in the right spirit, might do a kid a bit of good. None of us would have dreamed of beating a kid with a stick. Well, we might have dreamed of it but we wouldn't have actually done it.

We didn't, of course, get all our ideas from each other. They were the prevailing views—they were everywhere. In magazines and books, on the television screen. We all knew there were wrong ways to rear a child, but we had no idea that there were any other right ways.

A generation has passed—I'm a grandmother now—and mothers no longer have the time to sit around on weekday afternoons chatting with their neighbors. But it is still true that all the women who are in the same maternal support network are likely to hold similar views on child-rearing. The members of a parents' peer group are less likely today to be neighbors, but often they are. Often they've become friends because their children attend the same schools or day-care centers. If their children don't go to the same school, they have opportunities to play with each other outside of school. Thus, parents who belong to a given peer group are likely to have children who also share a peer group. To look at it from the other end, children who belong to a given peer group are likely to have parents who also share a peer group. The same is true in traditional societies. It has been true for millions of years.

This is how I believe culture is normally transmitted: from the parents' peer group to the children's. Not from parent to child but from group to group—parents' group to children's group.

When three-year-olds enter a peer group, most of them already have a culture in common. Most of them come from similar homes, homes that are typical for their neighborhood. If their parents are Americans of European origin, or second- or third-generation Americans whose ancestors came from somewhere else, it is safe to say that they all speak English, eat with a spoon and fork, and have bedtimes. They are dressed in similar clothes. They own many of the same toys, eat many of the same foods, celebrate many of the same holidays, know many of the same songs, watch many of the same TV shows.

There is no need for children who share a language to devise a new one. There is no need for children who share a culture to construct a new one from scratch. Children do construct their own cultures, but they don't usually have to construct them from scratch. Anything they have in common—anything shared or approved of by most of the children in the group—can be woven into the children's culture. The children's culture is a variant of the adult culture, and the adult culture they know best is the one they were exposed to at home. They bring that culture with them to the peer group, but they do it carefully, tentatively. They are alert to signs that there might be something wrong with it—that it might not be the culture of the Out-Theres. Alexander Portnoy, the fictional hero of *Portnoy's Complaint,* balked at using the word *spatula* in first grade because he thought it might be a word belonging to his home culture, not a word that could properly be used in school. I felt the same way, as a child, about using the word *pinky* to refer to my little finger.

Children in our society have to wonder if what they are learning at home is the right stuff, the same stuff their friends are learning. In tribal and small village societies, they don't have this worry: they know what is happening in their friends' homes. In traditional societies there is no privacy and children are exposed from infancy on to aspects of life that we, in developed societies, try to protect them from: birth and death, gore and gossip, sex and violence. There is, I assure you, as much sex and violence in traditional societies as there is in our own.

The difference is that in our own society most real-life scenes of sex and violence take place behind closed doors. So instead of learning about such things by watching their neighbors, today's children watch television. Television has become their window on society, their village square. They take what they see on the screen to be an indication of what life is like Out There and they incorporate it into their children's cultures. The *Sesame Street* characters, the superheroes and supervillains, are as much a part of the raw material of a children's culture as the language they learned at their mother's knee. Preventing an individual child from watching television would not protect that child against its influence, because television's impact is not on the individual child—it is on the group. Like all other aspects of the culture, what is portrayed on the television screen will affect an individual's behavior only if it is incorporated into the culture of the peer group. It often is.

The child whose home life is odd in some way, because he isn't permitted to watch television or because his parents are different from the other parents on the block, will nonetheless acquire the same culture as his peers. He gets it in the same place his peers get theirs: in the peer group. If his parents speak a foreign language, don't use spoons or forks, or believe in evil spells, he will nonetheless acquire the same language, customs, and beliefs as his peers. The only difference is that he has gotten them second hand. They have been passed to him, via the peer group, from the parents of his peers.

I know a woman who had many brothers and sisters and whose parents were unable to cope with the burdens of parenthood. No one ever told her, when she was a child, that she should take a bath. One day she noticed that her arms looked different from the arms of her classmates. She figured out what made them look different—hers were dirty—and began, on her own, to take baths.

Ah, you say, but many children who come from families like that never do catch on. True. But when parents who can't make a go of it have children who are similarly handicapped, I don't have to explain that—the behavioral geneticists have already explained it. Because some of children's psychological characteristics are inherited from their parents, heredity always gets in the way when it comes to explaining personality. That is why I like to look at language and accent, where heredity is not a factor.

The easiest way to tell who socialized a child—who gave the child her culture—is by listening to her. She got her language and accent in the same place where she got the other aspects of her culture. She got them from the children's peer group, which—in most cases but not all—got them from the parents' peer group.

Welcome to the Neighborhood

Psychologists and sociologists have long known that kids who grow up in neighborhoods where delinquency is endemic, or who associate with delinquent peers, are more likely to get into trouble. Thus, one way of rescuing a kid who is heading for trouble is to get him out of the neighborhood and away from his delinquent peers.

It worked for Larry Ayuso. At the age of 16, Larry was living in the South Bronx. His grades were too low to allow him to try out for the basketball team. Three of his friends had died in drug-related homicides. He was headed for high school dropout and a life (or death) of crime when he was rescued by a program that takes kids out of urban ghettos and puts them somewhere else—somewhere far away. Larry ended up in a small town in New Mexico, living with a middle-class white family. Two years later he was making A's and B's, averaging 28 points a game on his high school basketball team, and headed for college. When he went back to pay a visit to his old friends in the South Bronx, they stared at his clothes and said he talked funny. He didn't talk like them anymore. He didn't dress like them, he didn't act like them, and he didn't talk like them.

The *New York Times* reporter who wrote about Larry's metamorphosis is a product of our culture: a believer in the nurture assumption. He gave the credit to Larry's foster parents, the white couple in New Mexico. But kids like Larry can be rescued even if they aren't provided with new parents. Anything that takes them away from their delinquent peers has a good chance of succeeding. British studies have shown that when delinquent London boys move out of the city their delinquency rates decline—even if they move *with* their families. By living in one neighborhood rather than another, parents can raise or lower the chances that their children will commit crimes, drop out of school, use drugs, or get pregnant.

If the kids in one neighborhood are generally sensible and law-abiding, and those in another neighborhood are not, it isn't just because the well-behaved kids have rich parents and the other ones do not. It isn't just because they have educated parents and the other ones do not. The financial status and educational level of *their neighbors* also has an effect on the kids. The fact that children are like their parents isn't informative: it could be heredity, it could be environment, who knows? But the fact that children are like their *friends'* parents is very informative: it can only be environment. And since most kids don't spend a whole lot of time with their friends' parents, the environmental influence must be coming to them by way of their friends. It is delivered, according to group socialization theory, by the peer group.

From neighborhood to neighborhood, there are differences in the way adults behave and in the way they rear their kids. And from neighborhood to neighborhood, there are differences in the norms of the children's peer groups. In neighborhoods like the one where Larry Ayuso used to live, the norm is for kids to be aggressive and rebellious. Larry's former friends in the South Bronx are not "unsocialized": they have simply done what kids everywhere do, adapted their behavior and attitudes to those of their group. The fact that they behave and speak and dress differently from Larry's new friends in New Mexico does not mean they are less socialized: it means they were socialized by groups with different norms.

The kids in the South Bronx are aggressive for the same reason that the kids in the Mexican town of San Andres are aggressive: because that's how the other people in their communities behave. It's not because of the way their parents treat them. How do I know? Because you can move one of these families to a different neighborhood—a neighborhood where the parents don't fit in and aren't likely to become members of the local parents' peer group—and the behavior of their kids will change. The kids' behavior will become more like that of their new peer group.

Here is the conclusion of a recent study published in the *Journal of Quantitative Criminology:*

> When African American youths and white youths were compared without regard to neighborhood context, African American youths were more frequently and more seriously delinquent than white youths. When African American youths did *not* live in underclass neighborhoods, their delinquent behavior was similar to that of the white youths.

Another study looked at aggressive behavior in elementary school children. The researchers focused on children who were considered to be "high risk" on the basis of their family income (low), family composition (no father in the home), and race (African American). They found that children with these risk factors who lived in mostly black, lower-class neighborhoods were highly aggressive, but those who lived in mostly white, middle-class neighborhoods were "comparable in their level of aggression" to their middle-class peers. The researchers concluded that the middle-class neighborhoods "operated as a protective factor for reducing aggression among children from high-risk families."

Data Can Be Dangerous

"My son the doctor." A generation ago, before anyone ever heard of managed care, it was so common for Jewish parents to want their sons to become doctors, and so common for their sons to become doctors, that it got to be a joke. It was obvious to everyone, developmental psychologists included, that the sons applied to medical school because they had been brainwashed—oops, I mean socialized—by their parents to think of medicine as the most desirable of all professions.

But even before managed care, some voices failed to join the chorus. Did you hear the one about the Jewish parents who got mixed up and urged their son to be a *musician* instead of a *physician?* The punch line is that the kid decided to become a doctor anyway.

> Dr. Snyder's parents suggested that he go to a music conservatory after high school. "I didn't think being a musician was such a good job for a nice Jewish boy," he recalled. Many of his friends wanted to be doctors, and since, he said, "my major goal in life was to be like other boys," he decided to become a doctor too.

His parents got it wrong but it didn't matter. The idea that medicine is a desirable profession is transmitted in the same way as other cultural beliefs and attitudes: from the parents' peer group to the children's peer group, and thence to the individual child. The boy whose parents are listening to a different drummer nevertheless marches to the same tune as his peers.

Though the story of Dr. Snyder is a true one, it is only an anecdote, and as social scientists like to say, the plural of *anecdote* is not *data.* But I present this story precisely to show why data can be misleading. When you collect data you look at averages, overall effects. The exceptions get washed out, like the fluff on the lint filter. But in this case it is the exception that tells you what is really going on. The kid whose parents are atypical in some way, whose parents don't fit in, still ends up with the same attitudes as his peers.

There is another, more insidious way in which collecting data can produce misleading results, which I will illustrate by using my favorite example, language. If you look at kids who live in the same neighborhood and go to the same school, you will find that they all speak the same language and have the same accent. Most of their parents also speak that language and have the same accent. But because heredity is not a factor here, within a neighborhood you will find no correlation between the parents' language and accent and those of their children. That is what Derek Bickerton found in Hawaii: the parents spoke a bunch of different languages, but the second-generation Hawaiians of a given cohort all spoke the same version of creole. You couldn't tell, by listening to the children, which country their parents had come from.

Now let's say you decide to do an international study of language, collecting data on the way children talk all over the world. Your subjects include an upper-class British couple and their child, an Italian couple and their child, a Yanomamö couple and their child, and parent–child sets from a hundred other parts of the world. Hey, you've just found evidence for the nurture assumption! There is a strong correlation between the language the parents use and the language their kids use.

What has happened is that you've mistaken parents'-group-to-children's-group effects for parent-to-child effects. It is an easy mistake to make, and if we add heredity things become even more confusing. Let's say you want to show that harsh punishment by parents leads to aggressiveness in children and you decide to do your study in the Mexican town of San Andres. You find that almost all the parents beat their kids and almost all the kids are highly aggressive. But there are variations from family to family even in a homogeneous culture like that of San Andres. Because aggressiveness is to some extent genetic, and because the parent's behavior is to some extent a reaction to the child's, you find there is a tendency for the most punitive parents in San Andres to have the most aggressive children: there is a correlation between parental punishment and childhood aggressiveness. But it's a weak correlation. Darn, it's not statistically significant!

Calm down. All you have to do is add some subjects from La Paz, where parents hardly ever beat their children and children hardly ever punch their playmates. Put together all the data and voilà, you've found a strong correlation between parental punishment and children's aggressiveness. You've found that the parents who use harsh physical punishment tend to have aggressive kids, and kinder, gentler parents tend to have unaggressive kids. In fact, you've done the same thing that modern socialization researchers do when they make sure—with the best of intentions—to select their subjects from a wide assortment of ethnic groups and socioeconomic classes.

Depending on whether researchers look across cultural groups or within them, they can either find, or fail to find, correlations between parent and child. If they put together the data from several villages or tribes or neighborhoods, they are likely to find correlations that make it look like parents are influencing their kids, because the kids' behavior is more similar to that of their own parents than to that of the parents in some other place. Kids (as a group) tend to behave like the adults in their village or neighborhood. It isn't because individual kids are behaving like their own parents. If heredity is not involved, kids are as similar to their friends' parents as they are to their own.

When you see children behaving like their parents, it is easy to take it as evidence for the nurture assumption. But children and parents not only share genes: they also live in the same village or neighborhood and belong to the same ethnic group and socioeconomic class. In most cases the children's culture is similar to the adults' culture. Unless you pay attention to the exceptional cases in which the children's culture is *not* like the adults', it looks like the children learned to behave like that at home.

Seventy years ago, Hugh Hartshorne and Mark May carried out a study of what they called "character." The researchers presented children with temptations to lie, steal, or cheat in a variety of situations. They found that children who behaved morally in one situation would not necessarily behave morally in another. In particular, a kid who resisted the temptation to break rules at home, even when no one was watching, was as likely as anyone else to cheat on a test at school or in a game on the playground. The results implied that what children learn from their parents about morality doesn't go any further than the door of their home. Click. Click.

And yet—this was the mysterious thing—in a variety of situations, children tended to make the same moral (or immoral) choices as their friends and their siblings. This ceases to be a mystery when you consider that children who are friends or siblings usually live in the same neighborhood, go to the same school, and, at least in the case of friends, belong to the same peer group. They are participants in the same children's culture. Hartshorne and May concluded—this was in 1930, before the nurture assumption had clouded psychologists' minds—that "the normal unit for character education is the group or small community."

Cultural Creativity

When behavioral geneticists analyze the data from twin or adoption studies, they assume that any similarities between siblings not due to heredity must be due to growing up in the same home. "Shared environment," they call it. But in the long run it isn't the home environment that makes the difference. It is the environment shared by children who belong to the same peer group. It is the culture created by these children.

Children can create a culture almost from scratch but they usually don't. In traditional societies, the children's culture is very similar to the adults' culture because there are no handy alternatives and no need to seek them out. But even in traditional societies, the children's culture may contain elements the adults' culture lacks, such as the language of dirty words used by the Nyansongo kids. The children's culture persists for the same reason the adults' does: new members of the group learn it from the older ones.

It's a clever system—one that makes full use of the chief advantages kids have over adults, their flexibility and their imaginativeness. If the adults' culture seems to be working all right, the kids help themselves to whatever aspects of it they like. But if it seems stodgy and dated or fails to meet their needs, they are not constrained by it. They can make a new culture.

10 GENDER RULES

"It's the most awful thing I have ever done," the seven-year-old boy told the developmental psychologists. No, he hadn't killed his father or made love to his mother. He hadn't tossed his baby brother out the window or set their house afire. All he had done was to help the psychologists with an experiment by enacting a role in front of a video camera. He had followed their instructions; he had done what he was told. He had changed a doll's diaper.

The psychologists also asked a seven-year-old girl to let them film her playing with a toy truck, but she was made of tougher stuff. "My mommy would want me to play with this," she informed them, "but I don't want to."

What's wrong with these kids? We give them unisex names and dress them in unisex clothing. We tell our daughters that they can be truck drivers and our sons that it's all right to play with dolls. And we do our best to set them a good example. All over North America and Europe, fathers are changing diapers, mothers are shifting gears.

And yet our sons and our daughters still have these outdated notions. The grownups' ideas have been revised but the kids' have not. Over the past century the adult culture has grown steadily more egalitarian, but childhood is as sexist as ever.

I might as well admit it right off the bat: I don't believe that girls and boys are born alike. There are differences to begin with. But the differences we see between seven-year-old girls and seven-year-old boys are not just differences they were born with. Boys aren't born with an aversion to changing dolls' diapers; girls aren't born disliking trucks.

Sex differences increase over the first decade of life. So does overt hostility between the sexes. Boys put up "No Girls Allowed!!!" signs. Girls display their partisanship in equally unsubtle ways. Here's a song brought back from summer camp by the six-year-old daughter of a friend:

Boys go to Jupiter to get more stupider,
Girls go to college to get more knowledge.
Boys drink beer to get more queer,
Girls drink Pepsi to get more sexy.
Criss cross, apple sauce,
I HATE BOYS!

Such symptoms of sexism are commonly blamed on the parents, or the teachers, or the culture as a whole. But if the adults' society is less sexist than the children's, how can it be the adults who are having this effect on them? If you've followed me this far you already know my answer: it isn't the adults. It is the children themselves.

If you've followed me this far you know I am swimming against the stream: such is the power of the nurture assumption that neither the professor of psychology nor the person ahead of you in the checkout line is likely to agree with what I've said in the first nine chapters. But now we come to the development of maleness and femaleness and suddenly I find that I'm no longer swimming alone. When I say that a boy's masculinity and a girl's femininity are shaped by the environment they share with their peers rather than the environment they share with their parents, I am not saying anything new. Others before me—even professors of psychology—have come to a similar conclusion.

They came to that conclusion because efforts to blame this aspect of development on the parents have borne meager fruit. Do parents treat boys and girls differently? In the United States the answer is: not in any important way. They give boys and girls the same amount of attention and encouragement; they discipline them in the same way. The only consistent differences are in the chores they assign to boys and girls and the clothes and toys they buy for them. And these differences could be child-to-parent effects: reactions to, rather than causes of, the differences between the sons and daughters. Yes, parents buy trucks for their sons and dolls for their daughters, but maybe they have a good reason: maybe that's what the kids want.

Freud believed that a boy gets his ideas about how to behave by identifying with his father, a girl by identifying with her mother. The evidence does not support Freud's theory. A boy's masculinity and a girl's femininity are unrelated to those characteristics of their same-sex parent. Boys reared in fatherless homes are no less masculine, and girls reared in homes headed by lesbians are no less feminine, than boys and girls provided with a Dan Quayle–approved parental pair.

During the formative years of childhood, a girl becomes more similar to other girls and a boy becomes more similar to other boys. Rowdy girls become less rowdy; timid boys get bolder. And the differences between the sexes widen: the two overlapping bell curves pull apart so they don't overlap as much. It is the children themselves who are responsible for these changes. They don't identify with their parents: they identify with other children—others like themselves.

There Are Differences to Begin With

Of the forty-six chromosomes in the human genome, forty-five are unisex: both males and females have them. The forty-sixth is the Y chromosome, so called after its shape. The Y is found only in males. It is among the smallest of human chromosomes.

Nature is thrifty. If there is junk in our genome, it is there only because it is cheaper to leave it in than to winnow it out. We do not have multiple copies of essential genes because it is costly to do the processing that keeps them in working order. Thus, organisms are assembled the way that, according to Mozart, Salieri wrote music: with a lot of repeats. Bilaterally symmetrical organisms do not require a separate set of genes for each half—just a command to flip the instructions over and do the same thing on the other side.

Males and females have forty-five chromosomes in common because it is cheaper to duplicate than to vary. All the differences between them are encoded on, or triggered by, that one little Y chromosome; the rest of their genome contains the same instructions. Male kidneys and female kidneys, male eyes and female eyes, work the same way. Their bones form the same connections; their hemoglobin is concocted from the same recipe. Males have nipples, even though they don't need them, because it's easier to duplicate than to vary. Give a man estrogen and he will grow breasts.

Because nature is thrifty, only differences that made a difference were coded into our DNA. Only differences that made a difference in the environment in which our species evolved. These were things that, when present in males but not in females, increased the likelihood that the males would survive and reproduce, or that their close relatives would survive and reproduce. Or things that, when present in females but not in males, increased the likelihood that the females would survive and reproduce, or that their close relatives would survive and reproduce.

Boys and girls are alike in many more ways than they are different, but there are differences. One difference is obvious: it's the one the obstetrician (or the ultrasound specialist) looks at before making the traditional announcement, "It's a boy!" or "It's a girl!" Some differences are less obvious: at birth, on the average, boys are slightly larger and more muscular than girls. Some differences are not obvious at all, because they are inside the baby's head.

In a famous experiment of the 1970s, a pair of researchers showed college students a film of a nine-month-old baby wearing unisex clothing and playing with unisex toys. They told some of the students that the baby's name was Dana; others that it was David. Depending on whether they thought they were seeing a girl or a boy, the viewers of the film made different judgments about the baby. Dana was more likely to be seen as sensitive and timid, David as strong and bold. And yet it was the same baby.

This experiment was supposed to demonstrate that babies are really all the same and it is only because we give them names like Dana or David, and thereafter treat them differently, that they turn out the way they do. But sixteen years later another pair of researchers did a slightly different experiment: several babies were filmed, not just one, and the college students were asked to make judgments of all the babies. There were no indications in the films of the babies' real sex; none of them were given names. And yet, on the average, the female babies were judged to be more sensitive, the males to be stronger. If you borrowed a dozen healthy infants, dressed them in neutral clothing, called them things like "Jamie" and "Dale" and "Yan-Zhen," and asked passersby to guess their sex, I'll bet that well over half of the guesses would be correct.

In the first edition of my child development textbook, published in 1984, there was a sidebar called "The Case of the Opposite-Sex Identical Twins." It was based on a report by two psychologists at Johns Hopkins University, John Money and Anke Ehrhardt. Money and Ehrhardt were called in to counsel the parents of a pair of identical twin boys, one of whom had suffered a terrible accident. At the age of seven months, the child's penis had been destroyed in a botched circumcision. Now the parents—a young couple from a rural background with only a grade school education—had one intact son and one who was exactly like him in every way but one: he lacked a penis.

Doctors told the parents there was no satisfactory way to reconstruct a penis. The best alternative, they said, was to rear the injured twin as a girl. They recommended removing the child's testicles—thus eliminating the primary source of male hormones—and administering estrogen during puberty. The result would be a female-shaped body.

The parents agonized over the decision for several months and finally, when the child was seventeen months old, gave in. The boy was castrated and reconstructive surgery was performed, producing the outward appearance of female genitalia. He was given a girl's name and thereafter dressed and treated as a girl.

Judging from Money and Ehrhardt's report, the parents were wholehearted in their acceptance of their child's new gender. The psychologists heard from the mother several times over the next few years and she was always quite clear that one of her twins was a boy and the other a girl. In the sidebar in my textbook, I quoted the mother's words:

> She seems to be daintier [than her twin brother]. Maybe it's because I encourage it. . . . I've never seen a little girl so neat and tidy. . . . She just loves to have her hair set; she could sit under the drier all day long to have her hair set.

Although the child and the parents seemed to have made a good adjustment, Money and Ehrhardt did reveal some minor problems. "The girl had many tomboyish traits," they admitted, "such as abundant physical energy, a high level of activity, stubbornness, and being often the dominant one in the girls' group."

As I said in the first edition of my textbook, so what? Lots of little girls are tomboyish. Most of them nonetheless think of themselves as girls and have no doubts about their gender. I had my own history in mind when I wrote that, because I too had been something of a tomboy. Like the transformed twin, I had abundant physical energy and was stubborn. Unlike the transformed twin, I hated having my hair set and wasn't the least bit dainty. But I can't remember ever wishing to be a boy. I looked forward to becoming a mother and in the meantime lavished my maternal impulses on my pets and my dolls. Change a doll's diaper? Sure, no problem.

"The Case of the Opposite-Sex Identical Twins" made it through all three editions of my textbook, but by the final edition I was clearly having some misgivings. By then I was admitting that "there is a limit to how much social influences and learning can accomplish." But I still maintained that "if people consistently treat you like a girl, you will probably become a girl."

I no longer believe lots of things I said in that textbook, and one of them is the statement about becoming a girl if people treat you like a girl. Maybe it is true in some cases but certainly not in all and probably not in most. The opposite-sex identical twin did not, as it turned out, make a good adjustment to the change in gender. An article in a 1997 medical journal revealed the truth. The child had never fit in as a girl, never felt comfortable in the role of a female. And yet his parents and physicians kept telling him he was a girl. His anger and misery came to a head when he was fourteen; he felt his life was hopeless and contemplated suicide. At that point his parents finally revealed to him the secret of his past: that he had been born a boy. "All of a sudden everything clicked," he said. "For the first time things made sense and I understood who and what I was." He stopped trying to be a girl and became a boy again. The reverse metamorphosis was carried out in full view of his high school classmates; since his unfeminine behavior had already made him the butt of their jokes, his situation at school could hardly get worse. In fact, it got better. His peers found him more acceptable as a male than they had as a female. At the age of twenty-five he married a woman a few years older than himself and became, through adoption, the father of her children.

In a remote corner of the Dominican Republic, a mutation occasionally crops up that makes genetic males look like females at birth. At puberty their testosterone kicks in and masculine characteristics appear: the voice deepens, the shoulders broaden, and what appeared to be a large clitoris develops into a small penis. Researchers have studied eighteen of these people who were reared as girls. When their bodies became manlike in appearance, all but one of them chose to switch genders and abandon the female names and identities they grew up with. They married women; they took on men's jobs. The case of the opposite-sex identical twin differs from that of the Dominicans only because it wasn't due to a mistake of nature. It was due to a mistake made by a group of doctors and psychologists who thought that a little girl is a little boy minus a penis and testicles.

The idea that babies are born with the potential to become either male or female, and that the behaviors associated with the two sexes are entirely cultural, was popularized by the anthropologist Margaret Mead. It is another example of her tendency to see things through the lens of her prior beliefs. She described a New Guinea tribe—the Tchambuli—in which the men supposedly behaved like women and the women like men.

Submissive, anxious men; strong, bossy women. According to anthropologist Donald Brown, Mead got it wrong. In fact, among the Tchambuli, "polygyny was normal, wives were bought by men, men were stronger than women and could beat them, and men were considered by right to be in charge."

In every society we know of, the behavior of males and females differs. It differs far more in most societies than in our own. And the pattern of differences is the same all over the world. Males are more likely to be found in positions of power and influence. Females are more likely to be found tending to other people's needs. Males are the hunters and warriors. Females are the gatherers and nurturers. Boys are pressed into service as babysitters if a girl is not available, but girls are preferred for this job all over the world. Girls vie with each other to hold a baby; boys don't seem to find babies all that interesting. An Israeli researcher reported that in the homes she studied, lots of parents gave their sons dolls. But the dolls given to boys didn't get their diapers changed. The researcher saw their young owners "treading on them or beating them like hammers against pieces of furniture."

I don't think it's a coincidence that all over the world people have similar stereotypes for males and females. Social psychologists John Williams and Deborah Best gave questionnaires to college students in twenty-five assorted countries, asking them to check off adjectives that in their culture were associated more with one sex than the other. In all twenty-five countries, males were associated with adjectives like *aggressive, active, reckless,* and *tough.* Females were associated with adjectives like *affectionate, cautious, sensitive,* and *emotional.*

Stereotypes

To most people the word *stereotype* has a negative connotation: it implies prejudice. It implies making up your mind about someone too quickly and for the wrong reasons. But Williams and Best see stereotypes as "not essentially different from other generalizations." In their view, "stereotypes are simply generalizations about groups of people, not necessarily 'bad' generalizations." We have stereotypes not just about other groups but also about our own, and stereotypes about our own group are predominantly positive. This is a consequence of the tendency (described in Chapter 7) to favor our own group over others.

Humans—even very young ones—are excellent gatherers of statistics, excellent detectors of statistical differences. The human mind is built that way. Red fruits are, on the average, sweeter than green ones and it doesn't take long for kids to start preferring red foods to green ones. We mentally sort things into categories on the basis of their differences and then go on collecting more evidence on how they differ. Our minds carry out these jobs efficiently and automatically, usually without our conscious awareness.

Social psychologist Janet Swim did a study of the way males and females are stereotyped in American culture. She asked college students to estimate the differences between men and women on a number of dimensions, including the tendency to assume leadership in a group, performance on tests of mathematical aptitude, and the ability to interpret the body language and facial expressions of others. Then she compared these stereotypes with the actual results of studies in which sex differences were measured. She found that the stereotypes were surprisingly accurate —if anything, the college students were more likely to underestimate sex differences than to overestimate them.

Stereotypes are not always accurate; they are less likely to be accurate when they involve groups we don't know as well as we know men and women. But the real danger in stereotypes is not so much their inaccuracy as their inflexibility. We may be right when we see men as more apt to take on leadership roles and less adept at reading other people's feelings, but we are wrong if we think *all* men are like that. We are fairly good estimators of differences between means—the difference between the average member of group X and the average member of group Y—but we are poor estimators of the variability within groups. Categorization tends to make us see the members of social categories as more alike than they really are, and this is particularly true for the category we're not in.

The Social Categories *Girls* and *Boys*

During the first few years of life, little girls and boys collect statistics on several categories of people: *grownups* and *kids, women* and *men, girls* and *boys.* I have no formal data on which to base this statement, but I do not believe that young children have mental categories for *females* and *males.* I do not believe they have a mental category containing both girls and women, and another containing both boys and men. To children, grownups and kids belong to different species; it would be like lumping cows with hens and bulls with roosters. Children may know, in an intellectual sense, that boys eventually turn into men and girls into women, but this is something they had to be told or had to deduce—to them it's not obvious, it's not relevant, and it's only just barely believable. Because they have no pigeonhole labeled *males,* young boys put themselves in the pigeonhole labeled *boys,* and they tailor their behavior to that of boys, not of men. That is why a boy can see his own father changing diapers and still say that changing a doll's diaper was the most awful thing he had ever done. That is why a girl whose own mother was a physician could say that only boys can be doctors, girls have to be nurses.

So children collect statistics on the categories *girls* and *boys* and find statistical differences between them. They know, because they were told or they figured it out, which category they are in. Most find they like their own category best. Most find it more fun to play with the members of their own category—the members of their own sex—because those are the ones who usually want to do the same things they want to do. By the age of five or six, the majority of children in day-care centers and kindergartens are playing in small groups that are almost entirely single sex. They split up like this, if the adults permit it, whenever they have a choice of companions. Of course, as I've said before, when they have no choice they will play with anyone they can get.

The most important years for group socialization are the years of middle childhood, from six to twelve. During all that time, children in our society—a society that provides them with a plethora of potential companions—spend much of their free time with peers of their own sex. They are socialized—that is, they socialize each other, they socialize themselves—not just as children but as girls or as boys. This gendered socialization is not simply a consequence of spending time with other members of one's sex or even of liking the members of one's sex better: it is a consequence of self-categorization. A girl categorizes herself as a member of the category *girls* and a boy categorizes himself as a member of the category *boys,* and they get their ideas about how to behave from the data they've collected on those social categories. They've been collecting the data since the day they were born.

My evidence, as usual, comes from the exceptional cases. Consider the case of the opposite-sex identical twin: he was told he was a girl but he didn't *feel* like a girl. He wasn't interested in doing the things girls do. Here is his own description of his childhood:

> There were little things from early on. I began to see how different I felt and was, from what I was supposed to be. But I didn't know what it meant. I thought I was a freak or something. I looked at myself and said I don't

> like this type of clothing, I don't like the types of toys I was given. I like hanging around with the guys and climbing trees and stuff like that.

This was a genetic male whose male organs had been destroyed by a compound error of doctors; even after they started giving him estrogen and he started growing breasts, he still didn't feel like a girl. Then there are genetic males whose male organs are intact and who are reared as boys, and yet they do not feel like boys. The writer Jan Morris, born James Morris, was such a child.

> I was three or perhaps four years old when I realized that I had been born into the wrong body, and should really be a girl. I remember the moment well, and it is the earliest memory of my life.

Children like James Morris, children like "Joan" (the alias used for the opposite-sex identical twin during the years he lived as a female), are likely to be rejected by girls and boys alike. They are regarded, even by themselves, as "freaks"—as nails that will not or cannot be hammered down. Feminine boys have a particularly rough time of it: the other boys pick on them and, once they're past kindergarten, the girls don't want them either. Often they grow up friendless and alone. And yet they are socialized—they socialize themselves—and it is gendered socialization. James Morris categorized herself as a girl; therefore she was socialized as a girl, even though she was regarded by others as a boy. As an adult, Jan Morris voluntarily sought the same kind of surgery that had been perpetrated on Joan against his will, because it is very difficult to live in a male's body if inside you are female.

In an article in the journal *Child Development,* a researcher recounted a true story about a little boy named Jeremy, who one day decided to put barrettes in his hair and wear them to nursery school. Jeremy's parents evidently thought that was fine but his classmates were of a different opinion. One boy in particular kept teasing Jeremy about his new hairstyle by calling him a girl. To prove that he was not a girl, Jeremy finally pulled down his pants. "The boy was not impressed," reported the researcher. "He simply said, 'Everybody has a penis; only girls wear barrettes.' "

Jeremy's classmate was wrong in fact but right in theory: gender identity—the understanding that one is a boy or a girl—doesn't come like a label attached to the genitals. Nor is it something parents can give their child, try as they might. Milton Diamond, the psychologist who interviewed Joan after he became a male again, believes it comes from a process of comparing oneself to one's peers. Children, he says, compare themselves to the boys and girls they know and decide "I am the same" as one kind and "I am different" from the other. On the basis of how they feel inside—what their interests are, how they want to behave—they put themselves into one gender category or the other. And that is the category in which they are socialized.

Daja Meston, the boy who was reared in a Tibetan monastery (I told his story in Chapter 8), described himself as "a white body that houses a Tibetan." No surgeon can remedy that discrepancy. Daja was rejected by his peers because he was too tall and too white, but that didn't prevent him from categorizing himself as a boy like them and it didn't prevent him from being socialized as a Tibetan. In the same way, children such as Joan and James may categorize themselves as members of groups that reject them. You don't have to be liked by the members of your social category in order to feel that you are one of them. You don't even have to like *them.*

Gender Fences

Developmental psychologist Eleanor Maccoby—yes, the same Eleanor Maccoby who made a cameo appearance in Chapter 1 and played a leading role in Chapter 3—has described an experiment in which a pair of unacquainted children, between two and a half and three years old, were put together in a laboratory room strewn with toys. What happened next depended on whether the sexes were mixed or matched. Girls and boys were equally friendly when paired with a member of their own sex, but a disquieting asymmetry appeared when a girl was paired with a boy. The girl, instead of playing with her partner—the way she would have if he had been another girl—often became an onlooker. "When paired with boys," reported Maccoby, "girls frequently stood on the sidelines and let the boys monopolize the toys." These were little children, not yet three years old!

Playing with others involves cooperation, and cooperation means sometimes doing what the others ask of you. Bids for cooperation may take the form of suggestions or demands. Research has shown that as little girls get older they make more and more suggestions to their playmates, and their playmates—if they are other girls—become more and more amenable to following them. But, over the same period of time, little boys get less and less amenable to following suggestions, especially if the suggestions come from girls. They are more likely to listen to other boys, perhaps because such communications generally come in the form of demands rather than polite requests. These things are happening, mind you, at an age when there is hardly any difference in size or strength between the average boy and the average girl.

Perhaps this is why little girls start avoiding little boys: it is not much fun to play with people who won't listen to your suggestions and who grab your toys without even a by-your-leave. But soon little boys are also avoiding little girls, perhaps because it's more fun to play with people who want to do exciting things like making toy trucks go vroooom rather than dumb things like changing dolls' diapers. Or perhaps this mutual avoidance is the result of categorization into two contrasting groups, *girls* and *boys,* and the us-versus-them feelings that ensue.

For whichever reason, or for all three put together, segregation by sex gathers momentum over the years of childhood. The dividing line is sharpest just before puberty—just before it begins to fade. Even in parts of the world that are sparsely settled and where young children of both sexes play together, preadolescents form separate sex-segregated groups. They can do this because they are able to roam farther from home in their search for companions.

Much has been written about the differences between boys' groups and girls' groups in middle childhood. Eleanor Maccoby offers a succinct summary:

> The social structures that emerge in male and female peer groups are different. Male groups tend to be larger and more hierarchical. The modes of interaction occurring in boys' and girls' same-sex groups become progressively differentiated, and the different styles appear to reflect different agendas. Boys are more concerned with competition and dominance, with establishing and protecting turf, and with proving their toughness, and to these ends they are more given to confronting other boys directly, taking risks, issuing or accepting dares, making ego displays, and concealing weakness. Among boys, there is a certain amount of covert sexy (and sexist) talk, as well as the elaboration of homophobic themes. Girls, though of course concerned with achieving their own individual objectives, are more concerned than boys with maintaining group cohesion and cooperative, mutually supportive friendships. Their relationships are more intimate than those of boys.

Maccoby is talking, of course, about averages. There are exceptions to every rule,* and there are children who do not fit into these neat categorical descriptions. Some boys withdraw from the roughness and competi-

* There are exceptions even to the rule that there are exceptions to every rule.

tiveness of boys' groups; they are apt to become loners, at least in school. Some girls would rather play with the boys. If they are good enough at sports, they may be accepted.

It is unusual, however, for a girl to be accepted into a boys' game on the school playground. Most of the girls who play games with boys do it in neighborhood play groups, not in school. Neighborhoods offer fewer potential companions than schoolyards, so the children can't be as selective; this provides a handy excuse for the children who don't *want* to be as selective. In any event, neighborhood play groups often contain both sexes and a range of ages. The mixture of ages is what makes it possible for street games to be passed down from one generation of children to the next, from the older to the younger ones. The mixture of sexes is what makes it possible for so many women—more than 50 percent in some surveys—to say that they were tomboys in their youth and played with boys.

On school playgrounds and in coed summer camps, where there is no shortage of potential companions, girls and boys split up into warring camps of *us* versus *them.* Interactions between girls and boys on the playground often take the form of what sociologist Barrie Thorne calls "borderwork": interactions that deepen the division between the sexes, that make it more salient. Interactions that are hostile, at least on the surface—underneath they no doubt have more complicated meanings. Boys make forays into girls' games with the intention of breaking them up. They grab the girls' scarves or bookbags. They snap the bra-straps of the early developers. Girls are not always the victims in these skirmishes. I remember in my fifth-grade schoolyard some of the bolder girls (I was not one of them, having lost my boldness by then) used to chase after one of the boys—there was a cute red-headed boy who was frequently singled out—and threaten to kiss him. This was regarded, by the boy, as a fate worse than death and he usually managed to squirm away just in time. Men sometimes oppress women by kissing them against their will, but on the playground it is more often girls who use kissing as a weapon.

When group distinctions are salient, hostility between groups is most likely to emerge. Pressures on children to avoid showing any signs of friendliness to members of the opposite sex are most intense in the parts of the school where adults keep a low profile—the lunchroom and the playground. Boys, in particular, are taunted by their male peers if they play with girls or even sit down next to a girl. Adult influence *increases* the amount of friendly interaction between boys and girls. It is the kids themselves, not the grownups, who initiate and maintain the segregation of the sexes.

The parents I know are pleased if their kids have one or two friends of the opposite sex. Such friendships do exist, but if they begin in the preschool years, as many of them do, they generally go underground in middle childhood. The girl and the boy see each other only at home or in the neighborhood; at school they may fail to acknowledge each other even with a nod. Their parents are aware of the friendship but their peers are not. These are friendships I'm talking about, not romances. Underground romances between school-age children also exist, but many of them are one-way. The object of love may not even be aware of having been awarded the distinction.

Friendships and romances are personal relationships; they are not to be confused with groupness—the understanding that you are a member of a particular group and the feeling that you like your own group best. Groupness and personal relationships follow different rules, have different causes and effects. Sometimes they work at cross purposes, as when one finds oneself liking a member of a disfavored group. Sometimes they make conflicting demands and one has to choose between them. It has often been observed that men and women, when faced with this dilemma, tend to resolve it in different ways. A woman places high value on her personal relationships. A man pries himself loose from the arms of his beloved and goes off to fight a war. "I could not love thee, dear, so much," he assures her solemnly, "Lov'd I not honour more."* He tells her he's fighting for *her* but it's not true: he's really fighting for his group. In traditional societies it is the men who usually remain in the village they were born in, and fight for that village if need be; the women usually leave the village when they marry. Among chimpanzees, it is the males that form alliances with each other and go off together to kill the Kahamans.

I think groupness is stronger in males for evolutionary reasons: it is the males—larger and more muscular than the females, able to run faster and throw harder even in childhood, freer in adulthood to take physical risks because they don't get pregnant and don't have babies to lug around—who join with their fellows to defend the group and to launch attacks on other groups. Intergroup warfare was part of the environment in which our species evolved, and anything that gave us an edge over our adversaries was worth a bit of extra work for that little Y chromosome. The games boys like to play—the games they play all over the world—are very good preparation for warfare. As the writer Herman Melville once observed, "All wars are boyish, and are fought by boys."

* Richard Lovelace, "To Lucasta: Going to the Wars," 1649.

Many of the famous experiments in social psychology—the Robbers Cave study, the under- and overestimators—have used young males as subjects, and I suspect there was a reason: the results might not have been so clear-cut if female subjects had been included. The Robbers Cave researchers did a less well known experiment (I described the famous one in Chapter 7) in which the boys were allowed to make friends first and then the researchers divided them up into two competing groups, splitting up friendships. Split-up friendships ended; friends became enemies. I wonder what would have happened if the researchers had tried that with girls. "Oh please let Jessica switch with Claire so Jessica and I can both be Eagles!"

I do not mean to imply that groupness is absent in females. The male brain and the female brain both have a groupness module; they both have a relationship module. The difference, if there is one, is only in which takes precedence when they issue conflicting commands.

One Culture or Two?

Boys' groups tend to be hierarchical. There is a leader and he tells the others what to do. Boys vie with each other for status. They refrain from showing their weaknesses. They don't ask for directions because they don't want anyone to know they're lost.

Girls' relationships tend to be close and exclusive, though not necessarily lasting. Girls are less likely than boys to show hostility directly; they get back at their enemies by attempting to turn their friends against them. Leadership in a girls' group has its hazards: it can get you a reputation for being stuck-up or bossy. Girls don't believe in bossing their friends around—they believe in cooperation and taking turns.

When they're with their peers, boys strive to be tough; girls strive to be nice. I am not the first to point out these differences; nor am I the first to attribute many of the behavioral differences between adult men and women to the socialization they got, or the patterns of social interaction they learned, in their childhood peer groups. Eleanor Maccoby has said that girls and boys grow up in different cultures. Linguist Deborah Tannen, the author of *You Just Don't Understand,* has expressed a similar view.

Some writers disagree. Sociologist Barrie Thorne, who studied the ways boys and girls behave on school playgrounds, doesn't like the "different cultures" idea. She points out that girls and boys do interact in many social contexts: they interact with their siblings at home and with friends of both sexes in neighborhood play groups. In school classrooms the sexes often mingle peacefully in reading or study groups. Even on the playground, where awareness of the division between the sexes is most acute, girls and boys sometimes unite. Thorne tells about an incident she observed in which a boy named Don was unfairly punished by a playground aide and was very upset, and his classmates—girls and boys alike—came together in his support. Thorne feels that the behavioral differences and mutual avoidance of boys and girls are somehow transmitted to them by the adult culture. She doesn't say exactly how, and she admits that children are most sexist when they are most free from adult control, but she implies that calling children in a classroom "boys and girls" and putting sexist pictures on the wall must have something to do with it.

Although my own views on gender are much more compatible with those of Maccoby and Tannen, I do concede that Thorne has a point. Boys and girls don't really have separate cultures. Boys and girls of the same age and ethnicity, who live in the same neighborhood and attend the same school, are participants in the same children's culture. They have the same ideas about how boys and girls should behave, the same ideas about how men and women should behave. The different behaviors that are prescribed for people in different social categories are part of a culture. Boys and girls have different opinions on which way of behaving is *better*, but they agree pretty well on what boys and girls are supposed to do.

Different social categories, not different cultures. Social categories vary in salience depending on context, while the culture remains more or less the same. The way we categorize ourselves depends on where we are and who is with us, and even a very young child has a choice: she can categorize herself either as a *kid* or as a *girl*. If the age category is salient, the gender category automatically becomes less so. When a grownup is being conspicuously grownupish, like the playground aide who unfairly punished Don, it makes age categories come to the fore and gender recede into the background. That's why the girls and boys joined together in support of Don. If you give school-age children another way of dividing up—into ability-based reading groups, for instance—gender will become less salient to the degree that the reading groups become more so.

Two Sexes Or One?

Barrie Thorne used the fact that girls and boys interact in some contexts as an argument against the view that girls and boys are themselves responsible for the differences between them. But interaction doesn't prevent kids from developing notions of how girls are supposed to behave and how boys are supposed to behave. Interaction doesn't prevent them from categorizing themselves and their classmates as *girls* or as *boys,* and it doesn't reduce the salience of these categories.

What does reduce the salience of the gender categories is total lack of interaction: the absence of the opposite sex. When only one group is present, groupness weakens and self-categorization shifts in the direction of *me* and away from *us.* That's when you get within-group differentiation —that's when members of a group jockey for status, and choose or are chosen for different roles.

When there are no boys around, girls don't act so girlish. This was observed by some researchers who watched twelve-year-old girls playing dodgeball, a game played by children of both sexes. Two different groups of subjects took part in this study: middle-class African-American girls at a private school in Chicago, and Hopi Indian girls on a reservation in Arizona. The researchers purposely chose cultures that vary in the status accorded females: traditional Hopi culture is matrilineal and women have a good deal of social and economic power.

When no boys were present, both groups of girls were serious about playing dodgeball: they played competitively and some of them played very well. But as soon as some boys entered the game, the girls' manner of playing changed dramatically. Instead of standing in a way that indicated readiness to make a move, the Hopi girls stood with their legs crossed and their arms folded, looking shy and unathletic. The African-American girls, when boys were present, chatted with each other and teased the other players. Both groups of girls were completely unaware of the change in their behavior. When the researchers asked them why they thought the boys always won, they said the boys cheated. But the boys didn't cheat: they just played harder. They won even though, at this age, the average boy is noticeably shorter and lighter in weight than the average girl.

Boys and girls hold similar stereotypes of boys and girls: both think of boys as being more competitive than girls and better at sports. On the average, this is true. When the gender categories are salient, girls become more like their stereotype of a girl, boys become more like their stereotype of a boy, and the differences between them are exaggerated by contrast effects.

When there are no boys around, girls don't act so girlish. But when there are no girls around, boys still act boyish—at least in some ways. In some ways they do appear to be less masculine: to us rugged Americans, the graduates of the British all-male boarding schools, with their high voices and fastidious tastes, appear effete. But the hazing that goes on (or used to go on) in those schools is definitely a guy thing. Sir Anthony Glyn, the baronet's son, recalls his ungentle introduction to boarding school:

> A boy's first week at his preparatory school is likely to be the most traumatic experience of his life, one for which he is, at the age of eight, totally unprepared. Until that moment, he has not realized that there are so many people in the world who wish to hit him and hurt him and that they will be given ample opportunity to do so, both by day and by night.

The hitters and hurters are the other boys, the older boys. What has happened here is that the absence of girls has put the gender categories out of contention. As a result, age differences have become more salient and within-group vying for dominance has gone vroooming to the top. When there is no other group around, competition within a group increases; as the dodgeball players demonstrated, this is true for girls as well as boys. The domination of younger children by older ones is also found in both sexes. But girls' domination of younger children is of a different sort than boys': girls do it in a less aggressive way. It has been speculated that inhibition of aggression in females is a built-in (though imperfect) mechanism that evolved because those who lacked this brake were more likely to harm their own children.

Where children of both sexes go to school together—especially where they gather on the playground in dichotomous groups of girls and boys—the gender categories are highly salient and sexism reigns. Their fathers may change diapers and their mothers may drive trucks, but the sons play football and the daughters jump rope. The parents may sincerely believe that boys and girls are essentially alike—that a little girl is a little boy minus a penis and testicles—but the children know better.

Back to Our Roots

Oddly enough, girls and boys in modern egalitarian societies may be more stereotypically girlish and boyish than the children who lived in the hunter-gatherer bands of our ancestors. Among the few surviving populations of hunter-gatherers are a people called the Efe, who dwell in the Ituri forest of what used to be called Zaire. Here is a researcher's description of life among the Efe:

> Mau, an adolescent forager boy, sits in camp with his brother's 15-month-old daughter draped across his lap, lulled to sleep by the not-so-distant music of a finger piano. Mau reaches over to stir his pot of *sombe* as a group of young boys and girls play "shoot the fruit" using child-sized bows and arrows. The children come dangerously close to Mau's cooking fire, and he utters a disapproving "aa-ooh!" . . . As he scans the camp he notices a group

of women preparing for a fishing trip, while others lounge, smoking tobacco along with the men.

Because there are seldom enough children in a hunter-gatherer band to form separate play groups for boys and girls, the Efe boys and girls play together. Consequently the salient social categories for Efe children are not *boys* and *girls,* but *kids* and *grownups.* And the boys and the girls behave pretty much alike. Even among the adults, gender boundaries are less sharply defined than you might expect. In contrast, a neighboring tribe called the Lese, whose farming lifestyle allows for a greater population density, have a society that is highly differentiated by sex. The Lese live in settlements large enough to enable girls and boys to split up into separate groups.

Another group of traditional hunter-gatherers are the !Kung of Africa's Kalahari desert. Today they are farmers and herders, but twenty years ago some !Kung were still living in small nomadic bands. An anthropologist who studied them reported that among the nomadic !Kung, girls and boys played together and sex differences were minimal. Among the !Kung who had settled down and become producers of food, there were enough children for girls and boys to form separate groups and sex differences in their behavior were quite noticeable.

Girls and boys are more alike in behavior in places where they are too few in number to form separate groups, because in those places they categorize themselves as *kids.* They are alike because they are socialized in, and socialized by, the same peer group. The exaggerated sex differences we see today among children in our own society may indeed be a creation of our culture: it was the invention of agriculture, a cultural innovation only ten thousand years old, that made it possible for us to provide children with so many potential playmates.

A bit of advice to parents who want to rear androgynous children: join a nomadic hunter-gatherer group. Or move to a part of the world where there are just enough kids to form one play group and not quite enough to form two.

I'll Do It Your Way

Did you notice those Efe children running around with their little bows and arrows? Boys and girls were playing together, but they were playing a *boys'* game. And how about those neighborhood play groups in the American suburbs? The girls who participated in them became, by their own description, tomboys. You don't get much changing of dolls' diapers in those mixed-sex groups, not once the kids are past preschool age. If the girls want to play with the boys, they usually have to play by the boys' rules.

The urge to dominate their peers is detectable in human males at the tender age of two and a half. The greater aggressiveness of males—not just in humans but in almost all mammals—has been well documented. A stallion is more aggressive than a gelding (a castrated male horse), but it isn't just having testicles that does it. The opposite-sex identical twin, while living as a girl, was "often the dominant one in the girls' group" even though his testicles had been removed when he was seventeen months old. Girls who are born with a condition called congenital adrenal hyperplasia—a hormonal malfunction that can cause partial masculinization of the brain and genitals of a female fetus—tend to be assertive children even though the hormonal condition is medically rectified after they are born.

Most girls find out early in life that they don't have much influence on boys. They start avoiding boys before the boys start avoiding them. They would rather play with other girls because girls listen. Boys always want to do things *their* way.

So the girls form their separate groups, where they can do what *they* want to do. That works pretty well until adolescence. Then the sexes get back together again, driven by forces that—sorry—are outside the scope of this book. In adolescence, other ways of splitting up become more salient: you have the athletic cliques, the academic cliques, the delinquent cliques, and the none-of-the-aboves. Groups once again contain members of both genders. But on the whole they are run on the boys' terms. In mixed-sex groups it is males who do more of the talking and more of the cracking of jokes. The females do more of the listening and more of the laughing.

Downers

It has been alleged that girls' self-esteem plummets in early adolescence. Although this is not always found, and when found it is a smaller effect than the newspaper stories may have led you to believe, I accept that it is true on the average: for some girls, self-esteem does go down. What I don't accept is that it is the fault of the parents or teachers, or of a nebulous force called "the culture." It is due, I think, to the situation girls find themselves in at adolescence. By forming their own separate groups in childhood, they were able to avoid being dominated by boys. Then their biological clocks strike thirteen and suddenly they find themselves wanting to interact with a bunch of people who have been practicing the art of domination ever since they let go of Mommy's hand. It was bad enough when these people—the boys—were the same size or, for a brief time, a bit smaller. Now, to top it off, they are rapidly getting *bigger.*

For a teenage girl to have any sort of status in a group whose dominant members are boys, either she must be good at something they value or she must be pretty. If she has neither of these assets, chances are she will be ignored. These are not things she can acquire just by trying hard: she has little control over them. She may have had high status in the girls' groups of her childhood, but that is of no avail if in adolescence she turns out not to be pretty.

Two things that affect how a person feels about herself are status and mood. If her status in her group is low and there is nothing she can do to improve it, her self-esteem goes down. Her self-esteem also goes down if she's depressed. From early adolescence on, females are twice as likely as males to become clinically depressed.

The link between depression and low self-esteem is well-established. What is not so clear is which comes first—which is the cause and which the effect? Many clinical psychologists believe that low self-esteem causes depression, and no doubt this is true in some cases. But often the relationship works the other way around. If you know anyone with a bipolar mood disorder—manic depression, as it's more commonly called—you'll see what I mean. When people with this disorder are in a manic state, they think they can do anything, they think they're the best in the world. When they are depressed, they think they are worthless. Nothing has changed but their mood—they have the same history of good and bad experiences—but sometimes they feel good about themselves and sometimes they feel rotten.

Bipolar disorder occurs with equal frequency in both sexes but, starting in early puberty, unipolar depression (lows without highs) is more common in females. The drop in self-esteem that some girls experience at this age may be a symptom of depression, rather than the cause of it.

Why is depression more common in females than in males? No one knows for sure. My guess is that it's due to subtle differences in the brain—differences in the delicate balance between mechanisms that lead to action and mechanisms that inhibit action. When something goes wrong in the brain, males are more likely to tilt in the direction of too much action, and the result is violence. Females are more likely to tilt in the other direction, and the result is anxiety or depression. Manic depression, then, would mean that the balance between the two kinds of mechanisms is unstable.

To Hell with la Différence

Girls and boys are somewhat different when they are born. Over the next sixteen years, the differences increase. During childhood they increase because girls and boys identify, at least part of the time, with different groups. During adolescence they increase again, this time for physical reasons.

Nature is efficient; she is not kind. On the average, females are weaker and less aggressive than males, and in every human society—not excepting your noble hunter-gatherers—they run the risk of being battered. Female chimpanzees, too, are sometimes battered by their males. Things are better for women today than they have ever been over the past six million years. When I was a grad student at Harvard, there was still a professor in the psychology department who would say, in public, that women have no place in the laboratory. No college professor would dare say such a thing today.

Women are now permitted to play the games that used to be barred to them. The trouble is, they still have to play them by the guys' rules. What they learned in childhood puts men at an advantage and women at a disadvantage on the playing fields of contemporary societies.

But gendered socialization is not the only reason why people vary. The pressures from within and without to conform to the norms of one's group, the contrast effects that make these norms different, can only do so much. Psychological differences between the sexes are statistical differences: the distance between the twin peaks of the two bell curves. During childhood the curves pull a bit farther apart but they never part company: there is always an overlap. Some men are short; some women are tall. Some boys are gentle; some girls are tough. Even when they're in the company of their peers.

11 SCHOOLS OF CHILDREN

You probably remember how it was done. Maybe you even remember doing it yourself. The little ways that schoolchildren signal to their classmates—while still remaining within the letter of the classroom law—that they are not knuckling under to the teacher. Sociologist Sharon Carere, an ex-schoolteacher herself, has described some of the techniques children use for what she calls "playing the fine line"—defying the teacher in ways that the teacher has trouble objecting to. There is, for example, the wastebasket saunter:

> Students walked to the wastebasket in a saunter. Upon arriving, each dimension of disengaging from the refuse and allowing it to fall into the holding device was executed with painstaking care and precision, and was followed by a few seconds of simply watching it lie there.

And the split-body maneuver at the bookshelves:

> They positioned themselves at the structures either with book at hand, intently pouring over it to assess its adequacy for their reading desires and needs of the moment, or looking over the array of books, ostensibly searching for a title that caught their interest. What was noteworthy about this institutionally defined, directed behavior was that it was specific to only part of the students' bodies: normally the upper section appeared absorbed, while the lower half displayed social interaction and free-play concerns, including softly kicking the leg of the person next to them, using their feet to maneuver any object that might be on the floor near them, making a fist with the hand that was dangling from the arm not in use, and pok-

ing, usually gently (so as not to cause a commotion) the person next to them.

Half the fun is getting there. The trip to the wastebasket or the bookshelves can be livened up in many entertaining ways, such as "dancing up the aisles to an internal rhythm" or pretending to be a toy soldier, a tightrope walker, or a duck. For the real pro, "the action might even include a pause at the front of the room to deliver a special center-stage performance for the benefit of any fans who might be watching."

The fans, of course, are the other children in the classroom. The teacher is not a fan—she is one of *them,* the necessary foil without which the little acts of defiance would be pointless.

To children in school, the most important people in the classroom are the other children. It is their status among their peers that matters most to them—that makes the schoolday tolerable or turns it into a living hell. A large part of the teacher's power resides in her ability to put individual children in the spotlight, to make them the focus of their peers' attention. She can, if she is so inclined, hold up a child to public ridicule or public envy.

But a teacher can do much more than that. If, in this book, I seem to rob parents of much of their power and responsibility, I cannot be accused of perpetrating the same crime against teachers. Teachers have power and responsibility because they are in control of an entire group of children. They can influence the attitudes and behaviors of the entire group. And they exert this influence where it is likely to have long-term effects: in the world outside the home, the world where children will spend their adult lives.

Groupness in the Classroom

As they get older, children get better at navigating the complex array of social identities offered to people in modern societies. Without budging from her seat—without moving a muscle—a seven- or eight-year-old can switch back and forth among various self-categorizations. She can at one moment think of herself as a third-grade girl, at another as a third grader, at still another as a student in the Martin Luther King Elementary School. She can think of herself as a member of the highest reading group or as one of the smart kids in the class. (There is no need for her to have names for these social categories.) She can also slide back and forth along the me–us continuum: sometimes she feels like a member of a group, sometimes she is more concerned about her status as an individual.

Social categorization is always at play in the environment of the school.

Because there are so many children all in one place, there are many possibilities for forming subcategories. Big groups tend to fall apart into smaller groups unless there is something to hold them together.

Between parallel groups there are contrast effects. In the previous chapter I described the results of one such contrast: that between girls and boys. When children categorize themselves as *girls* or *boys,* and when these self-categorizations are salient, the differences between the sexes widen. Even if there hadn't been differences to begin with—and in this case there were differences to begin with—the mere existence of two dichotomous social categories is enough to produce them. The Rattlers and the Eagles taught us that.

Now you can see why ability grouping (or "tracking") has the effects it does. When teachers divide up children into good readers and not-so-good ones, the good readers tend to get better and the not-so-good ones to get worse. A group contrast effect at work. The two groups develop different group norms—different behaviors, different attitudes.

Groupness makes people like their own group best. You may wonder whether that can be true even of the members of not-so-good reading groups. Yes, I believe it is true even of them. They might recognize that they are not very good at reading but think they are better at other things—that they are nicer or handsomer or better at sports. They might recognize that they are not very good at reading but devalue the importance of reading. They might adopt an attitude that school sucks and anyone who does well at it is a nerd, a goody-goody, or a brown-nose. The Eagles looked down at the Rattlers for being dirty-mouthed; the Rattlers looked down at the Eagles for being wimps.

Attitudes such as those that I've posited for the not-so-good reading group—that reading is unimportant, that school sucks—have effects that compound themselves over the years. Being a poor reader may cause a child to categorize himself with the poorer students in the class even if the teacher doesn't formally acknowledge such groups. The child then adapts to the norms of that group and takes on its attitudes, and the attitudes are likely to be anti-school and anti-reading. The consequences are harmful and they are cumulative. Group contrast effects between quick learners and slow ones result in the slow learners adopting norms that make them dumber—or, more precisely, norms that cause them to avoid doing things that might have made them smarter.

Group contrast effects act like a wedge. They force themselves into any little crack between two groups—any little difference between them—and make it wider. Such effects have their origin in the deep-rooted tendency to be loyal to one's own group. I am one of *us,* not one of *them.* I don't want to be like (yuck) *them.*

In school, children's group alliances are often made on the basis of academic performance or motivation. Good readers versus poor ones. Nerds versus jocks. Brown-noses versus burnouts. It isn't until high school that such groups acquire labels and develop a stable membership, but there are cliques operating on similar principles even in elementary school. Kids who hang around with the good students in the classroom tend to have good attitudes toward schoolwork; those who hang around with the not-so-good ones tend to have poorer attitudes. And if a kid shifts from one group to another during the course of the school year—something that still can happen in elementary school—the kid's attitudes change to match those of his new group.

This is not a question of self-esteem; it is a question of acquiring skills by practicing them. The kids who have a poor attitude toward school simply do not do as much brainwork as the kids who think school is important. They don't have a poor attitude toward *themselves*—just toward school. They don't, as a rule, have lower self-esteem. African-American students, for example, who as a group perform less well in school than Americans of European or Asian descent, do not have lower self-esteem than children in other ethnic groups. Forget what you may have thought or read on this subject: on the average, the self-esteem of young African Americans is no lower than that of young European Americans. Self-esteem is a function of status *within* the group. People judge themselves on the basis of how they compare with the other members of their own group.

An Apple for Miss A

My textbook on child development was written before I saw the light and cast off my belief in the nurture assumption—before I understood the power of group socialization. In that book there is a sidebar entitled "An Apple for Miss A." It doesn't say anything I need to apologize for today, but when I wrote it I didn't fully understand what had happened in Miss A's classroom or why it happened. Now I think I do.

"Miss A" is what she was called in an article about her by educator Eigil Pedersen and his colleagues, published in the *Harvard Educational Review.* She was a first-grade teacher in the primary school Pedersen attended in the 1940s—an old school of the old school, built like a fortress, its windows reinforced with iron bars. An inner-city school surrounded by tenements and attended by the children of poor people and immigrants: two-thirds white, one-third black. A school that sent only a tiny minority of its alumni to college; most never graduated from high school. A school in which fights and behavior problems were rampant and were punished with the strap. There were two or three strappings a day. The good old days, huh?

Eigil Pedersen was one of the tiny minority of the school's alumni who made good. He graduated from high school and went on to college, and in the 1950s he returned to the school as a teacher. During the years he taught there he began to look into the school records for an explanation of why such a large proportion of the school's students never finished high school. But something he found in the records interested him so much that he abandoned his original research goal and concentrated instead on studying Miss A's effect on the students in her first-grade classes.

Miss A, Pedersen discovered, had had an extraordinary effect on her students. The fact that they made good grades in her class didn't prove anything—perhaps she was an easy marker—but Pedersen noticed that Miss A's students, on the average, made better grades the next year too, even though they were split up among several second-grade teachers. Following them through their school careers, Pedersen discovered that the academic superiority of Miss A's kids was still detectable in seventh grade. Intrigued, he extended his investigation to the world outside the school: he traced some of its alumni and interviewed them. He found that Miss A's ex-students were doing better in their adult lives than those who had been taught by other first-grade teachers. In terms of upward mobility, they had climbed higher than their schoolmates.

Judging by what her ex-students told Pedersen, Miss A is a strong candidate for sainthood. She never lost her temper. She would stay after school to help any of her students who were having trouble—they came from a variety of backgrounds but every last one of them learned to read. She would share her lunch with kids whose parents had forgotten (or couldn't afford) to provide them with one. She remembered their names twenty years after they left her classroom.

In the sidebar of my textbook, I attributed Miss A's long-lasting effect to the head start she gave her students in first grade. But head starts provided by programs like Head Start tend to peter out over time, even if they produce dramatic improvements in the short run. Why didn't the Miss-A effect peter out?

Here is a clue. Not one of Miss A's former students failed to correctly name her as their first-grade teacher when Pedersen interviewed them. But four people who *hadn't* been in her class incorrectly named her as their first-grade teacher. "Wishful thinking," Pedersen called it.

Did wishful thinking cause these people to construct memories of a classroom they had never set foot in? Memory is far less trustworthy than most people believe—yes, it can create as well as destroy—but I think something else was going on here.

To explain it I must digress for a moment and talk about leaders. Groups sometimes, but not always, have leaders. The leader isn't necessarily a member of the group; groups can be influenced from either the inside or the outside. A teacher is a leader who can influence a group even though she is not a member of it.

Leaders influence groups in three ways. First, a leader can influence the group's norms—the attitudes its members adopt and the behaviors they consider appropriate. To do this it is not necessary to influence every member of the group directly: influencing a majority of them is enough, or even just a few if they are dominant members, the ones at the top of the attention ladder. Cultural forces like television work the same way. According to group socialization theory, it is not necessary for every last boy in the group to watch a particular television show: as long as most of his peers watch it, the effect on the norms of an individual boy is the same, whether or not he watches it himself.

Second, a leader can define the boundaries of the group: who is *us* and who is *them.* This was something that Hitler, for example, was very good at.

Third, a leader can define the image—the stereotype—a group has of itself.

A truly gifted teacher can exert leadership in all three of these ways. A truly gifted teacher can prevent a classroom of diverse students from falling apart into separate groups and can turn the entire class into an *us* —an *us* that sees itself as scholars. An *us* that sees itself as capable and hard-working.

Don't ask me how they do it; I don't know. Jaime Escalante, an immigrant from Bolivia who taught calculus to a bunch of Mexican-American kids in an East Los Angeles high school (and who was immortalized in the movie *Stand and Deliver*), was a teacher of this sort. A biographer described Escalante's effect on his students this way: he made his students feel that they were "part of a brave corps on a secret, impossible mission." Another gifted leader is Jocelyn Rodriguez, a teacher at a middle school in the Bronx, New York. Rodriguez manages to form the students in her classes—mostly black and Hispanic—into a close-knit community. Each class thinks up a name for its community, designs a flag, and composes an anthem. "We're all really friends," one of her students explained to a reporter, "so we don't mind sitting close together."

One of the things that characterize these exceptional classrooms is the attitude the students adopt toward the slower learners among them. Instead of making fun of them, they cheer them on. There was a boy with reading problems in one of Rodriguez's classes and when he started making progress the whole class celebrated: "Every time he made a small step, the class would give him a round of applause."

You can see the same sort of thing in descriptions of schools in Asian countries—in Japan, for instance. Kids are criticized by their classmates for misbehaving and cheered for doing well. Misbehavior by one child is seen as a blot upon the entire class; one child's improvement is seen as a triumph for everyone. It's not because Japanese kids are *nicer*—out on the playground, bullying is as much of a problem there as it is in other countries. I don't know how the teachers do it—whether it is their pedagogical methods, the culture, or a combination of the two—but I think their we're-all-in-this-togetherness is a chief reason why Asian kids are ahead of Americans in many school subjects. With no group in the classroom adopting an anti-school, anti-intellectual attitude—with every kid working at maximum capacity—the teacher can go vroooming ahead.

Which brings us back to Miss A. I believe she possessed the mysterious ability to form the diverse bunch of kids in her classroom into a united group of motivated learners—an *us.* An *us* is a social category, whether or not it has a name. I think Miss A made her kids feel that they were in a special social category: "a brave corps on a secret, impossible mission." This self-categorization stuck with them even after they graduated from her classroom; it buffered them from anti-school attitudes and made them feel superior to the other kids in their grade. And the existence of this special social category must have been recognized even by those who hadn't been lucky enough to have Miss A as a teacher. That is why some of the people Pedersen interviewed claimed to have been in Miss A's class: they were, or had aspired to be, part of the group she created. Behind the barred windows of that old school, among the tenement kids who attended it, there was a group of motivated learners who thought of themselves as "Miss A's kids," even though some of them had never set foot in her classroom.

Perhaps Eigil Pedersen himself was part of that group. Perhaps that is how he managed to become one of the school's most successful alumni, despite the fact that his first-grade teacher was Miss B.

Long Division

There are many vicious circles in development—the child whose peers don't like him has fewer opportunities to develop his social skills, the overweight child avoids physical activity and gets fatter—but no circle is more vicious than the one having to do with intelligence. Children who may be only a little behind their peers to begin with tend to avoid doing those things that could have made them smarter. As a result they fall further and further behind. Meanwhile, the kids who started out a little ahead are doing push-ups with their brains.

Behavioral geneticists have found that the heritability of IQ increases across the lifespan—estimates for older adults are as high as .80, which seems to say that 80 percent of the variation in intelligence among the elderly can be chalked up to their genes. But putting it that way is misleading because not all the variation is due to the direct effect of the genes. Much of it is due to the choices people make in childhood and adulthood. Whether to watch TV or do homework. Whether to play ball or go to the library. Whether to remain in Brittany's circle or switch to Brianna's. Whether to go to college and what courses to take there. Whether to marry Roger or Rodney. The results of a lifetime of such choices show up in behavioral genetic studies as genetic influence on IQ—heritability—but in fact what the researchers are measuring (as I pointed out in Chapter 2) is a combination of direct and indirect genetic effects.

The increase in the heritability of IQ over the lifespan is due mostly to indirect genetic effects—the effects of the effects of the genes. What starts out as a small difference can balloon into a large one. IQ tests may actually underestimate the ballooning of the difference because they are graded on a curve: children are compared only to their agemates and at every age the same proportion of 130s and 100s and 70s are given out.

When children in a classroom split up into smaller groups on the basis of academic achievement, contrast effects cause the differences between the groups to widen. The effects tend to be more noticeable on the poorer achievers in the class than on the better ones because the better ones are already doing the best they can. I believe that group contrast effects of this sort are an important source of indirect genetic effects on IQ.

When children in a classroom split up into smaller groups on the basis of race or socioeconomic class, contrast effects again act to widen the differences between the groups—or to create differences if there were none to begin with. If you randomly divided children in a classroom into the Dolphins and the Porpoises, and if it happened that the Dolphins had one or two outstandingly good students or the Porpoises had one or two who couldn't keep up, the two groups might adopt group norms with contrasting attitudes toward schoolwork—even if the average IQs of the two groups started out the same. Now assume that over several years of schooling, the members of these two groups continue to identify themselves as Dolphins or Porpoises, associate mainly with their groupmates, and (depending on which group they belong to) either strive to do well in school or turn up their noses at schoolwork. What started out as a different attitude toward schoolwork might well end up as a difference in average IQ.

There is a book called *A Question of Intelligence,* by Daniel Seligman, which makes some of the same points as *The Bell Curve* but in a less inflammatory way. In one chapter Seligman talks about the black–white IQ difference and describes the efforts of social scientists to attribute it to differences in environment. He points out that differences in socioeconomic status, or differences in income, are not an adequate explanation: even if you look at kids in the same socioeconomic class, or at those whose parents bring in the same income, you still find a difference in average IQ. Seligman finds these results discouraging but he leaves the door open just a crack for a different environmental explanation:

> These details do not quite end the argument about environmental effects, however. In principal, it would still be possible that all or most of the black–white gap was attributable to other kinds of environmental factors —to factors not being captured in standard social-science data. A kind of last-ditch argument for the environment is sometimes made by positing an "X" factor. The X factor is something that nobody knows how to quantify or even describe very clearly, but—the argument goes—it comes with the experience of being black in America; it makes that experience unique and utterly noncomparable to the lives led by whites. In the process, it undermines the relevance of all those correlation coefficients that seem to show only limited environmental contributions to the gap. And in some way that nobody can make clear, the X factor works to suppress mental abilities.

I believe I know what the X factor is; I believe I can describe it quite clearly. Black kids and white kids identify with different groups that have different norms. The differences are exaggerated by group contrast effects and have consequences that compound themselves over the years. That's the X factor.

Around the age of three, children begin to notice that people can be categorized by race. Over the next few years, racial distinctions increase in salience and become one of the ways that children divide up into smaller groups. Whether they do or don't divide up this way depends partly on something as trivial as number: how many children there are at a given place and time. Just as girls and boys will play together if they don't have a wide choice of companions, and categorize themselves simply as *kids,* so will white and black children.

American children tend to learn more in classrooms that have fewer students. The reason may be that it is easier for the teacher to make a smaller class into a united group. The kids are less likely to divide up into contrasting groups with contrasting attitudes toward schoolwork if there aren't very many of them.

If the kids in a classroom vary both in socioeconomic class and in race or ethnicity, and if the two are linked in such a way that the members of one race or ethnic group are mostly middle-class and the members of the other are not, even the best teacher in the world might find it impossible to forge them into a single group.

Sociologist Janet Schofield spent several years studying sixth- and seventh-graders in a school she calls "Wexler." Wexler is a city school with a mixture of African-American and non-Hispanic white students in roughly equal proportions. The majority of the white children come from middle-class homes, the majority of the black children from working-class or low-income homes. Although the teachers and administrators are committed to the goal of promoting racial harmony, they haven't come close to achieving it. Black kids and white kids eye each other with a wary distrust that is only one notch short of the open hostility between the Rattlers and the Eagles. At Wexler it is rare for a black kid and a white kid to play together on the playground or sit together in the lunchroom.

The kids at Wexler come from different social classes but that's not what they notice: what they notice is a difference between two social categories defined in terms of race. Both the black kids and the white kids in this school see the whites as academic achievers, the blacks as academic resistors:

Sylvia (black): I guess they (blacks) don't care about learning. The white kids, when it's time to get their education, they can't wait.

Ann (white): The black kids don't really care what (grades) they get.

The differences between the groups are not just academic. Both the black kids and the white kids see the whites as soft and wimpy, the blacks as tough and aggressive. The white kids "just can't take it," a black girl told the sociologist. "They don't know how to fight." Attempts to cross the racial divide are likely to be met with disapproval by one's groupmates:

> *Lydia (black):* They (other black girls) get mad 'cause you've made a white friend. . . . They say that blacks are supposed to have black friends and whites are supposed to have white friends.

"For black students," Schofield observed, "succeeding academically often means leaving their friends behind and joining predominantly white groups within their classes." Black kids who do well academically are pressured by their peers not to work so hard. They are failing to conform to the norms of their group: they are "acting white." These kids do not get their anti-school attitudes from their parents. Parents of all racial and ethnic groups think education is important and hold high expectations for their children's academic success. Some researchers have found greater emphasis on education among black and Hispanic parents than among European Americans.

Schofield's work at the Wexler school dates from the late 1970s but things haven't changed much. A recent article in the *New York Times* quoted a teacher in the Bronx who said that some of her black students "would rather be paraded in handcuffs before television cameras than be caught reading a book." And "acting white" is still used as an insult among black kids.

The pressure on black kids to act black and on white kids to act white is of the same sort as the pressure on the Rattlers to refrain from crying and on the Eagles to refrain from cursing. It comes from within the group, not from outside, and it needn't be overt. Children seldom have to be urged to conform to the norms of their group. Nails that don't stick up need no hammering.

I've spoken here of black–white contrasts, but there are schools in which the contrast is between Asian Americans and European Americans or between two white groups or two black groups. In a school on Long Island, New York, the principal tells a journalist about tensions between Haitian immigrants and American-born black kids. The Haitians, who are also black, are the good students. A Haitian-born teenager complains about being taunted by the African Americans: "When we are nice and respectful of teachers, they say that we are trying to act 'white' and act as if we are better than them." In parts of Brooklyn and the Bronx, the children and grandchildren of black immigrants from Jamaica identify with groups that contrast themselves with other black kids. The Jamaicans are the academic achievers and they do very well indeed; their success stories are reminiscent of those of the children of Jewish immigrants a generation earlier. Colin Powell, the retired general who said no thanks when he was asked if he'd like to be president, is the child of Jamaican immigrants who settled in the Bronx.

There was a study done many years ago in Germany of children fathered by American servicemen and brought up by their German mothers. The researchers found no difference in IQ between the children of white fathers and those of African-American fathers, even though the biracial children were, by the conventional definition, "black." These were black kids who couldn't have had a group of their own because there weren't enough of them in any one school. They may have been rejected by their white schoolmates, as Daja Meston was by his Tibetan monasterymates, but evidently that didn't give them the idea that reading is unimportant and school sucks.

"Stereotype Threat"

Sticks and stones may break my bones but names can never harm me. That's not true, of course: names can hurt terribly. But the names that do the serious damage are the ones we call ourselves. The stereotypes we give ourselves are the ones that matter in the long run—not the ones imposed on us by other people. The power of other people's expectations to exert an insidious influence on our behavior, intelligence, or what have you, has been vastly overrated.

But the notion persists that when prophecies fulfill themselves, it must be the prophet's fault. "Stereotype threat" is what does the damage, according to social psychologist Claude Steele. It turns out that if you make a young woman who is good in math more aware of being female, she does less well on tests of mathematical ability, and if you make a young African American who is a good student more aware of being black she does less well on tests of academic ability. Steele found that all you have to do to lower the score of a bright black kid on a test of academic ability is to give her, before she takes the test, a short questionnaire that includes the question "Race?"

Self-categorizations are exquisitely sensitive to social context. What Steele is doing is evoking his subjects' groupness: he is increasing the salience of race or gender and thereby making it more likely that they will categorize themselves as *black* or *female.* Along with these self-categorizations come the norms associated with them. People feel uncomfortable about violating the norms of their group.

Steele attributes the discomfort associated with "stereotype threat" to fear of failure. It could just as easily be attributed to what, thirty years earlier, psychologist Matina Horner called "fear of success"—a hangup she detected in bright young women. I believe the discomfort is caused by a conflict between the desire to do well and the feeling that doing well would conflict with the norms of one's group. Horner herself, by the way, was evidently untroubled by any such ambivalence. When she was offered the presidency of Radcliffe College she didn't say no thanks.

As Claude Steele demonstrated, it is still possible to make some women feel they are violating the norms of their group if they do too well in math. He attributes these effects to injurious stereotypes that are held by the society as a whole. I attribute them to the stereotypes that groups have of themselves (which is not to deny that the society, too, might have stereotypes). In contexts where gender is less salient, girls and young women do better in science and math. Women's colleges produce a disproportionate number of outstanding female scientists. The women at these colleges live in the same society as the rest of us but they are less likely to categorize themselves as *women* and less likely to contrast themselves with men.

The society as a whole does not distinguish between African Americans whose parents came from Jamaica and those whose parents came from anywhere else. What has made the descendants of the Jamaicans so successful is that they have a different stereotype of themselves.

Intervention Programs

A recent issue of the American Psychological Society's *Observer* features an argument between two developmental psychologists: one a supporter of preschool enrichment programs like Head Start, the other a critic of them. The critic points out that Head Start was designed "to prevent school failure and improve adult outcomes among low income children" but there is little solid evidence it does so. The supporter is backed into a corner. She is forced to admit that Head Start produces no long-term gains in the academic achievement of African-American children and resorts to citing gains in "the accessibility of community-based services" for the families involved and the "higher rates of immunizations" for their children. Though these are worthy goals, they fall short of what the program was designed to do.

Most programs like Head Start have only temporary effects on the children they serve and some have no measurable effects at all. Interestingly enough, the ones that have no measurable effects at all tend to be those that try to change the *parents'* behavior. Programs that rely on visits by professionals to the children's homes can produce changes in the parents' behavior—a significant reduction in child abuse, for example. But they have no noticeable effect on how the children behave when they are not at home or on how well they do in school. The programs that get the parents involved produce no better results than the ones that leave the parents out. This is just what group socialization theory would predict.

For intervention programs to work, I believe they must modify the behavior and attitudes of a *group* of children. For such programs to have long-term effects, the children must remain in contact with each other so that they can continue to think of themselves as a group. Thus, I would predict that programs aimed at an entire schoolful of children should be more successful than those that pluck seventeen children from ten or twelve different schools.

An example of the sort of program I have in mind is one that was designed to reduce aggressive behavior and increase mutual helpfulness among school-age children. Training sessions were administered to all the children in selected target schools and resulted in a small but significant improvement in their behavior on the school playground and in the cafeteria. What had changed was the norms of the group. As my theory would predict, there was no detectable improvement in how the children behaved at home.

So far there has been no test of my prediction that intervention programs can have long-term effects if they focus on changing the norms of a group and if the members of the group retain their ties to it. The researchers who carry out long-term follow-ups of intervention programs never mention in their reports—and I presume they never notice—whether the children who took part in a group program kept in touch with each other after the program ended.

Language Lessons

One of the characters who made an appearance in Chapter 4, along with Cinderella, was a boy named Joseph—a real boy, though that is not his real name. When he was seven and a half years old, Joseph's parents immigrated from Poland to a rural area of Missouri. Neither Joseph nor his father could speak any English at all when they arrived in the United States. His mother had taken a six-week course in the language and could pronounce some English words.

Joseph's parents were unskilled workers. In Missouri, his father first found work as a laborer in a garden nursery and later as a custodian. His mother was not employed outside the home and seven years after immigration she still had very limited skills in English. I am telling you about his background so you won't think that Joseph had any sort of advantage—genetic or cultural—to make his transition easier. As far as I can tell from the report of the psycholinguists who studied him, he was an ordinary boy, the son of ordinary parents.

Joseph arrived in Missouri in May and had the summer to acquire some English-speaking friends and begin to learn their language. When school started at the end of August, the psycholinguists estimated his ability to speak English as about that of a two-year-old. The school provided him with no translator and no special classes for kids who can't speak English. He was put into a second-grade classroom with children of his own age, none of whom could speak Polish, and a teacher who couldn't speak Polish either. All his instruction was in English. It is a method sometimes referred to as "sink or swim."

For a while it looked like Joseph wasn't even trying to swim. During the first couple of months in his new school, he sank to the bottom and just stayed there, saying very little in class. But he was fully alert to what was going on around him, watching the other kids for clues to what the teacher was saying. When she told them, for instance, to take out their spelling books, Joseph looked around, saw the others taking out their spelling books, and took out his.

His progress was remarkably rapid. By the end of November he was producing sentences like this on the way to the playground: "Tony, I don't give you cars anymore if you don't let me play." Not perfect, but it got its point across to Tony.

Eleven months after he came to the United States, at the age of eight and a half, Joseph's use and understanding of English were rated as equivalent to that of an American-born six- or seven-year-old, though he still spoke with a Polish accent. After another year he had caught up with his agemates and his accent was barely detectable. The psycholinguists didn't check on Joseph again until he was fourteen; at that point his pronunciation was indistinguishable from that of his American-born peers, even though he continued to speak Polish at home. His performance in school showed a similar trend: he experienced some difficulty in reading in the early grades, but from the fifth grade on, his grades were average or a little above.

There was no group of Polish Americans in Joseph's school, no group of non-English-speaking kids with which he could identify. Like Daja Meston he was sui generis, and one isn't enough to make a group. So he categorized himself as just a kid, a second-grade boy, and adopted the norms of behavior appropriate for that social category. The norms included speaking English. If Joseph had been plunged, sink or swim, into a school for deaf children, the norms would have been quite different and Joseph would have learned to communicate with his hands instead of his tongue. A sociologist who visited a school for deaf children reported that it was "a place where one learned to be deaf." Here is a conversation between the sociologist and a veteran teacher at the school:

Sociologist: Have you seen any "deaf behavior"? What is it and what does it look like?

Teacher: I don't know that I can explain it but we've had kids come here who had a good bit of hearing and then later on they're acting more and more deaf . . . and it's not just the fact that they stop using their speech . . . which is a bad thing. That does happen, I hate to say.

Sociologist: Explain that a bit. I've heard that before. . . . If a kid came in here and could talk they (the students) make him stop talking, don't they?

Teacher: They stop talking.

Sociologist: Why? . . . Is there pressure on them to stop talking?

Teacher: From the other kids. And so they start acting deaf.

Now consider what would have happened if Joseph's parents had settled in an area where there were many Polish immigrants and if he had been one of several students in his class who knew little or no English. Let's say Joseph had gone to a school that offered a bilingual program for children who couldn't speak English. Would he have been better off?

Certainly he would have found the transition easier. Certainly the first months in his new school would have been less stressful. But would he have learned English as rapidly or as well?

This is a controversial question but by now you know that I am not one to shrink from controversy. The answer is no. Bilingual programs have been, in the words of one knowledgeable critic, "a dismal failure."

Group socialization theory can explain why these programs fail. They fail because they create a group of children with different norms—norms that permit them not to speak English, or not to speak it well. The fact that their teachers might speak grammatical, unaccented English is not enough. In the schools for the deaf, it's not the *teachers* who cause the children with "a good bit of hearing" to stop talking. Most of the teachers in those schools can hear.

Language is both a kind of social behavior and a kind of knowledge—something that can be taught. Teachers can transmit knowledge but they have only limited power to influence the behavioral norms of their students. Even an excellent teacher of English will be frustrated by the slowness of her students' progress unless she can convince them that speaking English is the norm for their group. The hard part is not keeping them afloat: it's persuading them to swim against the current.

In areas where there are many immigrant families, bilingual programs enable children to spend most of the school day in the company of other children with whom they share a native tongue. A teacher observed,

> The Russian students end up talking to each other in Russian, the Haitian kids talk in Creole, the Hispanic ones in Spanish. They stick together and create subcultures. They go to school together, they spend the day together.

If there aren't enough Russian kids to form a group of their own, programs designed to teach them English lump them together with other immigrant groups:

> One of the counselors, smiling, said that some of the Russian students speak English with a Spanish accent, while others have picked up a Jamaican accent.

If most of the kids in their group speak English with a Spanish accent, that's how they all will end up speaking it. The accent doesn't go away—why should it? It's normal for their group; it's the way they all talk. If they remain in that group through adolescence, that's how they will speak in adulthood. And if the language they use when they are together—the language they use in the lunchroom and the playground—is Spanish or Russian or Korean, English will never be more than a second language to them. They will think, they will dream, in Spanish or Russian or Korean.

The decision to leave their homeland is not the only hard choice immigrants must make. Once they arrive in their new country they face another decision. They must decide which is more important to them: to have their children retain the language and culture of their homeland or to have them become masters of their new language and culture. By settling in an area where there were no other Polish immigrants, Joseph's parents picked alternative 2. Their son became a "real American," indistinguishable from his native-born peers. But the Americanization of Joseph came at a price—a price reckoned in Polish. Though he learned it in the cradle, and though he continued to speak it at home, Polish became the language in which he felt like a fish out of water.

If Two's Company, How Many Does It Take to Make a Crowd?

Cultures are passed from the older generation to the younger one via the peer group, not at home. Children acquire the language and culture of their peers, not (if there's a discrepancy) those of their parents or teachers. If they don't have a culture in common they will create one. A culture designed by a committee of children is likely to be a pastiche, but if you're thinking "camel," forget it.

Most children don't have to create a culture: they can use the one they got from their parents, updating it a bit to suit their more enlightened tastes—or, now that television has become a major source of input for updating, their less enlightened tastes.

I do not deny that most children get their language and culture from their parents. If their parents speak English and so do most of their friends, there is no need for them to devise a new language or to learn English all over again. The same is true for the culture. This carryover—this agreement between parent and child—is one of the things that misled developmental psychologists. It is a false clue, a red herring. If we change nothing about the family but simply plunk it down in a place with a different language and culture, we get a completely different outcome for the children. They will, if they're still young, pick up the second language and culture as quickly and easily as they picked up the first. There seems to be no great advantage in having parents who can teach you the local customs before you venture outside. The chief advantage is that you are less embarrassed when, later on, you want to bring some friends home with you from school.

In the ordinary course of events, most children do end up with more or less the same language and culture as their parents, because most parents live in places where they share a language and a culture with their neighbors. When their children go to school they find themselves surrounded by other children who come from homes similar to theirs. All they have to do is to swim with the current.

But a large public school may serve several different neighborhoods, and these neighborhoods may have different cultures (subcultures, to be precise). Their inhabitants may speak with different accents and have different ideas about how to run a home, how to behave in public, how to lead a life. Remember peaceful La Paz and violent San Andres, the Mexican villages that have made several previous appearances in this book? Two neighborhoods in the United States, located within a few blocks of each other, can be as different as La Paz is from San Andres.

If there were a school halfway between La Paz and San Andres, attended by children from both villages, I imagine its atmosphere would be a lot like Wexler's, the school where sociologist Janet Schofield studied black–white relations. The kids from La Paz and those from San Andres would form separate groups; it would be rare for someone from one village to have a friend from the other. The San Andres kids would say that the ones from La Paz were soft and wimpy. "They don't know how to fight," they'd say. The La Paz kids would complain that the ones from San Andres were always pushing people around. Groupness would be salient. Children would feel pulled to conform to the norms of their own group. Contrast effects would exaggerate the differences between the groups.

Now imagine a slightly different scenario: the school is located closer to La Paz and most of the kids who attend it come from that village. But for some reason one boy from San Andres—I'll call him Miguel—also ends up in this school. What would happen? How would he behave?

Perhaps you're thinking that Miguel is going to be the terror of the playground, because what he learned in his village is going to make him a shark among the herrings. But I don't think a difference in culture—in behavioral norms—is what makes bullies. Every culture has its bullies; they are people who *violate* behavioral norms. It's a personality problem, not a cultural problem.

Assuming that Miguel is an average sort of boy, a boy like Joseph, what will happen (according to group socialization theory) is that he will learn to behave like the kids from La Paz while he is in school. This is because he's the only one from San Andres; he doesn't have a group. If Miguel commutes back and forth from his village to the school and has other friends at home, he will be bicultural: he will learn to swim with the sharks at home and with the herrings at school. But if all his friends are from La Paz—if these are the kids he plays with after school and on weekends too—he will, like Joseph, lose the culture of his native village.

He will acquire a new one, the culture of La Paz. He will adopt the behavioral norms of his new culture.

Number turns out not to be trivial. Whether a classroom of kids will split up into contrasting groups depends partly on how many kids there are in the classroom: bigger classes split up more readily than smaller ones. And whether the kids will form groups that differ in village of origin, or in race, ethnicity, religion, socioeconomic class, or academic ability, depends on how many there are in these social categories. You need a minimum number to form a group and I'm not sure what it is because there hasn't been much research on this question—not with children, anyway. In some cases two might be enough to form a group, but usually it takes more than two, perhaps more than three or four.

In a school where the majority of kids come from La Paz and a few are from San Andres, you will get mixed results. Some classrooms might have only one or two San Andres kids and in that case they would probably adopt the behavioral norms of the majority from La Paz. In other classrooms there might be four or five from San Andres and that might be enough for them to form their own group—a group in which the norm is to be aggressive.

In Chapter 9, I mentioned a study of African-American kids from "high risk" families—no fathers, low incomes. The ones who lived in low-income neighborhoods were more aggressive than their middle-class counterparts; aggressive behavior was the norm where they lived. But the ones who lived in mostly white, middle-class neighborhoods were not particularly aggressive. These black kids from fatherless, low-income homes were "comparable in their level of aggression" to the white, middle-class kids they went to school with. They had adopted the behavioral norms of the majority of their peers.

Number counts. I mean, number is important. A few students from a different socioeconomic class, ethnic group, or national background will be assimilated to the majority, but if there are enough of them to form their own group they are likely to remain different and contrast effects may cause the differences to increase. At intermediate numbers, things can go either way: two classes with the same number of majority and minority students may in one case split up into groups and in the other remain united. It will depend on chance events, on the characteristics of the individual children, and, crucially, on the teacher.

The teacher's job is most difficult, I think, when her students come from widely different socioeconomic classes. A child born into a home where the only reading material is on the back of the Cocoa Krispies box, and where the television set is turned on at dawn and left on till midnight, is going to arrive at school with a different attitude toward reading than one born into a home filled with books and magazines. A child born to college-educated parents is going to have a different view of the relevance of education—of the *normalness* of spending the first quarter of your life working your butt off in school—than one born to high school dropouts. The children will bring these attitudes with them to the peer group and if their attitudes are shared by the majority of their peers they will retain them. The atmosphere in the classroom is likely to be pro-reading in a school that serves a homogeneous neighborhood where all the homes are full of books and magazines. It's likely to be So what? Who cares? in a school that serves a homogeneous neighborhood where reading is something people do only out of necessity and never for pleasure. And a school that serves both kinds of neighborhoods is likely to split up into groups of kids with contrasting cultures.

According to a recent article in the journal *Science,* children do better in school if they come from homes that have a dictionary and a computer. The writer evidently thinks that it's the home that makes the difference. I think it's the culture, not the home. Homes that contain a dictionary and a computer are likely to be found in middle-class neighborhoods populated by college-educated parents. Such neighborhoods foster a pro-reading, pro-education culture. The kids bring this culture with them to the peer group and the peer group retains it because it is something they have in common.

Now you can see why kids who go to private schools and parochial schools do so well. These schools serve homogeneous populations: the children who go to them come from homes where the parents care enough about such things to actually *pay* for their kids' education. Throw a few scholarship students into these schools, sink or swim, and they take on the behaviors and attitudes of their classmates. They take on their culture. Margaret Thatcher, the former prime minister of Britain, was a scholarship student at a fancy private school.

And now, perhaps, you can see why it might not work to send a large number of kids from low-income neighborhoods to a private or parochial school. They might form a group of their own and retain the attitudes and behaviors they brought with them to the school.

The IQ Scores of Adopted Children

Short-term intervention programs usually have only short-term effects (if any) on a child's IQ. But what about a long-term intervention program? The most drastic intervention of all is adoption: giving a child a new family, usually of higher socioeconomic status than the one he or she was born into.

I just got e-mail from a colleague who asked the rhetorical question "Are parents important?" He immediately answered it in the affirmative. Adoption can raise a child's IQ, he said, and that shows that the child can gain from a better home environment.

Believers in the nurture assumption would like to attribute that rise in IQ to the family environment—to the adoptive parents. To the mobile over the crib, the books read out loud, the dictionary on the shelf, the computer on the desk. But the child raised in this home is raised in a middle-class neighborhood and goes to a middle-class school. His peers also come from homes in which mobiles are hung, books are read, and dictionaries and computers are purchased. This child is reared in a culture that considers reading and learning to be important, even enjoyable. He is part of a peer group that holds similar views. They look with favor on activities like reading books and using computers. They know the names of dinosaurs; they send each other e-mail.

It makes sense to me that adoption would raise a child's IQ, as long as the adoptive home is higher in socioeconomic status than the one his biological parents would have provided. If the adoptive parents are middle-class, it means they probably live in a middle-class neighborhood. If the adopted parents are unskilled laborers, they probably don't live in a middle-class neighborhood and neither I nor anyone else would predict, in this case, that adoption would raise the child's IQ. This is exactly what was found in a study carried out in France: adopted children reared by middle-class adoptive parents had higher IQs than those reared by working-class parents. In fact, there was a difference of twelve IQ points between the averages of the two groups.

Was it their experiences at home or their experiences at school and in the neighborhood that made the difference? The attitudes and activities of their adoptive parents or the attitudes and activities of their peers? My colleague would say "parents"; I would say "peers."

Unfortunately, this argument may turn out to be entirely rhetorical because right now it is not clear whether that twelve-point difference in IQ persists into adulthood (the French adoptees were tested at an average age of fourteen). Some evidence from behavioral genetic studies suggests that it does not persist. In childhood there is a modest correlation between the IQs of two adopted children raised in the same home—a correlation that I believe is due to their shared neighborhood and not to their shared home. But by the time these adoptive siblings reach adulthood, the correlation between their IQs has dwindled away to nothing. If these results are taken at face value, the implication is that neither the home nor the neighborhood has had any long-term effects on the adoptees' intelligence. However, the behavioral genetic studies probably underestimate the long-term effects of adoption, because the researchers didn't make a special effort (as the French researchers did) to find adoptees reared in homes that varied widely in socioeconomic status. Most adoptees are reared by middle-class parents in middle-class neighborhoods. Where there is little variation in environment, behavioral genetic methods cannot give an accurate estimate of environmental effects.

There is no question, though, that the effects of adoption on IQ do tend to fade in adolescence. I believe this is due to the fact that as kids get older they become freer to follow their own propensities. Teenagers sort themselves out into peer groups that vary in their attitudes toward intellectual achievement, and they can usually find anti-intellectual groups even in middle-class neighborhoods.

What is still unclear is how much the effects fade—how much of the increased IQ found in adopted kids reared by middle-class parents survives into adulthood. No one is sure because the answer depends on combining data from diverse—often incompatible—types of studies. Behavioral geneticist Matt McGue is probably the world's leading expert on adoption studies of IQ. His current guess is that the long-term benefits of adoption might amount to about seven IQ points.

Perhaps this finally closes the case on the boast John B. Watson made so long ago. "Give me a dozen healthy infants," he said, "and I'll guarantee to take any one at random and train him to become any type of specialist I might select—doctor, lawyer," you name it. An increase of seven IQ points is not to be sneezed at, but it is not enough to get a child of average genetic endowment into medical school.

Down with Group Contrast Effects

The neighborhood environment has effects during childhood because primary schools tend to be small and to serve homogeneous populations. One of the reasons these effects often fade in adolescence is that high schools tend to be larger. Number matters. Even if the population it serves is homogeneous, the larger enrollment in a high school permits the students to form more social categories and to divide up in more ways. Black or Asian kids reared in white neighborhoods, whose friends up till now had been white, might find a black or Asian peer group to identify with in high school. Kids who had trouble with their schoolwork in the early grades might get together and form anti-school—maybe antisocial—groups in high school. Once these groups form, whatever characteristics they started out with are exaggerated by group contrast effects.

Group contrast effects work like a teeter-totter: when someone goes up, someone else goes down. The average outcome is worse than neutral because it's so much easier to go down than up.

Once kids have split up into groups it is extremely difficult to put them back together again. It's better to discourage them from splitting up in the first place. There are ways that educators might be able to do this.

One way is to make the kids as homogeneous as possible. That is why—as paradoxical as it might seem—girls do better in math and science in all-girl schools, and why traditionally black colleges put out a disproportionate number of the nation's talented black scientists and mathematicians. It is why school uniforms just might work. I would be very interested in the outcome of an experiment that put primary school girls and boys into identical unisex uniforms.

Another way is to create new groups that cross-cut the other ones. It means giving kids harmless ways to split up—Dolphins versus Porpoises—rather than harmful ways—girls versus boys, rich versus poor, smart kids versus dummies. As the Eagles and the Rattlers demonstrated, this method has its risks. What starts out as a harmless way to split up can escalate to socks filled with stones. The trick is to keep the social categories in balance so that they cancel each other out. If a child can't decide whether she is a girl, a Dolphin, or a dummy, she may end up categorizing herself simply as a member of Ms. Rodriguez's sixth-grade class.

If all else fails, a surefire way of uniting people is to provide them with a common enemy. It works for chimpanzee groups; it works equally well for sports teams or, for that matter, chess teams. In my high school, Mexican-American and Anglo kids joined together to cheer for our school when Tucson High competed against Phoenix. The Robbers Cave researchers got the Rattlers and the Eagles to work together by telling them that vandals from outside had meddled with the camp's water system.

Leaders can bring people together or divide them up. Some of the things that teachers do nowadays, with the best of intentions, have the unintended result of making children more aware of the ways they can be sorted into social categories. I believe that a teacher's job is not to emphasize the cultural differences among the students (that can be done at home by the parents) but to downplay them. A teacher's job is to unite students by giving them a common goal.

12 GROWING UP

Except for the dog, I was alone in the house. I was sitting at my desk on a dark winter afternoon, reading an article about adolescent delinquency. It was January 20, 1994.

The article was by Terrie Moffitt, a developmental psychologist for whom I had, and still have, the greatest respect. In this article Moffitt reported that "illegal behavior" is so common during adolescence that it can be considered "a normal part of teen life." The news that teenagers routinely break the law hardly gave me pause. What stopped me in my tracks was Moffitt's explanation of this unendearing foible. "Delinquency," she said, "must be a social behavior that allows access to some desirable resource. I suggest that the resource is mature status, with its consequent power and privilege."

Wait a minute! I thought. Is she saying that teenagers commit illegal acts because they want to be like adults? That can't be right! If teenagers wanted to be like adults they wouldn't be shoplifting nailpolish from drugstores or hanging off overpasses to spray I LOVE YOU LIƧA on the arch. If they really aspired to "mature status" they would be doing boring adult things like sorting the laundry and figuring out their income taxes. Teenagers aren't trying to be like adults: they are trying to *distinguish* themselves from adults!

The thought blossomed like a magician's bouquet. Within a few minutes I had the basic outline of group socialization theory—the theory that children identify with a group consisting of their peers, that they tailor their behavior to the norms of their group, and that groups contrast themselves with other groups and adopt different norms. Only after I had gotten that far did I realize the full implications, and then I had to go back and reconsider the evidence before I was willing to accept the second half of my epiphany. Hey, it's not the parents! It's not the parents at all!

Everything fell into place. All the observations that didn't fit into the prevailing theories suddenly made sense.

I am not naive enough to believe that every cloud has a silver lining; some clouds are gray through and through. But if the Harvard Psychology Department hadn't kicked me out without a Ph.D., if health problems hadn't kept me from going back to graduate school and forced me to spend the past twenty years working at home, if I had had mentors and colleagues and students, it would never have happened. If I had gone through the routine brainwashing process and become a member-in-good-standing of the academic community, I would never have realized that the nurture assumption is just an assumption and an unwarranted one at that. I would never have written an article saying that parents count zilch and mailed it off to the same journal that published Terrie Moffitt's paper. I would not be writing this book and you, dear reader, would not be reading it.

It was adolescence that made me see the light because that is where it can be seen most clearly. Even true believers in the nurture assumption are willing to admit that teenagers—at least *some* teenagers—are influenced less by their parents than by their peers. But the true believers have convinced themselves that teenagers are different in this respect from younger kids—that some sort of madness overcomes them when their hormones hit the fan.

My position is that teenagers belong to the same species as the rest of us—that, despite all appearances to the contrary, they are members-in-good-standing of the human race. They are equipped with the same sort of brain, pushed and pulled by the same sticks and carrots. They want to be like the other members of their group, only better. They don't want to be like the members of other groups. These peculiarities do not pop out like a cuckoo when the clock strikes thirteen. These desires do not strut and fret their hour upon the stage and then are heard no more.

But one cannot help but wonder. If they are equipped with the same sort of brain as the rest of us, why do they so often give the impression of having forgotten how to use it? Why do they seem less socialized than younger children, even though they've been undergoing socialization for a longer time?

I confront some of the questions about adolescence in this chapter. It is titled "Growing Up" rather than "Adolescence" because it begins in childhood and ends in old age. If you are uninterested in teenagers and feel inclined to skip this chapter, I hope you will not skip its concluding section.

Why Do Children Grow Up?

A smart-aleck graduate student once pointed out to me* that there is a problem with my theory. If children tailor their behavior to the norms of their group, if the norms are determined by a majority-rules rule, and if (in societies like ours) peer groups consist of children of the same age, how do they ever grow up? Why do they stop acting like little kids and start acting like big kids? How do their norms ever change?

The traditional explanation—the one the grad student espoused—is that children emulate grownups. As they get older, they get better and better at pretending to be grownups. I reject that explanation for two reasons. First, as I said in Chapter 1, in most societies children who act like adults are considered impertinent. One of the first lessons children must learn is that they are not supposed to behave like grownups. Second, as I said in Chapter 9, a child's goal is not to become a successful adult, any more than a prisoner's goal is to become a successful guard. A child's goal is to be a successful child.

Among the Yanomamö of the Amazon rainforest, according to the anthropologist who studied them,

> A well-dressed man often sports nothing more than a string around his waist to which is tied the stretched-out foreskin of his penis. As a young boy matures, he starts to act masculine by tying his penis to his waist string, and the Yanomamö use this developmental phase to signify a boy's age: "My son is now tying up his penis." A certain amount of teasing takes place at that age, since an inexperienced youth will have trouble controlling his penis. It takes a while for the foreskin to stretch to the length required to keep it tied securely, and until then it is likely to slip out of the string, much to the embarrassment of its owner and the mirth of older boys and men.

We have the anthropologist's word on it that this style of, uh, dress is uncomfortable. The question is: What motivates the young boy to put up with the discomfort and the teasing and to start tying his penis to the

* I am still not a member-in-good-standing of the academic community. However, I now have colleagues who are, and they teach graduate students.

string around his waist? Is it because at some point he notices that that's how his father wears his? Anthropologists, developmental psychologists, and smart-aleck grad students think so. I think not. The test case would be a Yanomamö boy whose father for some reason failed to follow the local custom of tying up his penis. I've told you about kids like that—kids whose parents are atypical. They don't copy their atypical parents. This boy would do whatever the other boys were doing.

Children want to be like their peers. They want most of all to be like the kids who have high status in their peer group. Within children's groups that span a range of ages—as they do in the villages of tribal peoples like the Yanomamö—the kids with high status are the older kids. Younger ones look up to those a year or two ahead of them with admiration and envy.

In societies where education is compulsory, children rank "being left back in school" as the third most scary thing they can think of, beaten out only by "losing a parent" and "going blind." "Wetting my pants in school" comes in fourth. A Yanomamö boy with his penis not tied up is like an American child who has wet his pants in school: he is a boy who has been left back. It would be humiliating to walk around with a dangling penis when other boys his age or younger were already tying theirs up. When the Yanomamö boy ties his foreskin to the string around his waist, he's not pretending to be his father: he's concerned about maintaining his status among the other children in the village. It is the mirth of the older boys that provides the stick. It is the respect of the younger ones that provides the carrot.

In urbanized societies like our own, peer groups usually consist of children of about the same age. But even within same-age groups, children vary in physical and psychological maturity. In such groups, the more mature ones generally have higher status. It is the equating of maturity with status that makes little children want to behave, speak, and dress like bigger ones. Kids do not look to grownups for guidelines on how to behave, speak, or dress because kids and grownups belong to different social categories that have different rules. Wanting to have higher status—wanting to be like a bigger kid—goes on *within* the group, within the social category "kids." Grownups are a different kettle of fish. To a kid, grownups are not a superior version of *us:* grownups are *them.*

Do not be misled by the fact that among the Yanomamö both the boys and the men tie up their penises: it doesn't mean that the boys are trying to be like their fathers. Within a society there are numerous things that are common to more than one social category. Yanomamö men, women, and children all wear the same hairstyle, with a little bare patch shaved at the crown. American men, women, and children all eat with forks and spoons.

And do not be misled by the fact that sometimes a Yanomamö boy will pretend, in play, to be a grown-up man. The role he is playing is not that of his own father: it is a generic, idealized version of a man. In play, children can be anything they like—witches, horses, supermen, babies. They do not confuse these fantasies with reality. The American child who pretends to be a mommy in a game of House doesn't think she's a mommy in real life. The one who pretends to be a teacher in a game of School doesn't make the mistake of behaving like that in the classroom.

A child can get away with inappropriate behavior if it's clearly labeled "play," just as an adult can get away with an inappropriate remark if it's clearly labeled "joke." When they are not playing or joking, people are expected to behave, speak, and dress in a manner that is appropriate for their social category and the social context. This is true everywhere, at every age past toddlerhood. Yanomamö boys may tie up their penises like the grown-up men and wear their hair like the grown-up men and women, but they are expected to behave like boys.

Rites of Passage

The human mind wants to categorize. We put things into categories even when they fall along a continuum and not into convenient clumps. *Night* and *day* are as different as night and day, even though one fades imperceptibly into the other. The fact that the people they know span a continuum of ages doesn't prevent children from thinking of *kids* and *grownups* as separate social categories.

To make it easier for individuals to know which category they are in (and, therefore, how they are expected to behave), societies like that of the Yanomamö generally provide markers. For girls it is easy because nature provides her own marker—menarche, the first menstrual period. All the society has to do is endorse it, recognize it.

The coming of age of a Yanomamö girl is described in a remarkable book called *Yanoáma: The Narrative of a White Girl Kidnapped by Amazonian Indians.* It is the true story of a woman named Helena Valero, who was taken from her Brazilian parents when she was about eleven years old by a war party of Yanomamö men armed with poisoned arrows. She lived with the Yanomamö—she lived *as* a Yanomamö—for twenty years.

Among the Yanomamö, Helena explains, a girl experiencing her first menstrual period is said to be "of consequence."

> We all went back into the great *shapuno* [a ring of huts covered by a single round roof] where there were two girls of consequence. When girls are twelve to fifteen years old and are just about grown up, at the time when they begin, they are shut up in a cage made with *assai* palm branches and other branches of *mumbu-hena,* which I have seen only in those mountains. They tie all the branches with lianas, very tight, so that the girls cannot be seen. They leave only one little entrance. The men and boys must not even look that way.

The girl stays in the "cage" for about a week, with a fire burning all the time. Her food and water are severely restricted and she is not allowed to talk. Finally there is a brief ceremony involving the burning of dried banana leaves, and then comes the fun part.

> Then the mother, with the other women, accompanies her daughter into the woods to adorn her. . . . One woman begins to rub a little red *urucu* over all her body, which becomes pink. They then design wavy black lines, brown on her face and body; they make lovely designs. When she is completely painted, they push through the large hole in her ear those strips of young assai leaves. . . . Then they take coloured feathers and push them through the holes which they have at the corners of their mouths and in the middle of the lower lip. One woman also prepares a long, thin, white stick, very smooth, which she puts in the hole that they have between their nostrils. The young girl is really lovely, painted and decorated like this! The women say: "Now let's go." The girl walks ahead, and after her come the other women and the little girls.

The parade wends slowly through the center of the village so that everyone can admire the debutante. Though she is probably no more than fifteen years old (menarche comes later to girls in tribal societies), she is now considered old enough to marry. If her father has already promised her to someone she will take up residence with her new husband. She went into the cage a girl and came out a woman, as though a magician had waved his magic wand: Poof, you're a woman!

For boys it is a little different. Nature provides no convenient marker for the beginning of manhood, so most tribal societies make up for the lack by providing one of their own. Puberty rites are a favorite topic for anthropologists, and *male* puberty rites are the ones they most like to write about. Margaret Mead's colleague Ruth Benedict has provided a description of the initiation rites of the Zuñi Indians of New Mexico. Groups of Zuñi boys are initiated when they are about fourteen in a lengthy procedure that involves whippings by masked "scare kachinas."

> It is at this initiation that the kachina mask is put upon [the boy's] head, and it is revealed to him that the dancers, instead of being the supernaturals from the Sacred Lake, are in reality his neighbours and his relatives. After the final whipping, the four tallest boys are made to stand face to face with the scare kachinas who have whipped them. The priests lift the masks from their heads and place them upon the heads of the boys. It is the great revelation. The boys are terrified. The yucca whips are taken from the hands of the scare kachinas and put in the hands of the boys who face them, now with the masks upon their heads. They are commanded to whip the kachinas. It is their first object lesson in the truth that they, as mortals, must exercise all the functions which the uninitiated ascribe to the supernaturals themselves.

The details vary but male puberty rites in tribal societies tend to have much in common. Several boys are initiated together in a group. They are temporarily removed from the rest of the society. They go through an arduous preparation that usually involves the revelation of secret knowledge and often a great deal of terror and pain (Benedict mentions in passing a tribe that buries boys in hills of stinging ants). Once through the ordeal, they are reintroduced to the society and their new status is acknowledged. Perhaps they are not yet first-class adults; perhaps they remain adults-in-training until they have passed a further test, such as killing a man in battle or fathering a child. But they are no longer children.

Why, asks ethologist Irenäus Eibl-Eibesfeldt, are male puberty rites apt to be so harsh in tribal societies? Because, he says, the boy "must be emancipated from his family so that he can identify with the group on a new level. He must develop a group loyalty that extends beyond his loyalty to family." The initiation, according to Eibl-Eibesfeldt, removes the boy from "the sphere of the immediate family" and gives him to the group.

I agree with Eibl-Eibesfeldt about group loyalty but not about emancipating the boy from his family. The boy left "the sphere of his immediate family" when he graduated from his mother's arms to the children's play group at the age of three. The purpose of the puberty rite is to take him from the play group and put him, along with his childhood playmates, into a new social category, in which he is expected to take on the work and the responsibilities of a man. He must endure terror and pain and stand shoulder to shoulder with the other men of the village in defending it against its enemies. He is now "of consequence."

In contrast, the American or European fourteen-year-old is of little consequence to society, unless it is in the guise of a thorn in its side. At an age when a Yanomamö girl is considered old enough to marry and a Yanomamö boy old enough to give up his life defending his village, the American teenager is not considered old enough even to drop out of school.

Neither Fish nor Fowl

Human children have a peculiar pattern of growth that is not seen in most other mammals. They grow very rapidly in the first two or three years and then growth slows down and remains slow for about a decade. Then, in early adolescence, there is a sudden growth spurt and they quickly shoot up to adult size. It is as though nature were trying to keep children children as long as possible and then, as soon as the purposes of human childhood had been fulfilled, propel them into adulthood as rapidly as possible, thus shortening the period of uncertainty in which they are neither fish nor fowl.

It worked well for many thousands of years. When humans roamed around in bands of fifty or so, or lived in small villages, there were just two age groups: children and adults. You identified with one group or the other and took your cues on how to behave from your own age group. When young people shot up to adult size, they became adults. They worked and fought and had babies side by side with the other adults.

Now we live in complex times and two age groups are no longer enough: a person can be as big as an adult but not an adult. We've had to create new social categories to contain such people. One of these categories is called *teenagers.* During the 1960s an additional category came into existence, because our society contained a group of people who were older than teenagers but who refused to identify themselves as adults. They had their own category, though no ceremonies to mark the transitions. You entered it by leaving home to go to college or to join a roaming band; you left it upon reaching the upper boundary set by the members themselves. Never trust anyone over thirty, they said. They meant, anyone over thirty is *them.*

Today, with no Vietnam War to unite them, that age group has fallen apart into subgroups. Some of them are brown-nosing away in colleges and professional schools; some are having babies or programming computers or fixing cars or looking for jobs. The result is that there is no longer any buffer between teenagers and adults; the age group between them is, to all intents and purposes, missing. Teenagers nowadays tend not to see much of people in their late teens and early twenties: the "young adults" are off somewhere else. Which leaves the *real* adults—the parents and teachers and policemen who, God help them, are supposed to be in charge—to take the brunt of the teenagers' groupness.

We belong to a species that has a long evolutionary history of living in small groups and competing or warring with other such groups. The winners of these competitions were our ancestors, and it is to them we owe our inclination to identify with a group and to like our own group best. It is to them we owe our easily aroused hostility toward other groups.

In the hunter-gatherer or tribal society there were but two age groups, children and adults. Was there hostility between them? If so, it was subtle and muted. Children were designed by evolution to evoke nurturing from adults; they evolved that way because those who didn't have what it takes to make their parents love them were less likely to survive. Adults were designed by evolution to nurture children; they evolved that way because those who lacked this instinct—yes, instinct!—were less likely to succeed at rearing children to carry on their genes. The nurturing instinct is powerful in humans. It doesn't depend on the belief that you share genes with the small creature; a kitten or puppy can set it off quite as well as a human baby. I have even found myself thinking "Isn't that cute" about a sample size bottle of laundry detergent.

I believe that evolution gave us two independent systems, controlled by different mental modules, to make us want to take care of children. Evolutionary theorists, inspired by the idea of the "selfish gene," tend to talk about only one system, based on kinship: we love our kids because they carry our genes. This theory predicts that we should love the ones who resemble us more than the ones who don't, which turns out to be true. But it also predicts that we should love our older kids more than our younger ones, because the older ones are closer to being able to perpetuate our genes by producing grandchildren for us. Though the death of an eight-year-old does seem to hit parents harder than the death of a one-year-old, while both are alive it is the one-year-old who gets the attention and the kisses. The problem with a kinship-based view of parenting is that it puts all its eggs in one basket.

A two-baskets view of parenting is necessary to explain what happens in adolescence. Evolution provided us with two reasons to love our young children: because they carry our genes and because they're little and cute. Evolution gave us only one reason to love our teenage children: because they carry our genes. Once they balloon to adult size—once their faces lengthen and their noses grow and their sweat gets that gamy smell—adolescents no longer evoke our nurturing instinct. On their part, they no longer need us so much. They are capable of managing, at least in the kind of environment they were designed for, without their parents.

When the only age groups are children and adults, hostility between the groups is dampened by dependence on the one hand and nurturance on the other. But when teenagers have an age group of their own, hostility between age groups—between teenagers and adults—can emerge. Does emerge. It is mutual, I believe. The hostility is most visible when groupness is salient, because it is groupness that causes it. When groupness is not salient, it is perfectly possible for teenagers to have warm relationships with adults. Some of their best friends are grownups.

Now you can see why teenagers are so annoyed when adults take over their styles of dress or speech—why they are forced to invent new ones. They have attained an adult size and shape, more or less, but they don't want to be mistaken for grownups. They need ways of signaling their group identity and loyalty to the other members of their group. The big question of adolescent life—the unspoken question that teenagers are constantly asking each other and constantly answering—is: Are you one of *us* or one of *them?* If you're one of *us,* prove it. Prove it by showing you don't care about *their* rules. Prove it by doing something—a tattoo would be nice, a hole through your nose even better—that will mark you irrevocably as one of *us.*

You see the same sort of thing between warring villages in tribal societies: the creation of cultural differences and the use of visible markers—the more permanent the better—to trumpet the differences. If their counselors hadn't patched up things between them, perhaps the Eagles and the Rattlers would have done the same. The Eagles might have shaved a bare patch on the top of their heads, like monks in training. The Rattlers might have taken to painting their faces, like the bad boys in *Lord of the Flies.* Such markers have a practical value as well as a symbolic one: they make it easier to tell your friends from your enemies in the heat of battle. The distinctive uniforms worn by the members of professional sports teams are not only to remind the fans which side to cheer for.

A Mechanism for Social Change

Hostility toward adults doesn't pop out de novo in adolescence. Though it has been kept under wraps, it has been brewing for a long time, especially among boys. (Groupness, as I said in Chapter 10, appears to be stronger in males.) The foul language used by the Rattlers is typical. These boys came from respectable, church-going families. They learned the dirty words from older boys and from each other, not from their parents.

Sociologist Gary Fine spent three years observing the members of Little League baseball teams. He found out that boys who are "sweet, even considerate," with their parents can be remarkably nasty when they're with their teammates. Fine's prepubescents play pranks on adults and brag to each other about their sexual knowledge. They talk about girls in derogatory, sexually explicit terms and use "faggot" as a casual insult. Because four-letter words have lost their sting, boys from nice middle-class homes use the worst word they know, "nigger," and scrawl the worst kind of graffiti, a swastika. Their parents are not racists; their parents would be shocked. Which, of course, is exactly the point. It is a mistake to call the painting of swastikas by boys a "bias crime" and an even worse mistake to blame it on their parents. They paint swastikas because no one blinks an eye anymore if they paint "FUCK YOU."

But preadolescents are just toying with rebellion: they act this way only when their parents aren't looking. The in-your-face variety of rebellion awaits the moment when they balloon to adult size and become capable of managing—at least in the kind of environment they were designed for—without their parents. They might be immature but they aren't complete fools.

The in-your-face variety of rebellion that many teenagers indulge in today is a characteristic of societies that send adolescents to school. It isn't found, because it would be pointless, in societies that consider fourteen-year-old girls old enough to marry and fourteen-year-old boys old enough to shoulder the responsibilities and the weapons of men. Since these fourteen-year-olds are categorized (by themselves and by others) as grown-ups, they have no motivation to be different from adults. They may harbor resentment against particular adults—against the mother-in-law who works them like a slave or the father who competes with them for wives—but groupness doesn't play a role in these resentments. It doesn't play a role because, in most of these societies, teenagers have no opportunity to hang around with other teenagers. They have no concept of teenagehood. They have no groupness because they have no group.

Teenagers become a force to be reckoned with when they are gathered together in one place, as they are in the modern high school. As they were in an ancient high school, more than two thousand years ago. In Athens of the fourth and fifth centuries B.C., a series of Greek philosophers made their living by providing an education for the sons of rich Athenians. Philosophy proved to be a flimsy defense against the in-your-faceness of a bunch of teenage boys. Socrates grumbled that he don't get no respect: his pupils "fail to rise when their elders enter the room. They chatter before company, gobble up dainties at the table, and tyrannize over their teachers." Aristotle was similarly pissed off by his students' attitude: "They regard themselves as omniscient and are positive in their assertions; this is, in fact, the reason for their carrying everything too far." Their jokes left the philosopher unamused: "They are fond of laughter and consequently facetious, facetiousness being disciplined insolence."

They may have annoyed the living daylights out of their teachers* but they made fourth-century Athens the hot spot of the ancient world, the where-it's-at of its day. When you put together a group of people who are not children and not adults, what you have is a mechanism for rapid social change.

In a society that contains only two age categories, children and adults, a culture can be handed down virtually unchanged for hundreds of generations. Children are not changers of culture: they are still learning the ropes and are not sufficiently independent. Adults are not changers of culture: they are maintainers of the status quo. The changers of cultures are people in their teens or early twenties who have an age group of their own. Groupness motivates them to be different from the generation of their parents and teachers. They are so anxious to contrast themselves with the generation ahead of them that the differences don't even have to be improvements—indeed, they are often not improvements. They adopt different behaviors and different philosophies; they invent new words and new forms of adornment. And they take these behaviors, philosophies, et cetera, with them to adulthood. They leave to *their* children the burden of finding new ways to be different. Mom and Dad smoked pot? Yikes, we'll have to find something else to smoke!

Adolescents do not, of course, reject everything in their parents' philosophy. Sometimes the offspring of pot smokers do use pot. Though the

* According to Miss Manners, "Grown-ups have always lamented the appalling manners of the younger generations. It would be cheating the young of a source of satisfaction, if they did not."

choice of what to keep and what to chuck may be arbitrary, there are always some things that are kept. It wouldn't make sense for each generation to start all over from scratch.

Because the choice of what to keep and what to chuck is arbitrary, and because young people in developed societies tend to associate primarily with their agemates, each new cohort of high school or college students creates a culture of its own. Each new culture blends input it has received from the society as a whole—from the media, from what's going on in the world, from the cultures of previous cohorts—with something new, added by its creators as a way of distinguishing themselves from their predecessors.

The rapid turnover of cultures was especially noticeable during the late '60s and early '70s. A team of psychologists who studied adolescents during that period concluded that cohort membership was an important factor in the development of personality: each cohort seemed to exert distinctive pushes and pulls on the personalities of its members. For instance, the fourteen-year-olds in 1972 were more independent than fourteen-year-olds had been only a year or two earlier, but they scored lower in achievement and conscientiousness. Freedom mattered more to them than it did to their predecessors; doing well in school mattered less. The times they were a-changin'.

Groups within Groups

The social categories of younger children tend to be inclusive and based on straightforward demographic characteristics. A third-grade girl will identify herself as a third-grade girl, and this self-categorization doesn't depend on whether the other girls in her class like her or she likes them. If there are a lot of third-grade girls and nothing to hold them together, they might split up into subgroups based on other demographic characteristics such as race or socioeconomic class.

But schools contain groups within groups; even third-graders can choose from a menu of self-categorizations. Within the larger demographic groups are smaller ones—cliques—of kids who hang around together. The kids in these cliques generally have similar attitudes toward schoolwork, pro or con, and similar attitudes toward other things. In elementary school the cliques are still fluid; kids can move into and out of them. When they move, their attitudes shift to match those of their new friends.

In high school it is far more difficult to move into or out of a clique. By the time kids get to high school, most of them have been "typed" by their classmates and by themselves. The temporary cliques of earlier years have solidified into fairly rigid social categories based not on demographics alone: now they reflect the personalities, propensities, and abilities of the people who belong to them.

The other thing that has changed is the number of options available. High schools generally have larger enrollments than elementary schools and the students are freer to select their companions, so they are able to divide up more finely. You've heard, I'm sure, of some of the categories that high schools contain: the jocks, the brains, the nerds, the popular kids, the burnouts and delinquents. The larger the high school, the greater the choice of social categories. A big city high school is likely, for example, to contain a group of boys who have artistic or theatrical interests and who are not attracted to girls. Groups of this sort are seldom found in small rural high schools, which may be one of the reasons why male homosexuality is much less common in such settings. Having, or not having, a group to identify with could make all the difference to a kid who isn't sure what sort of person he is.

Birds of a feather flock together in high school, but they don't necessarily do it under their own wing power. Kids are often forced into social categories they would rather not belong to. No one chooses to be a nerd. In fact, in the typical American high school, no one chooses to be a brain. The kids pinned with that label are those who are not athletic or popular enough to get into one of the groups that have higher status. Among most European-American and African-American adolescents, braininess is not considered an asset. You might be able to get away with it, but only if you have other assets that are valued by your peers.

Perhaps braininess is not an asset because the kids who do well in school are seen as turncoats: too much under the influence of *them,* the parents and the teachers. Anthropologist Don Merten has described a similar social category in a junior high school in Illinois: its members are given the pejorative label *mels* (derived from the name *Melvin*). In this school, a boy who is a slow maturer, unathletic, and not particularly attractive can have his life ruined—or at least his adolescence—by being labeled a mel. Unlike a brain, a mel is not exceptionally smart or studious; like a brain, a mel is seen as being too much under the influence of adults. His failure to thumb his nose at adult standards makes him seem childish to his peers.

> Most early adolescents perceived the transition from elementary school to involve a dual set of changes—disengaging from one's childhood past and

> engaging one's adolescent future. In the eyes of their peers, mels failed both of these tasks, but especially the first. Once an individual was labeled a mel, he became a target for harassment.

Though it is difficult for a boy to shake off the label once it is pinned on him, it is not impossible if he is willing to resort to heroic measures. One of the subjects in Don Merten's study was a boy named William, who was teased and harassed in seventh grade but managed to kick off the traces of melness in the eighth. William went about it systematically. He disconnected himself from the other mels (the fact that they shared a social category doesn't mean they liked each other). He began to fight back when picked on and he stopped tattling on his persecutors. He intentionally broke school rules. The defining moment came when another kid took his pencil in the middle of an English class. William shouted loudly, "Screw you!" and got sent to the office by the teacher. Thus ended William's sojourn in the valley of the mels.

Some social categories in the high school are voluntary; some are assigned. The delinquent category is a mixed bag. Some of its members join voluntarily, drawn by the excitement and danger. Sensation seekers, psychologists call them. Others have no choice: none of the other groups will accept them. These are kids who were rejected by their peers in elementary school, often for being hyperactive, ill-tempered, or overly aggressive. By junior high they have found others like themselves and are egging each other on. The kids in adolescent peer groups are similar to begin with; groupness causes them to become more similar to each other and to contrast themselves with the members of other groups. The brains get brainier, the nerds get nerdier, and the delinquents get into real trouble.

Parents versus Peers

Most adolescents live in neighborhoods filled with people very like their parents; their peers grow up in homes very like their own. Kids bring what they learned at home to the peer group and retain much of what they have in common, which in homogeneous neighborhoods is quite a lot. If they grew up in a neighborhood where most of the boys planned on becoming doctors, like Dr. Snyder of Chapter 9, they do not necessarily abandon that plan the day their voice changes. In homogeneous neighborhoods, with kids who are doing well in school, adolescent rebellion may be a pro forma sort of thing, manifested in harmless though annoying ways. A girl dyes half her hair purple and becomes a vegetarian. A boy shaves off half his hair and listens to music his family can't stand. They do fill out their college applications, though. They might look foolish but they aren't complete fools.

High schools offer an assortment of peer groups, but in the kind of neighborhood I just described, most of these groups might be relatively benign from the parents' point of view. When the peer group and the parents have congruent goals and values, there is likely to be a minimum of trouble between teenagers and their parents.

Trouble is far more likely to occur when teenagers become members of groups with goals and values very different from those of their parents. The teenager who gets in with what her parents call a "bad crowd" is not going to have a serene home life. Her parents don't like her friends, they don't like the way she dresses or the way she acts, they don't like the reports they're getting from the school. They tell her to stop seeing her friends, but they can't control what she does when she's not at home so she sees them behind her parents' back and lies about it. The parents have two choices: they can get bossier and meaner in an attempt to regain control (see what I said about "Too Hard" parents in Chapter 3) or they can give up (see what I said about "Too Soft" parents).

Teenagers who are members of "nice" peer groups tend to get along well with their parents; teenagers who are members of delinquent groups tend to get along poorly with theirs. Developmental psychologists use this correlation as evidence of parental influence—evidence to support what they already believe to be true. Their view is that the nice teenagers are influenced by their parents because their parents use the right kind of child-rearing style; the bad teenagers are influenced by their peers and not by their parents, because their parents use the wrong kind of child-rearing style.

My view is that both groups of teenagers are equally influenced by their peers: it's just that they belong to different sorts of peer groups.

My husband and I had one teenager of each kind. Our daughters grew up in the same neighborhood and went to the same schools four years apart. In elementary school they belonged to similar sorts of peer groups; in high school they did not. The older one was a brain, the younger a burnout. Both turned out fine in the end (the older is a computer scientist, the younger a nurse), but one headed that way directly, the other took a more circuitous route.

Our two daughters were reared by the same parents but they were very different people, as siblings often are. The older one had little need for our guidance; she did what she wanted to do and it happened to be what we wanted her to do. The younger one had little use for our guidance; she rejected it out of hand. It conflicted with the goals and values of her peer group. We, her parents, were frustrated and angry, and she was often angry at us.

It is not surprising that a kid who belongs to one kind of peer group should get along well with her parents and a kid who belongs to a different kind of peer group should get along poorly with them. The question is: What made them become members of those peer groups? Was it something my husband and I did? Was it our fault? And if I say no, will you think I'm just trying to evade responsibility and get off scot-free?

But now I'm getting into issues that belong in the next chapter. I request a temporary adjournment. In the next chapter I will present my case and you will be the judge.

Why Adolescents Do Dumb Things and How to Stop Them

Sometimes—let's face it—they really are complete fools. They ignore our warnings and the warnings printed right on the package and get themselves addicted to tobacco. They have sex too early and too often and forget to use a condom. They drive too fast and drink too much, and—as Terrie Moffitt told us—breaking laws is a normal part of life for them.

My younger daughter was smoking cigarettes by the time she was thirteen, despite the steady diet of anti-smoking propaganda I fed her from the time she learned to talk. I had thought I was pretty clever about it: I emphasized the yuckiness and not the health risks. It didn't work. She belonged to a group—the burnouts—in which smoking was the thing to do. It was a group norm. Peer pressure, you're thinking? "A lot of bunk," according to the teenagers interviewed by psychologist Cynthia Lightfoot. Here is what one of them said about why they start drinking:

> You're trying very hard to show everyone what a great person you are, and the best way to do that is if everyone else is drinking therefore they think that's the thing to do, then you might do the same thing to prove to them that you have the same values that they do and therefore you're okay. At the same time, the idea of peer pressure is a lot of bunk. What I heard about peer pressure all the way through school is that someone is going to walk up to me and say "Here, drink this and you'll be cool." It wasn't like that at all.

As Lightfoot summed it up, "Peer pressure is less a push to conform than a desire to participate in experiences that are seen as relevant, or potentially relevant, to group identity." Teenagers seldom need to be pushed to conform to the norms of their group; that got settled a long time ago, in childhood.

Teenagers who smoke not only have peers who smoke: they often have parents who smoke. Most people, psychologists and nonpsychologists alike, assume that parental influence plays at least some role in teenage smoking. They assume that kids who see their parents smoking are more likely to think that smoking is a grownup thing to do and will therefore want to do it themselves. Earlier I attacked a similar assumption about why Yanomamö boys tie up their penises. Smoking turns out to be more complicated but it has one big advantage over penis-tying: we have drawers full of data on it.

In the past, the use of tobacco was an accepted part of the adults' culture in many American neighborhoods, and an accepted part of the kids' culture as well. Teenagers took it up because everyone their age was doing it. Parents had only mild objections or none at all. Smoking was passed on in the same way as other aspects of the culture—the same way that penis-tying is passed on among the Yanomamö.

It is not passed on that way anymore because it is rare to find an American neighborhood in which most of the adults smoke and rare to find parents who approve of kids' smoking, even if they themselves smoke. Nowadays smoking is more likely to be a signal of adolescent solidarity. It is a way to demonstrate your allegiance to a particular peer group within the high school, to show your disdain for other groups (the goody-goodies, the nerds), and to prove that you don't give a damn about adult concerns and adult rules. It's like wearing a certain kind of jacket to show which gang you belong to. It's like shaving a little bald patch on the top of your head to show which tribe you belong to.

Research has shown that the best predictor of whether a teenager will become a smoker is whether her friends smoke. This is a better predictor than whether her parents smoke. Teenagers who smoke are also more likely to engage in other kinds of "problem behavior": to drink, to use illegal drugs, to become sexually active at an early age, to cut classes or drop out of school, to break laws. They belong to peer groups in which such behaviors are considered normal.

But smoking, as I said, is complicated. The use of tobacco is addictive. People differ in how likely they are to experiment with addictive substances such as cocaine and nicotine and also in how likely they are to become addicted, and genetic factors are involved in both these differences. It turns out that smoking follows the same pattern that has been found for personality traits: two people who share genes are more likely to be alike—to both be smokers or both be nonsmokers—but sharing a home doesn't make this happy congruence any more likely. The reason that parents who smoke often have children who smoke is that smoking is partly genetic.

It took a behavioral geneticist—David Rowe of the University of Arizona—to disentangle the environmental influences from the genetic ones. The environment influences a teenager to smoke or not to smoke in only one way: she is more likely to smoke if her peers do. The genes exert their influence in two ways. First, via their effects on personality: an impulsive sensation seeker is more likely to end up in a peer group that favors smoking. Second, by making it more or less likely that she will become addicted to nicotine.

Exposure to peers who smoke is what determines whether or not a teenager will experiment with tobacco. Her genes determine whether or not she will get hooked.

Since we can't do anything about their genes, the only way to keep them from getting hooked is to keep them from experimenting with tobacco. Anyone who thinks this can be accomplished by putting "Danger! Poison!" on the cigarette pack needs to sign up for clue renewal. Humorist Dave Barry smoked his first cigarette the summer he turned fifteen, for reasons, he says, that were as compelling back then as they are for teenagers today:

> ARGUMENTS AGAINST SMOKING: It's a repulsive addiction that slowly but surely turns you into a gasping, gray-skinned, tumor-ridden invalid, hacking up brownish gobs of toxic waste from your one remaining lung.
>
> ARGUMENTS FOR SMOKING: Other teen-agers are doing it.
>
> Case closed! Let's light up!

Telling teenagers about the health risks of smoking—It will make you wrinkled! It will make you impotent! It will make you dead!—is useless. This is adult propaganda; these are adult arguments. It is *because* adults don't approve of smoking—*because* there is something dangerous and disreputable about it—that teenagers want to do it.

Telling them that smoking is yucky doesn't work either, as I learned to my displeasure. If adults think something is yucky, that makes it all the more appealing to an anti-adult.

Nor does recruiting a person their own age to lecture them about it. The lecturer is seen as a turncoat—a mel, a nerd, a goody-goody. A patsy of the adults.

Even making it harder for teenagers to get cigarettes doesn't do the job. When some towns in Massachusetts cracked down on stores that sold cigarettes to minors, the teenagers went right on smoking. The fact that it was more difficult to get cigarettes just made it more of a challenge.

Adults have limited power over adolescents. Teenagers create their own cultures, which vary by peer group, and we can neither guess nor determine which aspects of the adult culture they will keep and which they will chuck, or what new things they will think up on their own.

But our power isn't zero. Adults do control a major source of input to their cultures: the media. Media depictions of smokers as rebels and risk-takers—of smoking as a way of saying "I don't care"—make cigarettes attractive to teens. I see no way around this problem unless the makers of movies and TV shows voluntarily decide to stop filming actors (doesn't matter whether they're the heros or the villains) using tobacco.

Drastically raising the price of a pack of cigarettes might also help. At least it would cut down on the number of cigarettes used by the experimenters and thus cut down on the number who become addicted.

Anti-smoking ads? Very tricky. The best bet would be an ad campaign that gets across the idea that the promotion of smoking is a plot against teenagers by adults—by the fat cats of the tobacco industry. Show a covey of sleazy tobacco executives cackling gleefully each time a teenager buys a pack of cigarettes. Show them dreaming up ads designed to sell their products to the gullible teen—ads depicting smoking as cool and smokers as sexy. Show smoking as something *they* want us to do, not as something *we* want to do.

My younger daughter is no longer a teenager and she hasn't smoked in years. I don't know about Dave Barry.

Troublemakers

As Terrie Moffitt said in the article I started reading at the beginning of this chapter, breaking laws is a normal part of teenage life. Most of the people who commit criminal acts are people—especially males—in their teens or early twenties. Of a representative sample of teenage boys that Moffitt studied, only 7 percent of the eighteen-year-olds claimed never to have broken a law. Criminal behavior is uncommon in childhood and uncommon after the age of twenty-five or so. The troublemakers are people who are no longer children but not yet adults.

A large majority of the youthful lawbreakers were reasonably good children and will eventually become (if they live that long) reasonably law-abiding adults. Their delinquency is, as Moffitt puts it, "temporary and situational"—it depends on social context. Delinquency is not, by and large, something kids do on their own: it is something they do with their friends.

Their behavior may be antisocial but they are not unsocialized. They may be troublesome but they are not, in the majority of cases, "troubled." If they appear angry, it's probably because they've been caught. Most of them are normal kids who are behaving appropriately for their social context. They are conforming to the norms of their group (which may not happen to conform to the norms of yours), or doing what it takes to gain status in their group, or doing what it takes to avoid losing status. Want to change them? Then change the norms of their group. Lots of luck.

No, I'm being overly pessimistic. Adults do have some influence. The norms of teenagers' groups are based in part on the norms of adults' groups and are influenced by other cultural sources, especially the media. I believe the media's glamorization of violence—or, what might be even worse, their banalization of violence—is the source of much of the increase in criminal behavior over the past thirty years. The children of San Andres grew up thinking that aggressive behavior is normal because that's how a lot of the people in their village behaved. The children of North America and Europe grow up thinking that aggressive behavior is normal because that's how a lot of the people on their television screens behave. Kids bring these notions with them to the peer group and, since their peers live in the same village or watch the same shows, they incorporate them into the norms of their group. People in our society, they think, are *supposed* to behave that way.

They *are* supposed to behave that way in some societies. Yanomamö men, if they don't like the way their wife is behaving, hit her with a stick or shoot an arrow into some part of her anatomy they can do without. Ask Helena, the Brazilian girl who was kidnapped by the Yanomamö. When Helena came of age she was claimed by a Yanomamö headman, Fusiwe, who already had four wives. Fusiwe was a nice guy by Yanomamö standards—reader, she loved him!—but he got angry at her once for something that wasn't her fault and he broke her arm.

In such a society, it is the boy who *doesn't* behave aggressively who is out of step. Within the United States, there are differences from one subculture to another, and from one neighborhood to another, in tolerance of aggression and in attitudes regarding things like shoplifting and the use of drugs.

There are also differences from one peer group to another within a high school. As birds of a feather flock together, aggressive teens and those who are attracted to excitement and danger find others like themselves. Such personality characteristics are partly genetic, so when kids seek out other kids who are similar to themselves, to some extent they are seeking out those with similar genes.

Untangling the causes of delinquency will require an understanding of the four different factors that are involved: the culture, the age category within the culture, the peer group within the age category, and the individual. Some cultures foster impulsive, aggressive behavior. Within cultures that contain three or more age categories, there is apt to be trouble between teenagers and adults. Within schools that offer an array of peer groups, some groups pride themselves on being bad and contrast themselves with the goody-goodies. And where there is an array of peer groups, kids sort themselves out on the basis of their individual characteristics and gravitate to the group that provides the best fit.

Programs designed to cure delinquents of their delinquency have been notably unsuccessful. Usually the re-arrest rate of the kids who have been through the flavor-of-the-month program is almost as high as that of the kids who haven't. Sometimes it is higher. It is more likely to be higher when the delinquent kids are treated tough—sent to prison or to a modern version of what used to be called "reform school." In view of what I've told you, I hope you can see why putting kids who've committed crimes together with a bunch of other kids who've committed crimes is not likely to disabuse them of the notion that committing crimes is normal.

I will have more to say about criminal behavior in the next chapter.

From Childhood to Old Age

Adolescence is often described as an age of conformity—an age when people are most susceptible to the influence of their peers. But people are susceptible to the influence of their peers at all ages. I believe that childhood is a more conforming age than adolescence. Social psychologist Solomon Asch found that of all the subjects he tested, children under the age of ten were the most likely to yield to the majority in his famous test of group conformity (described in Chapter 7). Only a small fraction of his youngest subjects continued to make correct perceptual judgments when all the other children in the room were making wrong ones. Childhood is when the pressure to conform is the most severe; the nail that sticks up gets hammered down without mercy.

It is true that if you *ask* kids who influences them more—what they'd do if their parents and their friends gave conflicting advice—younger children are more likely to say they'd listen to their parents. But they are asked this question out of context and the one who's asking is a grownup. They may interpret the question as meaning "Whom do you love more?" and of course they love their parents more than they love their friends. The question has been answered by the relationship department of their mind but it is the group department that will, in the long run, determine how they will behave when they're not at home.

Childhood is a time of assimilation—a time when children learn to behave like the other members of their age and gender group. This is how they are socialized. In societies that have only two age groups, children and adults, fourteen years is time enough to produce a passable adult. In such societies it is pretty clear what a grownup man or woman is expected to do; there is not much choice in the matter.

But childhood is also a time of differentiation. Children learn what kind of people they are—plain or fancy, tough or soft, swift or slow—by comparing themselves, and by being compared, to the other members of their group, the other children of their age and sex. They bring this understanding with them when they move on to the next age category.

Adolescence, if the society provides one, is where they put it to use. In developed societies adults must specialize, and there is a wide variety of specialties to choose from. Adolescence is when the choices get made. When they sort themselves into groups, teenagers are defining themselves. They are choosing to travel in one direction rather than another. Such choices are not necessarily irrevocable—my younger daughter demonstrated that—but they do foreclose some options. A high school equivalency degree is not the same as a diploma. Going to college at twenty-eight is not the same as going at eighteen.

Like children, adults tailor their behavior to the social context. William James talked about the man who is tender with his children but stern with the soldiers under his command. But these temporary modifications of behavior no longer seem to have the power to produce long-term changes, the way they do in younger people. Childhood and adolescence are when people acquire the patterns of behavior, and the inner thoughts and feelings that accompany these patterns, that will serve them for the rest of their lives. The adult personality is quite resistant to change. "The character has set like plaster" is how James put it. From the grip of what, a century ago, he called "habit," an adult can "no more escape than his coat-sleeve can suddenly fall into a new set of folds."

The adult language is equally resistant to change; if anything, the hardening comes sooner. A person has only about thirteen years to acquire a language without an accent. Former Secretary of State Henry Kissinger immigrated to the United States as a teenager and never lost his German accent. His brother speaks unaccented American English. They both came to this country at the same time but the brother was a few years younger.

Childhood is when people learn to behave and to talk in a way that is appropriate for their society. The learning goes on at a deep level, ordinarily inaccessible to the conscious mind. Not until their parents complain (and possibly not even then) are children aware that they are bringing home the accents and behaviors of their peers. In adulthood, when people attempt to exert conscious control over the way they behave or the way they talk, they find it difficult or impossible to change them. These largely unconscious, largely involuntary patterns of behavior are what this book is about. They are what I believe we get from our peers and not from our parents.

Psychologists use the term *critical period* for a stage of life in which certain things must happen if they are to happen at all—imprinting in a duckling is the usual example. They use the term *sensitive period* for a stage of life in which certain things can be accomplished readily that are accomplished only with difficulty in other stages. Childhood is a sensitive period for the acquisition of a "native" language and the shaping of a "native" personality. These things may undergo further refinement in adolescence but the basic framework has been put in place.

The personality we acquire in our childhood and adolescent peer groups is the one that accompanies us through the rest of our lives. It is the "me" that continues to look out of our eyes even when our eyes require bifocals. This enduring, unchanging "me" is repeatedly surprised, often dismayed, and occasionally amused at the changes that take place in the physical container it inhabits. Old people fear (with good reason) that younger ones will not recognize them in their strange disguise. Some of them, now that the technology is available, try to halt or reverse the changes so that the outside doesn't get so out of step with what's inside.

I feel the mismatch just as keenly but haven't done anything to halt it.

Once in a while I'll catch a glimpse of myself in the mirror—the gray hair, the lines around the nose and mouth and eyes—and what I see strikes me, just for a moment, as ludicrous. It's "me" in a silly costume, dressed up for the role of Grandma in the high school play. White powder has been sprinkled in my hair, the lines are drawn with eyebrow pencil. Only they don't wash off.

Somewhere between the ages of seventeen and twenty-five, the "me" inside stops changing. Perhaps it stops changing because the brain has matured physically; if that's the case, then males (who mature more slowly) might remain plastic a bit longer than females. Perhaps it's because adults no longer have a peer group in the same sense they did when they were kids; if that's the case, then people who go to college might remain plastic a bit longer than those who don't. Or perhaps it's because the penalties for not conforming to group norms are so much milder in adulthood. If that's the case, there shouldn't be any systematic differences that depend on sex or education.

The personality shaped and polished in our childhood and adolescent peer groups is the one we take with us to the grave. My mother is dying of Alzheimer's disease and no longer talks at all, but she was still able to talk when she was eighty. On her eightieth birthday I asked her if she knew how old she was. She understood the question but had no memories left on which to base a reply. So she hazarded a guess.

"Twenty?" she said.

13 DYSFUNCTIONAL FAMILIES AND PROBLEM KIDS

According to the editorial in the *Journal of the American Medical Association,* Carl McElhinney was a child murderer. No, not a murderer of children, but a seven-year-old boy who had committed a murder. The editorial appeared a hundred years ago; it was reprinted in a recent issue of *JAMA* as a historical curiosity.

I cannot give you any details of Carl's crime because the focus of the editorial was not on the murderer himself but on his mother.

> Before Carl's birth Mrs. McElhinney was an assiduous reader of novels. Morning, noon and night her mind was preoccupied with imaginative crimes of the most bloody sort. Being a woman of fine and delicate perception, she appreciated to an extent almost equaling reality the extravagant miseries, motive, villainies set down in novels, so that her mind was miserably contorted weeks before the birth of her child Carl. The boy was an abnormal development of criminality. He has a delight in the inhuman. It takes intense horror to please this peculiar appetite. . . . I believe criminal record does not show a case so remarkable as this. As the boy matures these mental conditions will mature. He is dangerous to the community.

The cause of Carl's abnormal development, according to the editorialist, was the impression made on his mother's mind by the books she read while she was carrying him. Strong impressions on a woman's mind "may pervert or stop the growth, or cause defect in the child with which she is pregnant."

The editorial concluded, as editorials are wont to do, with a moral:

> We as scientific physicians . . . should teach our patrons how to care for our pregnant women, and the danger from maternal influences. The Spartans bred warriors, and I believe this generation can breed a better people. One of the future advances to help the generations to come, will be to teach them the power of maternal influences, with better care of our pregnant women.

The "better care of our pregnant women" would presumably include careful screening of the reading material permitted to them.

No doubt this sounds awfully silly to you. They were pretty dumb a hundred years ago, right? We know better now!

I ask you to consider the possibility that what the "experts" say today on the subject of why children sometimes turn out badly may be just as misguided as what they were saying—with, please note, exactly the same air of benevolent omniscience—a hundred years ago.

The idea of maternal influences—that what a pregnant woman does or sees or thinks can affect the child she is carrying—was not thought up by the physician who wrote the editorial. It is an ancient and pervasive idea, found in a great many cultures. I mentioned in Chapter 5 that parents in earlier times generally did not believe that the way they reared their children would have long-term effects on how the children turned out. And yet these people did realize that children are not all alike and that some turn out better than others. Since the same two parents can produce children with widely varying characteristics, it was not easy to see how heredity could account for the differences. And since many of the differences are present from birth (or at least from very early on), it was not unreasonable to attribute them to things that happened in the womb.

The result of this reasoning was that pregnant women in many traditional cultures were hedged in with rules: what they were allowed to do and see, what they were allowed to eat. Sometimes the prohibitions extended to the father as well. If the child turned out badly, the neighbors could blame it on the parents: they must have done something wrong while the mother was pregnant. They must not have followed the rules. You see, things haven't changed so much after all! The main difference is that in the old days the parents' period of culpability lasted only nine months.

Now it lasts forever. If you don't treat your kids right, not only will they turn out badly (according to the nurture assumption) but they will have "deficient parenting skills" as well, so *their* kids will also turn out badly, and *that's* your fault too.

I am going to try to get you off the hook by presenting evidence that maybe it's not your fault after all. But this is a two-way deal: I ask something from you in return. I ask you to promise not to go around telling people that I said it doesn't matter how you treat your kids. I do not say that; nor do I imply it; nor do I believe it. It is *not* all right to be cruel or neglectful to your children. It is not all right for a variety of reasons, but most of all because children are thinking, feeling, sensitive human beings who are completely dependent on the older people in their lives. We may not hold their tomorrows in our hands but we surely hold their todays, and we have the power to make their todays very miserable.

Let us not forget, though, that parents are also thinking, feeling, sensitive human beings, and that children also have power. Children can make their parents pretty miserable too.

Hand-Me-Downs

A cartoon strip that appeared on Father's Day shows cute, plump Cathy sitting with her parents, looking through the family photo album. "Here we are on Father's Day when I was just a year old, Dad," Cathy says. "You held my very first ice cream cone for me." In the next frame they're looking at a photo of Dad giving Cathy her first cotton candy. Two frames later it's a big box of chocolates, presented to Cathy by Dad to console her for her humiliation in the school play. French fries, caramel corn, and malted milks next appear, all thanks to Dad.

Now Mom speaks up:

> Ahah! Documented evidence! All fattening foods were introduced by your FATHER! All bad food habits came from your FATHER!! I am innocent! At last!! If you have a weight problem, it's all HIS fault!!

Alas, mothers don't get off the hook that easily. Cathy is not persuaded of Mom's innocence. And the cartoonist offers us only those two alternatives: either it's Mom's fault or it's Dad's.

Such is the power of the nurture assumption that it's the first thought that springs to everyone's mind: if Cathy has a weight problem—and, let's face it, she does—it must be the way her parents brought her up. Here is a newspaper columnist answering a question from the parent of an obese child by citing an "expert":

> The first thing adults can do, says pediatrician Nancy A. Held, is set an example. "If parents eat poorly and are sedentary, these are behaviors the child will copy."

The pediatrician is wrong and so is the cartoonist. The only thing Cathy's parents can be blamed for is giving their daughter their genes. Her parents, too, are cute and plump. Cathy came by her plumpness the same way she came by her cuteness.

I described in Chapter 2 how the effects of heredity and environment can be disentangled by means of behavioral genetic methods. The same methods used to study personality characteristics have also been used to study obesity, and with much the same results. Identical twins, whether they are reared together or apart, are usually very similar in weight—much more similar than fraternal twins. And adopted children do not resemble in fatness or thinness either their adoptive parents or their adoptive siblings.

Think of it: two adopted children are reared in the same home with the same parents. Their parents may be couch potatoes who nosh on caramel corn, or they may be dedicated broccoli eaters who work out daily in the gym. Both children are exposed to the same parental behaviors; both children are served the same meals and have access to the same pantry. And yet one child turns out lean and fit, the other is obese.

The heritability of fatness and thinness is somewhat higher than that of personality characteristics: about .70. But the important point is that the variation in weight that's *not* due to the genes—the part that's due to the environment—cannot be blamed on the *home* environment. There is no evidence that the parents' behavior has any long-term effects on their children's weight and very good evidence that it does not. And yet newspaper columnists and pediatricians go on telling parents, in tones that admit no uncertainty, that if they "set a good example" their children will be thin for life.

This is not merely an error: it is an injustice. If you have the misfortune to have a weight problem and your children have the same misfortune, you are not only blamed for your *own* presumably bad eating and exercising habits: you are also blamed for *theirs*. It's your fault you are overweight and it's your fault your kids are, too.

Forgive me for all the italics but this really gets my goat. The reason obese parents tend to have obese children is not because of the way they feed them or because of the bad example they set. Obesity is largely inherited.

A century ago a *JAMA* editorialist attributed the "abnormal development of criminality" in seven-year-old Carl McElhinney to the books his mother read while she was pregnant. Today a *JAMA* editorialist would no doubt attribute Carl's abnormalities to something else Mrs. McElhinney did wrong—something she did, or failed to do, *after* he was born. In neither case is attention paid to Carl's genetic heritage. Mrs. McElhinney is described as being obsessed with reading crime novels. "Morning, noon and night her mind was preoccupied with imaginative crimes of the most bloody sort." Carl and his mother share 50 percent of their genes and they both have a passion for crimes of the most bloody sort.

In Chapter 3 I recounted some stories of identical twins separated in infancy and reared in different homes. The Giggle Twins, both inordinately prone to laughter. The two Jims, who both bit their nails, enjoyed woodworking, and chose the same brands of cigarettes, beer, and cars. The pair who both read magazines back to front, flushed toilets before using them, and liked to sneeze in elevators. The pair who both became amateur firefighters. There was also a pair who, at the beach, would only go into the water backwards and only up to their knees. And a pair who were gunsmiths, and a pair who were fashion designers, and a pair who had each been married five times. These are not the imaginings of tabloid journalists; they were reported by reputable scientists in reputable journals. And there are too many of these stories for them all to be coincidences. Such spooky similarities are seldom found in the case histories of *fraternal* twins separated in infancy and reared apart.

Behavioral genetic studies have proved beyond a shadow of a doubt that heredity is responsible for a sizable portion of the variations in people's personalities. Some people are more hot-tempered or outgoing or meticulous than others, and these variations are a function of the genes they were born with as well as the experiences they had after they were born. The exact proportion—how much is due to the genes, how much to the experiences —is not important; the point is that heredity cannot be ignored.

But usually it is ignored. Consider the case of Amy, an adopted child. It wasn't a successful adoption; Amy's parents regarded her as a disappointment and favored their older child, a boy. Academic achievement was important to the parents, but Amy had a learning disability. Simplicity and emotional restraint were important to them, but Amy went in for florid role-playing and feigned illnesses. By the time she was ten she had a serious, though vague, psychological disorder. She was pathologically immature, socially inept, shallow of character, and extravagant of expression.

Well, naturally. Amy was a rejected child. What makes this case interesting is that Amy had an identical twin, Beth, who was adopted into a different family. Beth was not rejected—on the contrary, she was her mother's favorite. Her parents were not particularly concerned about education so the learning disability (which she shared with her twin) was no big deal. Beth's mother, unlike Amy's, was empathic, open, and cheerful. Nevertheless, Beth had the same personality problems that Amy did. The psychoanalyst who studied these girls admitted that if he had seen only one of them it would have been easy to fetch up some explanation in terms of the family environment. But there were two of them. Two, with matching symptoms but very different families.

Matching symptoms and matching genes: unlikely to be a coincidence. Something in the genes that Amy and Beth received from their biological parents—from the woman who gave them up for adoption and the man who got her pregnant—must have predisposed the twins to develop their unusual set of symptoms. If I say that Amy and Beth "inherited" this predisposition from their biological parents, don't misunderstand me: their biological parents may have had none of these symptoms. Slightly different combinations of genes can produce very different results, and only identical twins have exactly the same combination. Fraternal twins can be surprisingly dissimilar and the same is true of parents and children: a child can have characteristics seen in neither of her parents. But there is a statistical connection—a greater-than-chance likelihood that a person with psychological problems has a biological parent or a biological child with similar problems.

Heredity is one of the reasons that parents with problems often have children with problems. It is a simple, obvious, undeniable fact; and yet it is the most ignored fact in all of psychology. Judging from the lack of attention paid to heredity by developmental and clinical psychologists, you would think we were still in the days when John Watson was promising to turn a dozen babies into doctors, lawyers, beggarmen, and thieves.

Thieves. A good place to begin. Let's see if I can account for criminal behavior in the offspring without blaming it on the environment provided by the parents—on the parents' child-rearing methods or the lack thereof. Don't worry, I am not going to pin it all on heredity. But I can't do it without heredity, so if that bothers you, go and take a cold shower or something.

Criminal Behavior

How would you go about making a child into a thief? Fagin, of Charles Dickens's *Oliver Twist,* could have told Watson a thing or two. Take four or five hungry boys, make them into an *us,* give them a pep talk and a course in pocket-picking, and sic 'em on *them,* the rich folk. It's intergroup warfare, a tradition of our species, and the potential for it can be found in almost any normal human, particularly those of the male variety. Your schoolboy with his shining morning face is but a warrior in thin disguise.

But Fagin's method, which had worked flawlessly on the London slum children who were his other pupils, didn't work on Oliver. Dickens seemed to think it was because Oliver was born good, but there is another possibility: Oliver didn't identify with the other boys in Fagin's ring. They were Londoners and he was not. They spoke in a thieves' argot that was almost a foreign language to him. There were too many differences, and Oliver's run-in with the law came too soon to allow him to adapt to his new companions.

Oliver Twist was published in 1838, a time when it was still politically correct to believe that people could be born good or born bad—when it was still politically correct, in fact, to believe that badness could be predicted on the basis of one's racial or ethnic group membership. Dickens's other name for Fagin was "the Jew." It was by no means the worst of times, but it was certainly not the best of times.

Today both the individual explanation—that certain children are born bad—and the group explanation are held to be politically incorrect. Western culture has swung back to the view associated with the philosopher Rousseau: that all children are born good and it is society—their environment—that corrupts them. I'm not sure if this is optimism or pessimism, but it leaves too much unexplained. Even in the London slums of Dickens's time, not every child became an Artful Dodger. Even in the same family, one child may become a law-abiding citizen while another pursues a career as a criminal.

Though we no longer say that some children are born bad, the facts are such, unfortunately, that a euphemism is needed. Now psychologists say that some children are born with "difficult" temperaments—difficult for their parents to rear, difficult to socialize. I can list for you some of the things that make a child difficult to rear and difficult to socialize: a tendency to be active, impulsive, aggressive, and quick to anger; a tendency to get bored with routine activities and to seek excitement; a tendency to be unafraid of getting hurt; an insensitivity to the feelings of others; and, more often than not, a muscular build and an IQ a little lower than average. All of these characteristics have a significant genetic component.

Developmentalists have described how things go wrong when a child who is difficult to manage is born to a parent with poor management skills—something that happens, thanks to the unfairness of nature, more often than it would if genes were dealt out randomly to each new generation. The boy (usually it's a boy) and his mother (often there is no father) get into a vicious spiral in which bad leads to worse. The mother tells the boy to do something or not to do something; he ignores her; she tells him again; he gets mad; she gives up. Eventually she gets mad too, and punishes him harshly, but too late and too inconsistently for it to have any educational benefits. Anyway, this is a child who is not very afraid of getting hurt—at least it relieves his boredom.

The dysfunctional family. Oh yes, such families exist—there is no question about it! They are no fun to visit and you wouldn't want to live there. Even the biological father of this child doesn't want to live there. There's an old joke that goes like this:

> *Psychologist:* You should be kind to Johnny. He comes from a broken home.
> *Teacher:* I'm not surprised. Johnny could break any home.

Difficult for their parents to rear, difficult to socialize. For most psychologists these two phrases are virtually synonymous, because socialization is assumed to be the parents' job. For me they are two different things. It is true that there tends to be a correlation between them, due to the fact that children take their inherited characteristics with them wherever they go. But the correlation is not strong, because the social context within the home, where the rearing goes on, is very different from the social context outside the home, where the socializing goes on. Children who are obnoxious at home are not necessarily obnoxious outside the home. Johnny may be obnoxious everywhere he goes, but fortunately such kids are uncommon.

The word *socialization* is most often used to refer to the training in morality that children are presumed to get at home. Parents are held to be responsible for teaching their children not to steal, not to lie, not to cheat. But here again, there is little correlation between how children behave at home and how they behave elsewhere. Children who were observed to break rules at home when they thought no one was looking were not noticeably more likely than anyone else to cheat on a test at school or in a game on the playground. Morality, like other forms of learned social behavior, is tied to the context in which it is acquired. The Artful Dodger might have been as good as gold to his ol' mum, if he had had one.

It's harder to believe that Oliver might have been a thorn in his mother's side if she had lived. Oliver made friends wherever he went; women fell all over him. A sweet nature and a pretty face will do it every time. As Dickens described him, Oliver had precisely those traits that make a child easy to deal with. He was sensitive to the feelings of others and fearful of punishment and pain—timid, almost. He was bright, unimpulsive, and unaggressive.

Was Dickens right? Are some children born good? Let us do an experiment that John Watson would have approved of. Place in adoptive homes a bunch of infant boys whose biological parents had been convicted (or will later be convicted) of crimes, and a second bunch whose biological parents were, as far as anyone knows, honest. Mix them up: place some of each bunch in homes with honest adoptive parents and let others be reared by crooks. A dastardly experiment, you say? Well, that's what adoption agencies do. Of course, they don't purposely put babies in the homes of criminals, but sometimes it works out that way, and in places where careful records are kept both of adoptions and of criminal convictions—Denmark, for example—it's possible to study the results. Researchers were able to obtain background data on over four thousand Danish men who had been placed for adoption in infancy.

As it turned out, criminal convictions were numerous among the biological parents of the adoptees but infrequent among their adoptive parents. Thus, there were not many cases of boys who had honest biological parents being reared in the homes of crooks. Of this small group, 15 percent became criminals. But almost the same percentage of criminals (14 percent) was found among the adoptees whose biological parents were honest and whose adoptive parents were also honest. It seems that being reared in a criminal home does not make a criminal out of a boy who wasn't cut out for the job. Yet another blow to Watson, whose corpse is by now so thoroughly beaten that in all decency I should give it a rest.

The story is a little different for the boys whose biological parents were criminals. Of those who were reared by honest folk, 20 percent became criminals. And of the small group who came up unlucky both times —criminal biological fathers *and* criminal adoptive fathers—almost 25 percent went wrong. So it's not just heredity: it looks like the home environment does count for something after all. Try as you might, you can't make a criminal out of a kid like Oliver, but a kid like the Artful Dodger can go either way. Give him to a criminal family to raise and he is a little more likely to become a criminal.

Not so fast. It turns out that the ability of a criminal adoptive family to produce a criminal child—given suitable material to work with—depends on where the family happens to live. The increase in criminality among Danish adoptees reared in criminal homes was found only for a minority of the subjects in this study: those who grew up in or around Copenhagen. In small towns and rural areas, an adoptee reared in a criminal home was no more likely to become a criminal than one reared by honest adoptive parents.

It wasn't the criminal adoptive parents who made the biological son of criminals into a criminal: it was the *neighborhood in which they reared him.* Neighborhoods differ in rates of criminal behavior, and I would guess that neighborhoods with high rates of criminal behavior are exceedingly hard to find in rural areas of Denmark.

People generally live in places where they share a lifestyle and a set of values with their neighbors; this is due both to mutual influence and, especially in cities, to birds of a feather flocking together. Children grow up with other children who are the offspring of their parents' friends and neighbors. These are the children who form their peer group. This is the peer group in which they are socialized. If their own parents are criminals, their friends' parents may also be inclined in that direction. The children bring to the peer group the attitudes and behaviors they learned at home, and if these attitudes and behaviors are similar, in all probability the peer group will retain them.

I have told you about an adoption study of criminality; there are also twin and sibling studies. Behavioral genetic studies of twins or siblings usually show that the environment shared by children who grow up in the same home has little or no effect, but we've come to one of the exceptions. Twins or siblings who grow up in the same home are more likely to match in criminality—to both be criminals or both be honest. This correlation is often attributed to the home environment that the twins or siblings share—in other words, to the influence of the parents. But kids who share a home also share a neighborhood and, in some cases, a peer group. The likelihood that two siblings will match in criminality is higher if they are the same sex and close together in age. It is higher in twins (even if they're not identical) than in ordinary siblings, and higher in twins who spend a lot of time together outside the home than in those who lead separate lives.

The evidence shows that the environment has an effect on criminality but it doesn't show that the relevant environment is the home; in fact, it suggests a different explanation. When both twins or both siblings get into trouble, it is due to their influence on each other and to the influence of the peer group they belong to.

In the previous chapter I talked about Terrie Moffitt and her views on teenage delinquency. Moffitt distinguishes between two types of criminal behavior: the type that appears with the first pimple and is outgrown by the time the last tube of Clearasil hits the trash can, and the type that lasts a lifetime. Kids who were reasonably well-behaved in childhood and who will be reasonably law-abiding in adulthood often go through a phase in between where they are neither. As I said in the previous chapter, it's a group thing: a war between age groups. There is nothing psychologically wrong with most of these kids and it's not their parents' fault. They are socialized, all right—socialized by their peers.

The lifetime type of criminal behavior is far less common; it involves a small fraction of the population, mostly males. Their criminal behavior begins early—Carl McElhinney was a murderer at seven—and has the persistence of the Energizer bunny without its charm. Career criminals tend to have high levels of the characteristics I listed earlier: aggressiveness, lack of fear, lack of empathy, desire for excitement. Such people turn up from time to time in every society, even in those where their propensities are likely to lead to social ostracism or an early death. The members of an Eskimo group in northwest Alaska told an anthropologist that in the old days, when a man kept making trouble and nothing seemed to stop him, somebody would quietly push him off the ice. He was, as the *JAMA* editorialist said about Carl McElhinney, "dangerous to the community."

Are some people born bad? A better way of putting it is that some people are born with characteristics that make them poor fits for most of the honest jobs available in most societies, and so far we haven't learned how to deal with them. We are at risk of becoming their victims but they are victims too—victims of the evolutionary history of our species. No process is perfect, not even evolution. Evolution gave us big heads, but sometimes a baby has a head so big it can't fit through the birth canal. In earlier times these babies invariably died, as did their mothers. In the same way, evolution selected for other characteristics that sometimes overshoot their mark and become liabilities rather than assets. Almost all the characteristics of the "born criminal" would be, in slightly watered-down form, useful to a male in a hunter-gatherer society and useful to his group. His lack of fear, desire for excitement, and impulsiveness make him a formidable weapon against rival groups. His aggressiveness, strength, and lack of compassion enable him to dominate his groupmates and give him first shot at hunter-gatherer perks.

Unlike the successful hunter-gatherer, however, the career criminal tends to be below average in intelligence. I take this to be a hopeful sign: it suggests that temperament can be overridden by reason. Those individuals born with the other characteristics on the list but who also have above average intelligence are evidently smart enough to figure out that crime does not pay and to find other ways of gratifying their desire for excitement.

Where's Daddy?

In a hunter-gatherer or tribal society, children who lose their father are in danger of losing their lives. Where life hangs on threads, all it takes is one snip. In some societies they don't even wait for the fatherless ones to die of natural causes. According to evolutionary psychologist David Buss,

> Even today, among the Ache Indians of Paraguay, when a man dies in a club fight,* the other villagers often make a mutual decision to kill his children, even when the children have a living mother. In one case reported by the anthropologist Kim Hill, a boy of thirteen was killed after his father had died in a club fight. Overall, Ache children whose fathers die suffer a death rate more than 10 percent higher than children whose fathers remain alive. Such are the hostile forces of nature among the Ache.

In traditional societies, fathers defend their children against these so-called "hostile forces of nature," and a man who is dominant in his group can defend his children better than one who is lower on the totem pole. In industrialized nations you can still hear little boys—the sons of men who've never been in a fistfight, much less a club fight—telling each other, "My daddy can beat up your daddy." "My daddy can *sue* your daddy" would be more like it but that's not what they say (at least not until they're much older), because this is about power, not money. The message being conveyed here is, "You can't pick on me, because if you do, my daddy will beat you up, and he can do that without fear of being beaten up by your daddy." Among chimpanzees it is the mother, not the father, who comes running to the rescue, and when two young chimpanzees play together the one with the dominant mother is likely to have the

* For the Ache, a "club fight" means fighting with clubs, not fighting at a club.

upper hand. If the play gets too rough, his mother can wallop his playmate without fear of reprisal from the playmate's mother.

In a society where "My daddy can beat up your daddy" is still a credible threat, having a strong father versus a weak one, or having a father versus not having one, can have important repercussions on a child's status in the peer group and therefore (according to group socialization theory) can have long-term effects on a child's personality. But in societies like ours, where parents and peers are kept in separate compartments of a child's life, the parents' status no longer serves as a shield. The exception is when a parent has so much power or prominence that even the kid's peers cannot help but be aware of it. This is not necessarily a good thing—it can easily backfire, especially if the child doesn't have the other characteristics that would lead to high status in the group.

Having a father or not having one: How much difference does it make for an ordinary child in a developed society? I will not deny that children are generally happier if they have two parents; I will not deny that they are happier if they have evidence that both parents care about them and think well of them. But happiness today does not inoculate a child against unhappiness tomorrow, and (as I said in Chapter 8) there is no law of nature that says misery has to have sequelae. This book is about the long-term consequences of what happens while you're growing up. Do children with fathers turn out better in the long run than children without fathers? And if they do turn out better, is it *because* they had a father?

Most people think so. In 1992, Vice President Dan Quayle administered a tongue-lashing to Murphy Brown—a fictional character on a TV show—for having a baby without a husband. Characters on TV shows have unprotected sex all the time,* so I don't think *that* was what was bothering Quayle: it was the thought of this poor innocent (fictional) child growing up without a father. Two years later, sociologists Sara McLanahan and Gary Sandefur lent support to Quayle's apotheosis of fatherhood by writing a book called *Growing Up with a Single Parent* and asserting in italics, right on page 1,

> *Children who grow up in a household with only one biological parent are worse off, on average, than children who grow up in a household with both of their biological parents, regardless of the parents' race or educational background,*

* At least they certainly give that impression. On the other hand, their activities result in remarkably few pregnancies. Though this phenomenon deserves further investigation, a discussion of the fertility of fictional characters is beyond the scope of this book.

regardless of whether the parents are married when the child is born, and regardless of whether the resident parent remarries.

How are such children worse off? McLanahan and Sandefur settled on three indicators. Adolescents who do not live with both their biological parents are more likely to drop out of high school and more likely to be "idle" (neither working nor going to school), and the girls are more likely to become mothers while still in their teens. Father absence is not, of course, the only factor associated with these problems, but McLanahan and Sandefur believe it is an important one—important enough that "parents need to be informed about the possible consequences to their children of a decision to live apart."

The possible *consequences* to their children of a decision to live apart. Clearly, McLanahan and Sandefur believe that the parents' living apart is the *cause* of the kids' problems—that at least some of the worse-off kids would have managed to graduate from high school, get a job, and remain (unlike Murphy Brown) unimpregnated, if only Daddy had been around.

But the graphs and tables in McLanahan and Sandefur's book contain some curious findings: a lot of things you'd think would matter turn out not to matter. The presence of a stepfather in the home doesn't improve the kids' chances at all. Nor does contact with the biological father outside the home: "Studies based on large nationally representative surveys indicate that frequent father contact has *no* detectable benefits for children." Nor does having another biological relative living in the home: the presence of a grandmother doesn't help. In homes with live-in grandmothers, kids are left alone less often than in homes with two biological parents, yet that doesn't stop them from dropping out of school or getting pregnant. In homes with stepfathers, kids are given as much supervision as in homes with biological fathers—they are as likely to have their whereabouts monitored or their homework checked—yet that doesn't stop them from dropping out of school or getting pregnant. The number of years the kids spend in a single-parent family also doesn't matter: those whose fathers stuck around until they were on the brink of adolescence are no better off than the ones whose fathers went bye-bye when they were babies or, for that matter, fetuses.

The fatherless ones who *are* better off—and this is curious too—are the ones whose fathers have died. "Children who grow up with widowed mothers," McLanahan says, "fare better than children in other types of single-parent families." In some studies, in fact, they fare as well as children who grow up with two living biological parents. Researchers have had to grasp at straws to account for the different "consequences" of missing fathers and dead fathers. Widows are more financially secure than other single mothers? But remarried mothers are also more financially secure and having a stepfather doesn't help. The death of a parent is less stressful than a parental divorce? Among the more common causes of the premature death of a parent are suicide, homicide, cancer, and AIDS, and none of these strikes me as particularly stress-free.

"Consequences" is the word the researchers like to use, and even when they virtuously refrain from using it you can tell that's what they're thinking. But the data they use to support their beliefs do not show causes and consequences: the data are entirely correlational. They show only that certain things tend to go together with certain other things. If the epidemiological researchers I talked about in Chapter 2 had discovered that broccoli eaters are, on average, wealthier than broccoli shunners—and quite possibly they are—it would be rash to assume that if you start eating broccoli your income will go up or that if you stop eating it you will lose all your money. It would be equally rash to assume that if you win the lottery you will develop a taste for broccoli. The daughter of a married couple is, on average, more likely than the daughter of a single parent to graduate from high school and to avoid getting pregnant: that is a correlation. To conclude from it that the daughter of the married couple will drop out of school and have a baby if her parents split up is no better than concluding that if you stop eating broccoli you'll lose all your money. It could be true but the data do not prove it.

When the biological father is living but not living with his kids, you have a family situation that is statistically associated with unfavorable outcomes for the kids. Let me show you how it might be possible to account for the unfavorable outcomes without reference to the children's experiences in the home or to the quality of parenting they receive there.

Most single mothers are nothing like Murphy Brown: most of them are poor. Half of all homes headed by women are below the poverty level. Divorce usually leads to a drastic decline in a family's standard of living—that is, in the standard of living of the ex-wife and the children in her custody.

The loss of income impacts the kids in several ways. For one thing, it can affect their status in the peer group. Being deprived of luxuries such as expensive clothing and sporting equipment, dermatologists and orthodontists, can lower kids' standing among their peers. Money is also going to play a role in whether the kids can think about going to college. If it's out of the question, then they may be less motivated to graduate from high school and to avoid getting pregnant.

But by far the most important thing that money can do for kids is to determine the neighborhood they grow up in and the school they attend. Most single mothers cannot afford to rear their children in the kind of neighborhood where I reared mine—the kind where almost all the kids graduate from high school and hardly any have babies. Poverty forces many single mothers to rear their children in neighborhoods where there are many other single mothers and where there are high rates of unemployment, school dropout, teen pregnancy, and crime.

Why do so many kids in these neighborhoods drop out, get pregnant, and commit crimes? Is it because they don't have fathers? That is a popular explanation, but I considered the question in Chapter 9 and came to other conclusions. Neighborhoods have different cultures and the cultures tend to be self-perpetuating; they are passed down from the parents' peer group to the children's peer group. The medium through which the cultures are passed down cannot be the family, because if you pluck the family out of the neighborhood and plunk it down somewhere else, the children's behavior will change to conform with that of their peers in their new neighborhood.

It's the neighborhood, not the family. If you look at kids *within* a given neighborhood, the presence or absence of a father doesn't make much difference. Researchers collected data on 254 African-American teenage boys from an inner city in the northeast United States. Most of the boys lived in households headed by a single mother; others lived with both biological parents, a mother and a stepfather, or in other kinds of family arrangements. Here are the researchers' conclusions:

> Adolescent males in this sample who lived in single-mother households did not differ from youth living in other family constellations in their alcohol and substance use, delinquency, school dropout, or psychological distress.

Within an economically disadvantaged inner-city neighborhood, the kids who live with both parents are no better off than those who live with only one. But within a neighborhood like this, the majority of families are headed by single mothers, because mothers with partners generally can afford to live somewhere else. The higher income of a family that includes an adult male means that kids with two parents are more likely to live in a neighborhood with a middle-class culture and, therefore, more likely to conform to middle-class norms.

But why doesn't the higher family income help the kids reared in stepfather families? The answer is that these kids have another problem: too many moves. They've been shifted around from one residence to another more often than kids in any other kind of family setup, and each time they moved they lost their peer group and had to start all over again with a new one. Each time they moved there was a new set of peer group norms to adapt to and a new social hierarchy to climb, and each time they had to start all over again from the bottom.

Moving is rough on kids. Kids who have been moved around a lot—whether or not they have a father—are more likely to be rejected by their peers; they have more behavioral problems and more academic problems than those who have stayed put. McLanahan and Sandefur found that changes of residence could account for more than half of the increased risk of high school dropout, teen births, and idleness among adolescents being reared without their fathers. Together, changes of residence plus low family income could account for most of the differences between kids with dads and kids without them.

Both of these disadvantages can be explained in terms of things that happen outside the family. Changes of residence jeopardize a kid's standing in the peer group and interfere with socialization because it's difficult to adapt to group norms when the norms keep changing. Family income determines what kind of neighborhood the kid will live in and what kind of norms the local peer group is likely to have. Too many moves and low income increase the risk that a kid will drop out of school and/or have a baby.

But dropping out of school and having babies are things we already knew were susceptible to peer group influence. To convince you I'll have to take on a broader topic—the effects of divorce. The effects on the children's personalities, on their psychological health, and on the stability of their own marriages. Does the parents' divorce really do terrible things to the kids? And if not, how come everybody thinks it does?

Divorce

The most famous—and most pessimistic—study of the children of divorce is the one by clinical psychologist Judith Wallerstein. Wallerstein found a very high rate of emotional disturbance among the children of middle-class divorced couples. Her books sold a lot of copies but as science they are useless: all the families she studied had sought counseling and all were getting divorced. There was no control group of intact or self-sufficient families with which to compare the children of her patients and no way to filter out her professional biases. A study done shortly before Wallerstein wrote her first book demonstrated how easily professionals can be swayed by their preconceptions. The researchers showed some schoolteachers a videotape of an eight-year-old boy and told them that the boy's parents were divorced. These teachers judged him to be less well-adjusted than did other teachers who saw the same videotape but thought the boy came from an intact home.

A properly controlled recent study of the children of divorce gives a more optimistic picture than the one Wallerstein painted. The subjects were part of a massive British survey of children born in a single week in 1958; they were twenty-three years old at the time this study was done. They were asked to check off answers to questions about their mental health, for example: "Do you often feel miserable and depressed?" "Do you often suddenly get scared for no good reason?" "Do people often annoy and irritate you?" "Do you wear yourself out worrying about your health?" High scores on this test—lots of yeses—were taken to indicate a high level of psychological distress.

Parental divorce increased the chances that a subject's score on this test would fall above an arbitrary cutoff, but not by much: 11 percent of the people from divorced families scored above the cutoff, versus 8 percent of the people from intact families. The difference in the average number of yes responses was only half a test item.

There is a difference, but it's a small one. I hinted all along that it was going to turn out this way. I said that within a given neighborhood, the presence or absence of a father doesn't make *much* difference. I said that changes of residence plus low family income can account for *most* of the differences between kids with dads and kids without them. There are differences I haven't accounted for yet, and now they have cropped up again in the British study of the children of divorce. It is time to stop sweeping them under the rug.

Nowadays, studies of the effects of divorce or fatherlessness are generally carried out by researchers who know enough to control for a wide variety of potentially confounding factors. They control, for example, for socioeconomic class. Divorce and fatherlessness are more common in lower-income, less educated sectors of society, and this has to be taken into account. The researchers also control for racial and ethnic group, because different groups have different norms regarding marriage.

What they don't control for—because they have no way of doing it in this kind of study—is heredity. They are searching for effects of the child's environment with a method I scoffed at in Chapter 2: comparing foxhounds reared in kennels with poodles reared in apartments. The researchers look at one child per family. The child is, in most cases, the biological offspring of the parents. The parents provide the child's genes and also provide, or fail to provide, the child's environment, and there is no way of distinguishing the effects of one from the effects of the other. To distinguish them it is necessary to use behavioral genetic methods and to study adopted children or pairs of twins or siblings.

Relax, it has already been done, and done well, for a wide variety of psychological characteristics. Within the populations that have been studied—mostly American and European, preponderantly middle-class—almost all the characteristics show the same pattern. Heredity accounts for about half of the variation among the individuals who participated in these studies. The other half is environmental in origin but, as I explained in Chapter 3, it cannot be attributed to any environmental influence that is shared by two children growing up in the same home. In fact, any features of the environment that are shared by two children growing up in the same home are pretty much ruled out as important influences on what they will be like as grownups.

Within the populations that have been studied are many families that have been broken by divorce. Of the subjects who took part in these studies, some sizable fraction must have been reared by a divorced mother, or by a mother and a stepfather, or in some other arrangement that wouldn't pass muster with Dan Quayle. Sorry, Dan, but there is no evidence that it made any difference. If the parents' presence or absence in the home or the relationship between them—quarreling incessantly or writing each other little love notes—had any lasting effect on the kids, we should see it in the behavioral genetic data, and we do not.

More precisely, if the parents' presence or absence had any lasting effect on the kids, it must have been a different effect for each kid. Unfortunately, this doesn't bolster the position of researchers who say things like "Parents need to be informed about the possible consequences to their children of a decision to live apart." What consequences? If you can't say what the consequences are—if the parents' decision to live apart makes one child shyer and the other bolder, or makes one laugh more and the other laugh less, and if there are no overall trends—what were you planning to inform them about?

In the studies that produce the little differences I am now trying to account for—the studies that fill up developmental psychology journals and occasionally make their way into newspapers and magazines—consequences are reported all the time. But the consequences, or differences, are found only when the researchers fail to control for heredity. The home environment is revealed to be ineffective—that is, to have no predictable or consistent effects on the kids—only after genetic influences are siphoned off. If the research method does not provide a siphon, then genetic influences cannot be eliminated and they are invariably mistaken for evidence of the influence of the home environment. Cordial, competent parents tend to have cordial, competent kids, and most researchers simply take it for granted it's because of the warm and orderly home life these parents provide for their kids.

The best example of mistaken conclusions is divorce itself. It is well known—and also, as it happens, true—that children reared in broken homes are more likely to fail in their own marriages. Why are the sins of the parents visited on the children? Is it the anxieties the children drag along with them to adulthood, left over from their exposure to years of parental conflict? The repressed anger that has festered ever since Daddy moved out? Judith Wallerstein would have us think so.

But a twin study of divorce offers a different explanation. More than 1500 pairs of adult identical and fraternal twins answered questions about their own marital histories and those of their parents. The divorce rate was 19 percent among the twins whose parents had remained married. Among those whose parents were divorced, the chances of getting divorced were considerably higher: 29 percent. The chances were just as high—30 percent—for those with a divorced fraternal twin, and they were higher still—45 percent—for those with a divorced identical twin. The analysis churned out by the researchers' computer was boringly similar to those of other behavioral genetic studies: about half of the variation in the risk of divorce could be attributed to genetic influences—to genes shared with twins or parents. The other half was due to environmental causes. But *none* of the variation could be blamed on the home the twins grew up in. Any similarities in their marital histories could be fully accounted for by the genes they share. Their shared experiences—experienced at the same age, since they were twins—of parental harmony or conflict, of parental togetherness or apartness, had no detectable effect.

Heredity, not their experiences in their childhood home, is what makes the children of divorce more likely to fail in their own marriages. But don't bother to tiptoe through the chromosomes in search of a divorce gene. There is no divorce gene. Instead there is an assortment of personality characteristics, each roughhewn by a complex of genes and shaped and sanded by the environment, that together increase the chances that a person will marry unhappily.

Don't look for a divorce gene. Look instead for traits that increase the risk of almost any kind of unfavorable outcome in life. Traits that make people harder to get along with—aggressiveness, insensitivity to the feelings of others. Traits that increase the chances they will make unwise choices—impulsiveness, a tendency to be easily bored. Does this list sound familiar? Yes, it is similar to the list of characteristics that are often found in criminals. The same traits that make some kids good candidates for Fagin's school also lower their chances of a happy marriage. In childhood, individuals with these traits may be diagnosed with what psychiatrists call "conduct disorder." The adult form is called "antisocial personality disorder" and research has shown that it can be inherited.

The children of parents who will later get divorced sometimes start acting troublesome years before the parents actually split up. This observation has been taken to show that it isn't the divorce itself that causes problems in the kids—it's the family conflict that precedes it. But the finding that conflict-prone parents tend to have troublesome kids may be due to the genes they share rather than the home they share. A group of researchers at the University of Georgia discovered that what predicted conduct disorder in children was not parental divorce but parental personality: parents with antisocial personality disorder were more likely to have children with conduct disorder.

The links between divorce, personality problems in the parents, and troublesome behavior in the children are complex: the effects go every which way. People with personality problems are difficult to live with so they're more likely to get divorced; the same people are more likely, for genetic reasons, to have difficult kids. There might even be a child-to-parent effect: a difficult kid can put a real strain on a marriage. Earlier in the chapter I quoted the joke about Johnny, the kid who could break any home, but it is not funny if you have a kid like Johnny. Some children make every member of the family wish they could get out. Judith Wallerstein talks about the heavy load of guilt the children of divorcing parents are burdened with—the kids think their parents' divorce was *their* fault. What Wallerstein doesn't consider is that sometimes there may be an element of truth in what the kids think. Divorce occurs less often in families that contain a son than in those with only daughters. The presence of that boy either makes the parents happier or makes the father more reluctant to walk out. But what if the boy is not a satisfying kid? What if he is nothing but trouble?

Of course, most people who get divorced do not have serious personality problems and most children of divorce do not have conduct disorder. Most children of divorce do fine in the long run—the British study demonstrated that. The twenty-three-year-old children of divorced parents were only slightly more likely to say yes to questions about depression, anxiety, and anger.

Then why are clinical psychologists like Judith Wallerstein so certain that parental divorce is bad for children? Because, as social psychologist David G. Myers has pointed out, it *is* bad for them—only it's not bad for the reasons Wallerstein gave or in the ways she assumed.

Divorce is bad for children in several ways. First, it comes with a heavy financial penalty: the children of divorced parents usually experience a severe decline in standard of living. Their financial status will determine where they live, and where they live will make a difference. Second, it's bad for them because they often have to move to a new residence. Sometimes they have to move several times. Third, it increases the risk that they will suffer physical abuse. Children living in homes with stepparents are far more likely to be abused than those living with two biological parents. Fourth, it's bad for them because it disrupts their personal relationships.

In Chapter 8, I made a distinction between groupness and personal relationships. Groupness, I said, is what enables children to be socialized. The roughhewn personality we are born with has to be shaped and sanded into something more suitable for the culture we grow up in, and this happens during childhood through adaptation to a group—usually a group of other children. Long-term modifications of personality and ingrained patterns of social behavior are handled by the groupness department of the mind.

The department that oversees personal relationships doesn't produce long-term modifications of personality, but that doesn't make it unimportant. In our thoughts and emotions, the relationship department is much closer to the surface—much more accessible to the conscious mind—than the department that produces the long-term modifications. Relationships can dominate our moment-to-moment feelings and actions and leave residues in our memories like stacks of old love letters in the attic.

Relationships are important; they've always been important to the members of our species. That is why evolution endowed us with the motivation to form them and the motivation, if they're going reasonably well, to have them continue. Strong emotions like love and grief provide the power; Steven Pinker explains how they do it in his book *How the Mind Works.*

Divorce and the parental conflict that surrounds it make children unhappy. It disrupts their relationships with their parents and messes up their home life. This unhappiness, the disrupted relationships and the messed-up home life, are what the clinical psychologists and developmentalists are seeing when they study the effects of divorce on children. In studies of divorce, the children are usually interviewed at home or in a setting they've gone to with their parents. Or, what's worse, researchers rely on the parents' reports of the children's behavior, though even at the best of times—even when the parents are not in the midst of a divorce—what they say about their children agrees poorly with reports by neutral observers.

When life at home is disrupted, the child's behavior at home is of course disrupted, and so are the emotions associated with the home. These are the changes the researchers are seeing. If they want to find out how the child's life outside the home is affected by the parents' divorce, the researchers will have to collect their data outside the home, and if they want to do it right they will have to use unbiased observers—observers who are unaware of the child's family situation. What the researchers will find under these conditions, judging from the behavioral genetic data I mentioned earlier, is that parental divorce has no lasting effects on the way children behave when they're not at home, and no lasting effects on their personalities.

Physical Punishment and Child Abuse

I've come now to a topic I approach with trepidation. I have no fear that *you* will misunderstand me but I worry about those who do not read the book and only hear about it second-hand. Words can be misquoted or taken out of context; people are denounced for opinions they do not hold and never expressed. If I am going to be denounced, I'd rather have it be for opinions I do hold, so let me state them clearly here and now.

First, I do not think it is okay to beat children or to do anything to them that causes injury or long-lasting pain. Second, I don't think an occasional smack, at an appropriate time and on an appropriate part of the anatomy, does a kid any harm.

Physical punishment is used by parents around the world and in the vast majority of American homes. It is also found in other species. I believe it is part of the built-in repertoire of parental behaviors. One of my purposes in writing this book is to relieve parents of the guilt that has been imposed upon them by the professional givers of advice on child-rearing. If you have occasionally lost your temper and hit your children, it is unlikely that you've caused them any lasting harm. On the other hand, it is possible that you've harmed your relationship with them. If you have been unjust and they are old enough to realize it, they will think less of you. You don't get off the hook entirely.

But it isn't because your children will think less of you that the professional advice-givers warn you against hitting them. The trouble with hitting children, they tell you, is that it will make them more aggressive.

The logic is persuasive. By spanking your child, you are providing him or her with a model of aggressive behavior. You are teaching your child that it's okay to hurt people in order to make them do what you want.

For many years I believed this story and, in good faith, passed it on to the readers of my child development textbooks. I didn't notice that we also provide children with models for many other things that we don't want them to do and that they don't in fact do, such as leaving the house whenever they feel like it. And models for many things that we *want* them to do but they don't do, such as eating broccoli.

Child-rearing styles can change with dizzying speed, as one generation of advice-givers is replaced by the next. If the new ones didn't tell you something different from their predecessors, they couldn't stay in business. But advice-givers are not heeded equally by all segments of the population. Countries like the United States contain many subcultures, and your views on child-rearing will depend in part on which one you belong to. Asian Americans and African Americans tend to pay less attention to European-American advice-givers and hence are less averse to spanking a child. It is middle-class European Americans who currently abjure the use of spankings and who favor instead the use of time-outs. Last week a little boy with light brown hair was running wild through the aisles of our local supermarket. Running half an aisle behind him was his father, shouting, "Matthew! You're going to get a time-out!"

Black parents tend to be unenthusiastic about this method of enforcing discipline. "Time-outs are for white people," they explain to interviewers.

Perhaps the white people are too credulous. Most of the research on punishment—the research on which the advice-givers base their advice—is as worthless as Judith Wallerstein's study of the children of divorce. One of the reasons it is worthless is that researchers often fail to take into account the subcultural differences in child-rearing styles.

There is ample evidence that parents in minority ethnic groups and in low-income neighborhoods administer more spankings. In some—though not all—of these groups, the children tend to behave more aggressively and to get into more trouble. It is easy to mistake these subcultural differences for the "consequences" the researchers are looking for. Middle-class white kids are spanked less and also tend to be less aggressive, so if a study lumps together kids from middle-class white neighborhoods with kids from low-income black neighborhoods, the researchers are almost guaranteed to find a correlation between spanking and aggressiveness. Their hopes will be dashed, however, if they include too many Asian Americans among their subjects, because these parents do use physical punishment but they don't have aggressive kids.

The other problem with most studies of punishment is that they provide no way of distinguishing causes from effects. Within any ethnic group or social class, some kids are more aggressive than others and some get spanked more than others. If aggressive kids get spanked more, is the kids' aggressiveness caused by the spankings or are the parents doing a lot of spanking because they don't like the way the kids are behaving? Impossible to tell in most cases.

One way that researchers deal with the cause-or-effect problem is by following children over a period of years. The August 1997 issue of *Archives of Pediatrics and Adolescent Medicine* contains a study of this sort by psychologist Murray Straus and his colleagues. The researchers controlled for the initial level of antisocial behavior in the children by looking for *changes* in their behavior over time. If a mother administers more than the average number of spankings when the kid is six years old, is the kid even more troublesome when he is eight? Yes he is, the researchers concluded. Over the two-year period of the study, the children who received frequent spankings became more troublesome and more aggressive. "When parents use corporal punishment to reduce antisocial behavior," the researchers asserted, "the long-term effect tends to be the opposite."

The study made news. It was picked up by the Associated Press and reported in newspapers and magazines around the country; an abstract of it appeared in *JAMA.* Neither the Associated Press nor *JAMA* mentioned another study, by psychologists Marjorie Gunnoe and Carrie Mariner, that appeared in the same issue of *Archives of Pediatrics and Adolescent Medicine.* The topic was the same and the method was similar but the results were different. "For most children," Gunnoe and Mariner concluded, "claims that spanking teaches aggression seem unfounded." For black children of any age, and for the younger children in the study regardless of race, these researchers found that spanking actually led to a *decrease* in aggressive behavior.

Darn it, this sort of thing happens all too often in psychology. Effects are flimsy; results are evanescent. Throw the whole issue of *Archives of Pediatrics and Adolescent Medicine* in the recycling bin and forget about it.

No, wait. Fish it out and look more closely at the methods the researchers used. Ah ha, there *is* a difference. In the first study, the researchers assessed the children's behavior by asking their mothers—the same mothers who were administering the spankings. The mothers' replies were based on how the kid was acting *at home.* In the second, it was the children themselves who were questioned: the researchers asked them how many fights they got into *at school.* Kids who suffered many spankings at home were no more likely to report an increase in fighting in school than kids who were not spanked.

Spanking at home may make kids behave worse at home or it may just be an indication that the mother–child relationship, or the mother's life in general, is not going well (the kid may not be behaving as badly as the mother thinks he is). In any case, the evidence indicates that being spanked at home does *not* make kids more aggressive when they're not at home. The conclusion of the first group of researchers, that if parents stopped hitting their kids it could "reduce the level of violence in American society," seems a tad overblown.

However, I've been speaking of physical punishment within the normal range—an ordinary spanking from time to time. Surely I'm not crazy enough to tell you that punishment beyond the normal range—child abuse—has no lasting psychological effects on its victims?

Not quite that crazy. For one thing, abuse can damage the brain if the child is hit on the head or shaken violently. For another, there is such a thing as post-traumatic stress disorder. In extreme cases, prolonged abuse may even lead—at least according to some researchers—to multiple personality disorder, the Three Faces of Eve phenomenon.

But we are looking here at a very wide range of parental behaviors. For abuse not severe enough to produce any of the results I just enumerated, it is not clear to me that there are any psychological effects that children take with them when they leave home. There may be, but there is no conclusive evidence of it.

There are, of course, lots of studies. Abused children, according to the reports, have all sorts of problems. Aside from being more aggressive than kids who have not been abused (a well-established finding), they also have trouble making and keeping friends and trouble with their schoolwork. When they grow up they are more likely to abuse their own children. "The intergenerational transmission of child abuse," psychologists call it. They mean transmission via experience and learning—transmission by environmental means. They aren't talking about genes.

They hardly ever do; I don't know why. If you backed them into a corner, very few of them would deny that psychological characteristics are in part inherited, which means passed from parents to children. But somehow they are able to block this understanding from their minds when they do their research and write it up for publication. They are nowadays willing to admit that children's behavior affects the way their parents act toward them and that usually there is no way to distinguish a child's effect on a parent from the parent's effect on the child. But only the behavioral geneticists mention the possibility that some of the observed correlations between parental behavior and child behavior might be due to heredity. The others do not mention it at all, except to discount it. They discount it even though their research methods provide them with no way of eliminating it as a possibility.

Why would a parent abuse a child? One reason might be mental illness. Mental illnesses are in part inherited; they run in families. They run in families in which the members are biological relatives, not in adoptive families.

Probably only a minority of abusing parents are mentally ill. But it is likely that many more have personality characteristics that by now will sound familiar. People who are aggressive, impulsive, quick to anger, easily bored, insensitive to the feelings of others, and not very good at managing their own affairs are unlikely to be good at managing their children's. The unlucky offspring of such people are dealt a double whammy: a miserable home life and a genetic endowment that decreases their chances of success in the world outside their home.

Cinderella had a miserable home life but she didn't inherit any genes from the stepmother who abused her. The hidden message of the folktale is that you will turn out okay—you will triumph over adversity—if you were lucky enough to inherit good genes. *Oliver Twist* conveys the same message. The villain in the novel turns out to be Oliver's wicked half-brother, the son of a wicked mother. Oliver had a different mother—she was nice, like Oliver. Such stories are no longer politically correct; they don't seem fair. They aren't fair.

It isn't fair that in an abusing family, one child is often picked out to be the chief victim. If this child is removed from the abusing home and placed in foster care, sometimes he or she is victimized again. Certain characteristics, such as an unattractive face or a troublesome disposition, increase the risk that a child will become the victim of abuse. It is also possible that the victim may *lack* certain characteristics. The mystery is not why some children are abused: it is why most children are not. Children are so much trouble! They can be so infuriating! But most parents do not harm their children and most children do not suffer harm —even the children of people who themselves were abused in childhood. Evolution provided children with features and signals that defuse our anger—that make us feel protective toward them and, if they are ours, to love them. Some children, through no fault of their own, may lack these protective devices, or have them in a form that is too weak to do the job.

Even more unfair is the fact that children who are victimized at home also tend to be unpopular with their peers. There are children who are victims wherever they go. If they do not turn out well, should we blame it on their experiences at home or their experiences on the playground? Psychologists don't know and they don't ask; they just assume the home must be more important.

One researcher who has challenged that assumption is sociologist Anne-Marie Ambert, of York University in Canada. Ambert asked her students at York to write autobiographical accounts of their pre-college lives; to guide them she provided a list of questions. One of the questions was "What above all else made you unhappy?" She was surprised by the way her students responded. Only 9 percent described unfavorable treatment or attitudes on the part of their parents. But 37 percent described experiences in which they had been treated badly by their peers —experiences they felt had had lasting ill effects on them. Ambert came to the conclusion that "peer abuse," as she calls it, is a serious problem that has not received adequate attention.

> There is far more negative treatment by peers than by parents in these autobiographies. . . . This result, corroborated by other researchers, is startling considering the often single-minded focus of child-welfare professionals on parents while neglecting what is perhaps becoming the most salient source of psychological misery among youth—peer conflict and mistreatment. . . . In these autobiographies, one reads accounts of students who had been happy and well adjusted, but quite rapidly began deteriorating psychologically, sometimes to the point of becoming physically ill and incompetent in school, after experiences such as being rejected by peers,

excluded, talked about, racially discriminated against, laughed at, bullied, sexually harassed, taunted, chased, or beaten.

One last thing that may be involved in the unhappy lives of abused children is frequent changes of residence. Too many moves. Even when they remain with their parents, these children are moved from place to place more often than those in happier families. But in many cases they do not remain with their parents: when a child is diagnosed as abused, usually he or she is removed from the abusing home and put into a foster home. And, if that doesn't work out, into a second foster home and perhaps a third. It has been assumed that the harmful effects of foster care are due to the repeated loss of parents and parent substitutes, but repeated moves also deprive a child of a stable peer group. Even unfriendly peers may be better than nothing, because the lack of a stable peer group disrupts the child's socialization.

Babies undoubtedly need parents or parent substitutes. I believe that familiar caregivers are an aspect of the environment, like light and pattern, that a baby's brain requires in order to develop normally. But parents or parent substitutes may not be necessary for children over the age of five or six (see what I said in Chapter 8 about children raised in orphanages). For older children a stable peer group may be more important. The theory behind foster care is that kids need families. I think they need a stable peer group more than they need families. By trying to provide them with families—trying, in some cases, over and over again—well-meaning agencies rob them of peers.

Abused kids, as I said, have all sorts of problems. On the average they are more aggressive than other children, but that could be due to heredity: their abusive parents are aggressive, too. Their other problems could be due to peer abuse rather than to parental abuse, or to being moved from place to place too often. We just don't know. The right kinds of studies haven't been done yet (see Appendix 2).

The Kid Gets into Trouble and the Parent Gets the Blame

I see it in the news all the time; it always makes me angry. The Smith kid gets into trouble and the judge threatens to throw his parents in jail. The Jones kid burglarizes a house and his parents are fined for their failure to "exercise reasonable control" over his activities. The Williams kid gets pregnant and her parents are criticized for not keeping track of where she was and what she was doing. One set of parents, when they found it impossible to keep their teenage daughter out of trouble, chained her to the radiator. They were arrested for child abuse.

Blaming the parents is easy if you've never been in their shoes. Sometimes chaining the kid to the radiator is the only thing they haven't tried. The parents of reasonably well-behaved teenagers don't realize how crucially their ability to monitor their kid's activities depends upon the willing cooperation of the kid. An unwilling teenager cannot be monitored; my husband and I found that out. Kids can always outwit you if they really want to. If you try to enforce your rules by grounding them, they don't come home at all. If you stop giving them an allowance, they mooch off their friends or steal. The adolescents who can be monitored are the ones who are willing to be monitored, and they are the ones who need it least. Parents have remarkably little power to maintain control over the adolescents who need it most.

The adolescents who need it most are the ones who belong to a peer group their parents don't approve of. The parents don't want their kid associating with this group, but what can they do? These are the kid's friends and she will see them whether they want her to or not. All normal teenagers would rather spend time with their friends than with their parents; that's why parents impose curfews. The curfews are a tacit acknowledgment that the teenager would rather be somewhere else than at home. Parents tolerate this preference—and joke about it with their own friends—if they have no objections to their kid's friends. If they do have objections it is no joke.

Sometimes teenagers join delinquent peer groups because they live in a neighborhood where those attitudes and behaviors are normal. But even in nice middle-class neighborhoods like the one where I raised my kids, there are delinquent peer groups. Some kids join these groups because they have been rejected by the other groups; some join them out of choice. Kids identify with a group because they feel it consists of others "like me." The parents think the group is having a bad influence on their kid and they may be right, because whatever the members have in common tends to be exaggerated by their influence on each other and by contrast effects with other groups. But the influence is mutual and the kids had a lot in common to begin with.

Is it the parents' fault when a teenager becomes a member of a delinquent peer group? Socialization researchers who study styles of parenting maintain that parents who use an "authoritative" style—not too hard, not too soft, but just right—are less likely to have a teen who joins the wrong kind of peer group. Less likely to have a teen who gets into trouble. But this claim rests on data of doubtful validity.

The originator of the style-of-parenting research is developmental psychologist Diana Baumrind. Baumrind started off by studying preschoolers. She did one study in which she showed that kids with Just Right parents had fewer social and behavioral problems than kids with Too Hard or Too Soft parents. Her study had no controls for genetic influences, of course, and no way of distinguishing the effects of the child on the parents from the effects of the parents on the child, and the results were different for girls and boys (see what I said about "divide and conquer" in Chapter 2), but hardly anybody complained. Baumrind's work gets cited in every textbook of child development.

Nowadays, Baumrind's followers do not do research on preschoolers: they concentrate on adolescents. The advantage is that adolescents can fill out lengthy questionnaires. You can ask them how their parents treat them—whether their parents are too hard, too soft, or just right—and ask the same adolescents how many fights they've gotten into, how many joints they've smoked, and how they did on the algebra exam. The correlations these researchers are looking for are correlations between what the adolescents say about their parents and what they say about themselves.

There are still no controls for genetic influences, of course, and no way of distinguishing the kid's effects on the parents from the parents' effects on the kid, and the results are different for different ethnic groups. But now there is an additional source of confusion: the fact that the same teenagers are supplying both kinds of data. They are the source of the data on their parents and the source of the data on themselves. I noted a similar problem with Murray Straus's study of the effects of punishment: the same mother who told the researchers how often she spanked her kid was also telling them how the kid was behaving.

Whenever you ask the same people to answer two different kinds of questions, you are likely to find correlations between their replies to the first kind of question and their replies to the second kind. The correlations result from, or are inflated by, something statisticians call "shared method variance." People have response biases that bias their answers to all the questions you ask them. A happy person tends to check off upbeat answers to all the questions: Yes, my parents are good to me; yes, I'm doing fine. A person who cares about presenting a socially acceptable face to the world checks off socially acceptable responses: Yes, my parents are good to me; no, I haven't been in any fights or smoked anything illegal. A person who is angry or depressed checks off angry or depressed responses: My parents are jerks and I flunked the algebra test and to hell with your questionnaire.

What teenagers tell researchers about how their parents act toward them—whether their parents are too hard, too soft, or just right—agrees poorly with what the parents themselves say. A recent study that used multiple sources of information to find out what the parents were doing, instead of relying on what the kids said, failed to find a significant advantage of Just Right parenting, even though the researchers stacked the deck in their favor by eliminating in advance all the parents that didn't fit neatly into the types defined by Baumrind. They eliminated almost half the families they started out with!

But I am getting carried away; you are not interested in abstruse critiques of research methods. You want to know why I had so much trouble with my kid. You want to know what mistakes I made so you can be sure not to make them.

My kid turned out okay in the end; like most teenagers who cause their parents grief, she calmed down and wised up as she got older. She became an adult and quite a nice one. I've asked her what her father and I did wrong; I've asked her what we could have done differently. She doesn't know. She has a daughter of her own now and would like to know, but she doesn't. I note, however, that she has chosen to rear her own daughter in a neighborhood very like the one in which she was reared. A neighborhood that, when she was a teenager, she couldn't wait to get out of.

My husband and I didn't treat our two daughters alike: they *weren't* alike. It would have been impossible to use the same tactics on both of them and foolish to try. Of the mistakes made by the style-of-parenting researchers, the most serious is to assume that a parenting style is a characteristic of a parent. It is a characteristic of the relationship between a parent and a child. Both parties contribute to it.

Truth and Consequences

"Parents need to be informed about the possible consequences to their children of a decision to live apart," said sociologists Sara McLanahan and Gary Sandefur earlier in this chapter. If the parents do decide to live apart, and if their kid subsequently drops out of school or has a baby, McLanahan and Sandefur are prepared to blame the kid's problems on the parents' decision. McLanahan and Sandefur are making a mistake that is made too often in psychology and sociology—a mistake that is made all the time, despite the fact that students are repeatedly warned against it from the day of their first lecture in Intro Psych. The mistake is to confuse correlation with causation.

Good things tend to go together. So do bad things. These are correlations. Educational psychologist Howard Gardner would have us believe that there are several different "intelligences" and that someone who was stinted on one might have gotten a generous helping of another. But the fact is that people who score low on tests of one kind of intelligence are also likely to score low on tests of other kinds. We are pleased when we hear about a child who is mentally retarded in most respects but who is a whiz at drawing or calculating: it appeals to our sense of fairness. But such cases are uncommon. Far more commonly, nature is unfair to mentally retarded children by giving them no talents and making them physically clumsy as well. That is why they compete in the Special Olympics instead of the regular Olympics.

Good things tend to go together. People who score high on tests of one kind of intelligence are also likely to score high on tests of other kinds. The high score on one test doesn't *cause* the high test on the others but there is a correlation between them. No one knows for sure why they correlate.

"Everything is related to everything else," said a psychologist whose specialty was statistics. He told the story of a pair of researchers who collected data on 57,000 high school students in Minnesota. The researchers asked the kids about their leisure activities and educational plans, whether they liked school, and how many siblings they had. They asked about their fathers' occupation, their mothers' and fathers' education, and their families' attitudes toward college. There were fifteen items in all and 105 possible correlations between pairs of items.* All 105 yielded significant correlations, most at levels of significance that would be expected by chance less than .000001 of the time.

Everything is related to everything else but not in random fashion: good things tend to go together. People who eat healthy diets are also likely to get more exercise, to have physical checkups from time to time, and to live longer. Successful people tend to be taller than less successful people and to have higher IQs; if they get married they are more likely to

* Fifteen times fifteen is 225. But fifteen of those 225 are correlations between an item and itself, so those don't count, and half the remainder are correlations between the same two items in reverse order. A correlation between *A* and *B* is the same as a correlation between *B* and *A*.

stay married. Teachers and parents have higher expectations for kids who have done well in the past; these kids are also likely to do well in the future. Kids who do well in school are less likely to smoke and less likely to break laws. Kids who receive lots of hugs tend to have nicer dispositions than kids who receive lots of spankings.

Correlations come with no built-in arrows to distinguish causes from effects. If they did, some of the arrows would have points on both ends because the effects go in both directions, and some would have no points at all because the causes are something the researchers did not measure.

A study by psychologist Michael Resnick and a dozen of his colleagues, published in a September 1997 issue of *JAMA,* was entitled "Protecting Adolescents from Harm: Findings from the National Longitudinal Study on Adolescent Health." The researchers asked lots of adolescents lots of questions and found lots of correlations among the replies, but the headline that got into the newspapers was "Study Links Parental Bond to Teenage Well-Being." The researchers called it "parent-family connectedness"; they said it was "protective" against almost every kind of adolescent "health-risk behavior" they could name. What they meant was that adolescents who had more parent-family connectedness were less likely to smoke cigarettes, use illegal drugs, or have sex while still in junior high.

What the researchers actually found was that adolescents who *said* they got along well with their parents, and who *said* that their parents loved them and had high expectations for them, were less likely to *say* that they had smoked something or slept around. The researchers' conclusions were based entirely on the adolescents' answers to their questions—the same error made by the style-of-parenting researchers. *JAMA* would turn down a medical article if the physicians who tested a new drug knew which patients had received the drug and which got the placebo: administration of the drug has to be kept separate from the judgment of its effects. And yet the journal published a study in which the adolescents answering the questions were the only source of information about both the "protective factors" in their lives and their presumed effects.*

The nurture assumption is a powerful thing; it opens doors. According to *Time,* the *JAMA* study cost the federal government 25 million dollars. The *Time* essayist who reported this news, herself the mother of a teenager, was inclined to be skeptical:

* The parents were questioned too, but their responses were not included in the analyses reported in the *JAMA* article.

The opus, paid for by 18 federal agencies, probably got the attention it did because it offers so much comfort to parents whose little Mary doesn't make a move without calling her pal Molly, while treating Mom like a potted plant. "The power and the importance of parents continue to persist, even into late adolescence," says University of Minnesota professor Michael Resnick, the lead author of the survey. A reassuring finding: although your child may seem to ignore you, she is living off the remnants of the bond built during the years before getting her ears pierced was the most important thing in her life.

Perhaps she is. Despite my criticism of the researchers' methods, I have no doubt that some kids—and I'm not ruling out the essayist's little Mary—continue to get on reasonably well with their parents even after their biological clocks strike thirteen, and that these kids are less likely to do dumb things like use drugs or engage in risky sex. Perhaps what misled those eighteen federal agencies into thinking they were getting their 25 million dollars' worth was the positive way the researchers phrased their findings: *good* relationships with parents exert a *protective* effect. Expressed in a different (but equally accurate) way, the results sound less interesting: adolescents who don't get along well with their parents are more likely to use drugs or engage in risky sex. The results sound still less interesting expressed this way: adolescents who use drugs or engage in risky sex don't get along well with their parents.

What we've got here is one of those correlations where the imaginary arrow has no points at all because the cause is something the researchers didn't measure. The missing link is personality—the subjects' personality characteristics. People with certain types of personality are more likely to engage in risky behavior, and these same people are more likely to have problematic relationships, not just with their parents but with everyone.

A study done in New Zealand supplies the missing link. It was carried out by Avshalom Caspi and his colleagues and published in a psychology journal a couple of months after the *JAMA* study appeared. *Time* took no notice of it.

The New Zealand researchers gave personality tests to about a thousand young people and found that certain traits were good predictors of risky behavior. Eighteen-year-olds who are impulsive and quick to anger, who aren't afraid of danger and who seek excitement, are more likely to drink too much, drive too fast, and engage in risky sex. These same young people also tend to have difficulty establishing and maintaining close personal relationships.

As the researchers pointed out, these disadvantageous personality traits are heritable to the same degree as benign ones: genetic influences account for about 50 percent of the variations among individuals. And the traits show up early: the researchers were able to see signs of them when their subjects were only three years old. That's right, they had data on these same subjects—ratings of their behavior by trained examiners—made when they were three years old. The three-year-olds who were more impulsive and quicker to anger than others of their age, and who had more trouble staying focused on a task, tended to remain that way, and those individuals tended to engage in more "health-risk behavior" when they got older.

Admittedly, this sounds more discouraging than the results of the *JAMA* study. But in order to find a solution to a problem, we must first understand what's going on. Biology is not destiny; the fact that heredity plays a role in determining people's characteristics doesn't mean they can't be changed. We've just got to figure out how to do it. If we haven't done so yet, it may be because psychology's faith in the nurture assumption has gotten in the way.

Why Pop Psychology Blames Mom and Pop

On the shelves of my local library are many books by people like John Bradshaw, who writes about "dysfunctional families," and Susan Forward, who writes about "toxic parents." When I want a book that takes a more scientific approach, like McLanahan and Sandefur's *Growing Up with a Single Parent,* I have to fill out a request form and the librarian obtains it for me from a university library. I suspect, therefore, that I've spent far too much time railing against the McLanahans and Sandefurs and not nearly enough denouncing the Bradshaws and Forwards. Though I'm not planning to even up that imbalance—I don't have the stomach for it, to tell the truth—I do need to say something about the books that fill up my library's shelves. Why are clinicians like Forward and Bradshaw so certain that their clients' problems are the fault of their clients' parents, and why do I believe they're wrong?

I've mentioned many times the behavioral genetic finding that children reared in the same home by the same parents don't turn out alike. This is not a problem for the Bradshaws and Forwards of the world because they don't expect children to turn out alike. They *expect* dysfunctional parents to exert their toxic effects on each child individually, because each child is cast in a different role or is born at a different time or resembles a different grandparent. The Bradshaws and the Forwards are not going to lose any sleep over the behavioral genetic data. For that matter, they're not likely to lose sleep over any kind of data; their theories are stretchy enough to swallow anything I can throw at them. Theories that are not based on scientific methods or results are difficult to overturn with scientific arguments.

What I can do, however, is to show you why they came to the conclusions they did and how it's possible to look at the same things and see them in a different light. It is not their observations I doubt: it is the way they interpreted them.

Typically, a patient comes into the psychotherapist's office and complains that she (it's more often a woman) is miserable. She talks to the therapist for a while and he decides it's all the fault of the patient's parents. They belittled her or smothered her or didn't give her enough autonomy or made her feel guilty or sexually abused her. The therapist convinces the patient that whatever is wrong with her is not her fault, it's the fault of her parents, and after a while she says, "Thank you very much, Doctor, I feel much better now."

The question that interests me is not the one about why the patient got better or if she really did get better; I'll leave that to other writers. The question for me is: Why is the therapist so convinced it's the parents' fault? What does he see that makes him so sure?

He sees that dysfunctional patients have dysfunctional parents. He sees that parents treat their children differently, casting them in different family roles. The overburdened child or the family scapegoat or the baby of the family whose parents won't let go—they all end up in his waiting room. He sees that people who are unhappy had unhappy childhoods.

Of course, he doesn't see these things directly: mostly he sees them from the point of view of the patient. What he knows is what his patient tells him. However, sometimes he interviews her parents too, and they usually come across as even worse than the patient depicted them. He also sees how the patient acts when her parents are present. She tends to become a younger, sicker version of herself. The therapist concludes that the patient's problems are the result of how her parents treated her while she was growing up.

What alternative explanations has he failed to consider? What mistakes might he be making? I have thought of nine.

First is the possibility that dysfunctional parents pass on their dysfunctional traits genetically. Psychotherapists don't like this idea, perhaps because they think it means their patients' problems are incurable. Not at all. Many things caused by biology can be fixed; many things caused by the environment cannot be. And what if our destinies *were* written in our genes? If it were true—which it is not—what purpose would it serve to deny it?

Second is the possibility that the patient was cast in a particular family role because that's the role that suited her: it was typecasting. The parents may have been responding to characteristics she already had, rather than causing her to have them.

Third is the possibility that other people—people outside the family—responded to her in the same way. If she had characteristics that made her the family scapegoat, then maybe she was scapegoated on the playground too. And maybe the experiences on the playground were responsible for her current problems.

Fourth, maybe her parents did have problems that had an impact on her life, but the impact could have been on her social environment outside the home. If her father was an alcoholic, maybe he was unable to hold a job and they lived in poverty. If her parents got divorced, maybe she was moved from place to place too often.

The fifth has to do with the way she acts when her parents are present. People, regardless of their age, do behave differently in the presence of their parents. A mistake made by psychologists of every stripe is to assume that the way people behave with their parents is somehow more meaningful, more important, more lasting than the way they behave in other contexts. It is not. The evidence I've presented in this book shows that if anything, the way people behave with their parents is *less* important, *less* lasting, than the way they behave in contexts that are not associated with the parents. Children bring their outside-the-home behaviors home; they do not ordinarily bring their home behaviors out with them. What we're seeing, when the patient's parents are present, is her home personality, which does indeed reflect the way she was treated at home but does not have the importance the therapist attributes to it.

The sixth has to do with the way the *parents* act in his office. Before you judge these people, I'd like you to walk a block or two in their moccasins. They are the defendants in a jury-rigged trial. Only there is no jury and no counsel for the defense, either—just a prosecuting attorney, and he's on the patient's side. The parents are being tried for the crime of producing a dysfunctional child. They were convicted before they walked in the door and they know it. How do you expect them to behave?

The seventh asks the question: Who is the witness against the parents?

The answer: their dysfunctional child. Her presence here in the therapist's office signifies that she's unhappy. And, as you would expect, she remembers her childhood as unhappy. But her unhappy childhood may not be what is making her unhappy—it may be the other way around. Her current unhappiness may be causing her to *remember* her childhood as unhappy. Memory is not the accurate recording device we like to think it is. Depending on how we feel when we are doing the remembering, we pull out happy memories or sad ones from the storeroom, or neutral ones that we color to fit our mood. Depressed people are more likely to remember that their parents weren't good to them. When they stop being depressed, their memories of their parents improve. The childhood memories of identical twins are surprisingly similar, even those who were reared in different homes. They come up with similar memories partly because they are likely to be similarly happy or unhappy as adults. Yes, there are genetic influences on happiness, too.

Eighth is the fact that things that cause us distress or pleasure do not necessarily have the power to change our personalities or to make us mentally ill. Relationships mean a lot to us; parents are, without a doubt, important people in our lives. We care what they think of us. But that doesn't make us putty in their hands. The fact that the patient feels strong emotions when she thinks about her parents is not evidence that they are responsible for whatever's wrong with her. If you deprived her of food she might feel just as strongly about cheeseburgers, but no one thinks her hunger is the cheeseburgers' fault.

That brings us to the ninth and last thing the therapist overlooked: the pervasive influence of the nurture assumption. Both the therapist and the patient are participants in a culture that has, as one of its cherished myths, the belief that parents have the power to turn their children into happy and successful adults or to mess up their lives very badly. The belief that if anything goes wrong, it must be the parents' fault.

It is a harmless myth of our culture that children are born innocent and good, blank slates for their parents to write upon. The other side of the myth—that if children do not turn out as we hoped it must be their parents' fault—is not so harmless. We exonerate the child only by putting the burden of blame on the parents.

Clinical psychologists are quite sure that children can be, and often are, permanently messed up by the mistakes their parents made in rearing them. The physician who wrote the *JAMA* editorial was just as sure that Mrs. McElhinney made her son Carl into a murderer by reading too many murder mysteries before he was born.

14 WHAT PARENTS CAN DO

If you thought you would find the title of this chapter centered above an empty page, you have either overestimated my sense of humor or underestimated my chutzpah. It does take a lot of nerve to put those four words at the top of the page, after what I've said about advice-givers in the previous thirteen chapters. But it wouldn't be fair—and it wouldn't be accurate—to leave you with the impression that parents are wallpaper.

On the other hand, I don't want to raise false hopes. So let me begin with a true story my colleague David Lykken tells about a pair of reared-apart twins—one of the pairs studied at the University of Minnesota by the research team of which he is a member.

They are identical twins separated in infancy; they grew up in different adoptive homes. One became a concert pianist, talented enough to have performed as a soloist with the Minnesota Orchestra. The other cannot play a note.

Since these women have the same genes, the disparity must be due to a difference in their environments. Sure enough, one of the adoptive mothers was a music teacher who gave piano lessons in her home. The parents who adopted the other twin were not musical at all.

Only it was the nonmusical parents who produced the concert pianist and the piano teacher whose daughter cannot play a note.

What Children Learn at Home

David Lykken, who began his career as a clinical psychologist and who has made significant contributions in several diverse areas of psychology, has retained his faith in the power of parents to shape their children's lives. He explains the paradox of the mismatched twins this way:

> The piano-teacher mother offered lessons but did not insist, whereas the other adoptive mother, not musical herself, was determined that her daughter would have piano lessons and determined also that she would make the most of them; she shaped her daughter's early environment with a firm, consistent hand.

The nonmusical mother insisted that her daughter take lessons and made sure she practiced. Of course, the child must have had some innate talent to begin with—not everyone with a determined mother becomes a concert pianist. But without the determined mother the child's talent would have gone to waste. The twin with the wishy-washy mother cannot play a note.

I give you, as a counterexample, a daughter of my own. My older daughter never played with the Minnesota Orchestra but she was proficient enough to be the accompanist for her high school chorus and to perform in public many times. Like the wishy-washy mother, I offered my daughter piano lessons (from a teacher in our community) but did not insist. Unlike the determined mother, I never made her practice: she did it on her own. My daughter is certain that if I had bugged her about practicing it never would have worked—she would have quit. Recently I asked her what gave her the motivation to keep at it. She replied, "I enjoyed playing and wanted to play better, and I got better when I practiced." Virtuosity is its own reward.

Although I did not force my daughter to take piano lessons or to practice, or even urge her to do so, I did provide her with a mildly musical home. I sang in a chorus during most of her childhood and we sometimes had rehearsals at my house. Today my daughter plays the piano mainly to accompany herself; in her spare time she studies voice and sings in a chorus.

Yes, in some ways parents do have an influence. The case of the nonmusical twin is an exception to which I will shortly return. More often than not, musical parents have musical kids. The sons and daughters of physicians often become physicians. It would be foolish to deny that parents can influence their children's choice of a profession or their leisure-time activities. I do not deny it.

Parents influence the way their children behave at home. They also supply knowledge and training that their children can take with them when they go out the door and that may prove useful out there. A child who learned to speak English at home does not have to learn it all over again in order to converse with her peers—assuming, of course, that her peers speak English. The same is true for other behaviors, skills, and knowledge. Children bring to the peer group much of what they learned at home, and if it agrees with what the other kids learned at home they are likely to retain it.

Children also learn things at home that they do *not* bring to the peer group, and these may be retained even if they are different from what their peers learned. Some things just don't come up in the context of the peer group. This is true nowadays of religion. Unless they attend a religious school, practicing a religion is something children don't do with their peers: they do it with their parents. That is why parents still have some power to give their kids their religion. Parents have some power to impart any aspect of their culture that involves things done in the home; cooking is a good example. Anything learned at home and kept at home—not scrutinized by the peer group—may be passed on from parents to their kids. Maybe even how to run a home. The games of House that children play in nursery school give them the basic outlines of how family life is normally arranged in their community, but a lot of the details are left out.

Furthermore, some things that are learned at home may be retained even if they *are* brought to the peer group—even if they are different—because groups demand conformity only up to a point. There are behaviors that are obligatory and those that are optional, and which is which depends upon the group you're in. Language is one of the obligatory ones in any children's group: a kid who comes to the group with a different language or accent is expected to change, and does change. In boys' groups during middle childhood, it is obligatory to behave in a "masculine" manner: tough, unemotional, concerned about status. Girls' groups grant more leeway to deviate from the "feminine" pattern of behavior. This difference in how strictly the patterns are enforced may reflect a sex difference: groupness appears to be stronger in males (see Chapter 10).

What is obligatory can also vary over time. Patriotism is obligatory to the members of a group during times of war but may be optional in peacetime. As a result of changes in the adult culture, it is possible that boys' groups will become more permissive about the range of behaviors they allow their members. So far, however, developmentalists have seen no signs of such a shift.

If knowledge, skills, or opinions acquired at home are in an area that the peer group regards as optional—an area where conformity is not enforced, where differences may even be appreciated—the child may retain them. Most children's peer groups permit their members to vary in their talents, hobbies, political preferences, and future career plans. The kid who knows how to play the piano is not a nail that needs hammering down.

Children learn how to play the piano at home. They learn what it's like to be a doctor or why it's best to be a Democrat or how to wrap the corn husks around the tamales. What they don't learn at home is how to behave in public and what sort of people they are. These are things they learn in the peer group.

Can the Family Be a Group?

Near the end of Chapter 7, I talked about the reasons why families do not usually function as groups. In the privacy of the modern North American or European home, I said, the family is not a salient social category because it's the only one around. There are no competing groups to bring out the family's groupness and so it falls apart into a bunch of individuals—each with his own agenda, her own patch of turf to defend. Self-categorizations are on the *me* end of the continuum; *us* seldom makes an appearance in the home.

It may be different in Asian cultures, where people seem to identify more closely with their families and there is less emphasis on individual achievement and autonomy. In precolonial China, if a man committed a serious crime his whole family—his parents and children, brothers and sisters—were executed along with him. The idea was that the whole family shared in the responsibility. Perhaps Asian children do categorize themselves as "a Wang" or "a Nakamura" even when they're at home. Perhaps Asian families can assimilate as well as differentiate.

Under the right conditions it can happen in Western families too. Watch the members of an American family when they're traveling together in an unfamiliar place, a place where there are other people but the kids don't have to worry about being spotted by their classmates. Out of its familiar territory the family draws together and becomes a group. The little rivalries between siblings evaporate like puddles on the sidewalk in Tucson. But the respite is temporary. As soon as the parents and children get back into their car and are alone together again, groupness wanes and rivalry waxes. They revert to being a bunch of individuals—each with his own agenda, her own patch of turf to defend. Mom, he's putting his foot on my side!

Where groupness is weak or absent, differentiation triumphs over assimilation. The members of a family diversify, each seeking something to specialize in, a personal niche to fill. This niche-picking widens the family's assortment of skills and reduces head-to-head competition between siblings. But parents, too, occupy family niches and can, from the child's point of view, fill them up. Perhaps that is why the twin with the piano-teacher mother never learned to play: her family already *had* a piano player. The daughter would have had to compete with her mother if she took up the same instrument. What a pity her parents didn't encourage her to take up the tuba! My daughter had no competition at all in her family: neither of her parents could play the piano and her little sister was too young.

Family niche-picking can have lasting effects when it involves cultivating different talents or interests. The piano-playing twin found herself a career for life; her twin sister, even if she tries to make up for missed opportunities by taking lessons as an adult, can never hope to be more than a competent amateur. Choices made in childhood—made *at home* —about careers or politics or religion can have repercussions that echo through a lifetime. They may be brought to the peer group but they are not modified by the peer group because the kids don't notice or don't care.

When it comes to personality and social behavior, however, it's a different story. The evidence indicates that within-the-family niche-picking or typecasting leaves no permanent marks on the personality. One of the ways that children get typecast is by birth order: the oldest child is seen by his parents as more responsible, sensitive, and dependent than his younger siblings; he's seen by his siblings as bossy. But consistent differences that depend on birth order generally do not show up in personality tests given to adults. Nor do researchers find consistent differences in personality between only children and children with siblings. (See Chapters 3 and 4 and Appendix 1.)

Can a Parent Be a Leader?

Leaders, as I said in Chapter 11, can influence the behavioral norms of a group. They can define the group stereotype that the members have of themselves and the boundaries of the group—who is *us* and who is *them.* Can a parent ever be a leader of this sort? Can he or she form the family into a cohesive group and delineate its goals?

Yes. But it happens rarely in Western societies, perhaps because Western families tend to be small and it may require a family group of a minimum size. The other requirement is a strong and determined parent.

One such family that springs to mind is the Kennedys. But I would rather tell you about a different family, one you probably never heard of. They flourished in Long Branch, New Jersey, not far from where I live. The parents, both dead now, were Donald Thornton, who worked all his life as a laborer, and his wife Tass, who before her marriage was a chambermaid in a hotel. Both were African Americans from poor families. Donald dropped out of high school at fourteen; Tass briefly attended a teacher's college in the South.

Donald and Tass had five daughters, close together in age. Later they took in a foster child, a girl close in age to the others. According to Yvonne, the third oldest, there was no reason to expect anything unusual from these six children.

> As little girls, there had been nothing special about us, nothing to set us apart from the other black children in Long Branch, New Jersey. By any ordinary expectation, we should have grown up to graduate from high school and get factory or clerking jobs, that is, if we had been lucky enough to avoid getting pregnant and becoming high school dropouts, perhaps single mothers living on welfare and having an illegitimate child every other year or so.

Except that Donald Thornton had other ideas. He was determined that all his daughters were going to be women of accomplishment and he dedicated his life to that goal. As Yvonne tells it in her book *The Ditchdigger's Daughters,* here's how it began:

> The idea had not come out of pride and ambition; it started as a joke. Daddy dug ditches at Fort Monmouth, New Jersey, and when Mommy gave birth to a fourth and then a fifth girl, his fellow ditchdiggers kidded him about having nothing but female offspring. What kind of man is it, they teased, who can't produce even one son for himself? . . . "You won't be laughin'," he predicted, "when my little girls grow up to be doctors."

Many parents make similar boasts but few have Donald Thornton's single-mindedness and strength of character. Somehow he formed his daughters into a group. He gave them an image of themselves: You're better than the other kids in the neighborhood. You may not be any smarter but you work harder. He gave them a goal: You're going to be doctors. And he defined the boundaries of the group:

> "I don't want no one dilutin' the message," he told Mommy, who had more of a sense of us as children and would have let us go to the playground to rollerskate and play ball. Daddy would have none of it. "There's five of them," he argued. "They can play with each other. What do they need to go outside the family for? . . . If we stick together . . . there's nothin' this family can't do."

Like Jaime Escalante, one of the teachers who appeared in Chapter 11, Donald Thornton made his kids feel that they were "a brave corps on a secret, impossible mission." He was aided by the fact that the Thornton girls not only were bright and diligent like their parents but were musical as well, like their mother. When the girls weren't studying they were practicing; they had no time to associate with other kids or to get into trouble. The Thornton Sisters became a successful band that played at the Apollo Theater and on college campuses up and down the East Coast. The band earned enough to cover their educational expenses.

Donald didn't turn all his daughters into doctors but his fellow ditch-diggers stopped laughing long ago. Two daughters became physicians (one has a Ph.D. as well as an M.D.). The others are an oral surgeon, a lawyer, and a court stenographer. The foster daughter is a nurse. As Yvonne put it, she and her sisters are "women of accomplishment, independent women, women capable of taking care of themselves."

It doesn't happen often, but sometimes a family can be a group. Sometimes a parent can be its leader.

And sometimes parents can lead their children astray. I know of another New Jersey family in which the parents didn't want their children to play with the other kids in the neighborhood and insisted that they do nothing but schoolwork and practice. In this case the parents were well off and well educated. There were only three kids—two boys and a girl—and perhaps that made the difference. Perhaps you need a minimum number of the same sex to create a sense of groupness. The family settled in an out-of-the-way place; the children went to school but they were discouraged from having friends outside the family. The girl was so unhappy at home that she asked to go to boarding school—the only child I've ever heard of to make this request (which was granted). The middle child was very bright and eventually graduated from a topnotch university, but he was socially inept and got into trouble with the law over a hacker stunt that went wrong. The youngest child dropped out of college and went to work for a tree-trimming service.

Another kind of parent-leader is the one who devotes his life to making his kid into a prodigy. The father of the golf player Tiger Woods and the mother of the actress Brooke Shields are two examples; others can be found in the cheering sections of many outstanding gymnasts, figure skaters, and chess players. Such parents are awarded, in the popular press, part of the credit for their kid's success and most of the blame for the kid's hangups, and to some extent they deserve both. But you can't make just any kid into a star: these parents had to have the raw material to start with. Where did they get it? They bred it. They produced an offspring with half their genes. Tiger Woods and his dad both have the same single-mindedness I described in Donald Thornton—the same ability to focus on a goal and to work with great persistence to achieve it. Heredity, which plays a role in all personality characteristics, must play a role in this one.

The prodigy is an interesting case; many of these kids seem to come with their own built-in motivation. If it isn't there to begin with, I doubt a parent could provide it. In fact, often it is the child who is the prime mover and the parent who becomes the servant of the child's consuming interest. Intellectually gifted children receive certain things from their parents that less gifted children do not get—books, computers, trips to the museum—but they get them because they demand them. It is not the parents who are pushing: it is the child.

The danger in raising a prodigy is that many of these kids lack a peer group—they miss out on normal relationships with other kids their age. Children who do not have normal peer relationships are at risk of turning out peculiar. Though garden-variety gifted children generally fare very well, the true prodigies—the ones who are off the chart—have more than their share of psychological problems. Sometimes there is not much a parent can do: some kids are so intellectually advanced that they have nothing in common with their agemates. Some kids really don't want to do anything except practice golf or gymnastics or chess. But if parents were more aware of the importance of peers, perhaps they would try harder to see that their kid had some.

The Power of Parents to Choose Their Child's Peers

It is the one power that nearly all parents have—the one way that they can determine the course of their child's life. At least in the early years, they can determine who their child's peers are. When Joseph's parents plucked him out of his school in Poland and plunked him down in Missouri, they not only changed his childhood: they set him on a new road with a different destination. Joseph is an American now, with all the pluses and minuses that go along with it. He is no longer Polish, not even in his dreams. Though it wasn't his parents who taught him to be an American, he has them to thank or blame: by moving him to this country they gave him American peers.

You don't have to do anything quite that drastic to have an effect on your child's life. Just by moving to a different neighborhood, just by choosing your child's school, you can change the course of a life. It's a little scary, isn't it? Especially since it is so hard to predict what effect your decision will have. On the whole, children learn more in schools that contain a higher proportion of smart kids; on the whole, children are less likely to get into trouble in neighborhoods where delinquency rates are low. But a kid with below-average intelligence might be rejected by his peers in a school where everyone else is above average. A kid from a poor home might be shunned in a place where everyone else is well off.

Not that being rejected by one's peers is the end of the world. It hurts like hell while it's happening and it does leave permanent scars, but it doesn't keep a kid from being socialized (you can identify even with a group that rejects you), and I've noticed that many interesting people went through a period of rejection during childhood. Or got moved around a lot, which has similar effects. I was moved around a lot as a child and went through four years of rejection, and there is no doubt that I would have been a different person if it hadn't happened. A more sociable person, but more superficial. Certainly not a writer of books—a job that has as its first requirement the willingness to spend a good deal of time alone. The biologist and author E. O. Wilson recalls his childhood this way:

> I was an only child whose family moved around quite a bit in southern Alabama and northwestern Florida. I attended 14 different schools in 11 years. So it was, perhaps, inevitable that I grew up as something of a solitary and found nature my most reliable companion. In the beginning, nature provided adventure; later, it was the source of much deeper emotional and aesthetic pleasure.

If it were up to me, I would take the risk that my child might be rejected and put her or him into the best school I could find—a school with smart, hard-working kids. A school where no one makes fun of the one who reads books and makes A's. Such schools do exist. There is an overcrowded old school in Brooklyn, New York, called Midwood High. Half of its four thousand students are from the neighborhood, half have won admission on the basis of their test scores and grades in junior high. This is a "magnet school"—kids compete with each other to get into it. According to the *New York Times,*

> Once in the school, the 2,000 magnet students mix with another 2,000 students from the surrounding neighborhood in Flatbush, sharing many of the same classes. The high expectations are contagious, said Midwood's principal, Lewis Frohlich. More than 70 percent of the students earn Regents diplomas, compared with nearly 25 percent citywide, the dropout rate is less than 2 percent, and about 99 percent of those who graduate go on to college.

The principal is right: attitudes are contagious, if a group contains enough carriers of the contagion and if it remains intact and doesn't split up into subgroups. The magnet students—the ones who competed to get in—are not the only ones who do well at Midwood High. Almost everyone does well. The *Times* reporter interviewed some of the students—semifinalists in the Westinghouse Science Talent Search—and asked them if their classmates gave them a hard time about being "nerdy science nuts." The question startled them, she said. "At Midwood, being a science nut, apparently, is a good way to make friends, and being ambitious is far from shameful." Many of the students in this school are the children of immigrants. They bring with them to the peer group their parents' belief in the power of education and they do not lose it, perhaps because so many of their peers have parents with the same belief. The kids at Midwood don't divide up into two opposing groups, pro-school and anti-school. Schools like this should be studied carefully to find out why they work so well. I can't give you the answer.

The contagiousness of attitudes has its down side: bad attitudes are as catching as good ones. Many parents fear that their kid has gotten in with a "bad crowd" and that these companions are having an unwholesome influence on her. Often they are right, though in all likelihood their kid is as much an influencer as an influencee. Whichever way the wind is blowing, kids with delinquent tendencies do get into more trouble in the company of other kids with the same tendencies. Your kid would probably be better off without those friends.

Unfortunately, your power to influence your children's friendships shrinks as they get larger. With little children, parents have almost complete control over who their friends are, at least when they are not in school. But once they turn ten, all bets are off. If you forbid an older kid to see her friends, and if she is the kind of kid who is attracted to the kind of friends you don't want her to see, there's a good chance she will see them behind your back and lie about it. And lying quickly becomes a habit, if it isn't already.

Your options are limited. I don't recommend chaining her to the radiator, though I understand why you might be tempted. You could switch her to a different school or you could move. Neither is a perfect solution. If she is the kind of kid who is attracted to the kind of friends you don't want her to see, switching schools or neighborhoods might not help: she might seek out new friends just as bad as the old ones.

But sometimes a change of venue can work wonders. I once had an interesting conversation on a WordPerfect help line with a woman I'll call Marion. Marion lived in Provo, Utah; she had eleven children, ranging in age from teens to early thirties. When she heard that I was a writer of textbooks on child development (which I was at that time), she told me the story of her next-to-youngest child. All the others were doing fine but this one kid got in with some bad companions, she said, and started talking about dropping out of high school. "I got him out of there so fast his head was spinning," Marion told me. She sent him to live with his oldest sister in a small town in a remote corner of the state. Draconian, but it worked. The boy graduated from high school and was making plans to go to college.

There is one other circumstance in which it might be worth considering a move: if your kid is getting picked on all the time. If my kid were at the very bottom of the local totem pole, and if all the higher-ups were beating on him, I would want to get him out of there. Victims are victimized partly because they get a reputation for being victimizable, and it is extremely difficult to change the peer group's mind about things like that. Usually, moving is a disadvantage for a kid because he loses his peer group and whatever status he had in it. But if the peer group is making his life miserable and his status is zero, he doesn't have much to lose.

The final drastic solution is home schooling. This wouldn't work for teenagers and is risky for younger kids, unless you have several close together in age or can assemble a viable group of the offspring of your friends or neighbors. Although you are protecting your children from the malign influence of the kids in the school they would otherwise attend, you may end up producing misfits, poorly suited for the world in which they will eventually have to live.

Self-Esteem and Status

According to the advice-givers, self-esteem is the most valuable thing a parent can give a child. "Parents play the largest and most important role in shaping their children's sense of themselves," declares science writer Jane Brody in the pages of the *New York Times.* If the parents do a good job of shaping, the child will end up with an adequate supply of self-esteem. If they do not, the child has a one-way ticket to failure. "Lack of self-esteem leads so many youth to fall by the wayside," moans physician Liana Clark in an essay in *JAMA.* "Girls have sex and become mothers. Boys turn to drugs and guns. All of these tragedies occur because they do not believe in their abilities."

These writers may be putting the cart before the horse—mistaking an effect for a cause. According to psychologist Robyn Dawes, trying to elevate people's self-esteem is futile because this strategy "ignores the simple principle that much of our feeling results from what we do rather than causing us to do it." There is no solid evidence, says Dawes, that low self-esteem is "an important causal variable in behavior." The approach promoted by the feel-good gurus could even have a negative effect: "What these beliefs do is discourage people from attempting to craft a decent life for themselves and instead encourage them to do whatever is necessary to *feel* good—about themselves."

Feeling *very* good about yourself may, in fact, be counterproductive. The problem is that people with high self-esteem tend to think they are invulnerable. There is a theory that violence is caused by low self-esteem but a recent review concludes the opposite: "Violence appears to be most commonly a result of threatened egotism—that is, highly favorable views of self that are disputed by some person or circumstance." The reviewers point out that violence is a risky business and thus is more likely to appeal to people who have no doubts about their physical prowess, cleverness, and good luck. There is also evidence that people with high self-esteem are more apt to drive under the influence of alcohol or exceed the speed limit. A study of college women found that those with high self-esteem underestimated their chances of getting pregnant: they considered unprotected sex to be less risky than did those with lower self-esteem. These are women who do not want to get pregnant but their self-esteem makes them think "It can't happen to me."

I have to admit, though, that having low self-esteem isn't so hot either. This is the problem of many of the people who end up in the offices of psychiatrists and clinical psychologists: they are the "internalizers"—the ones who berate themselves instead of going out and shooting someone. The goal of traditional psychotherapy is to get them to stop blaming themselves and start blaming their parents, and occasionally it works. Because these patients are inclined to be depressed—the low self-esteem is as likely to be a symptom of depression as the cause of it—they tend to dredge up unhappy memories of childhood. It is pretty easy to convince them that their problems are all the fault of Mom & Dad.

According to the advice-givers, you can arm your children against a hostile world by making them feel good about themselves. I don't believe it. You cannot coat a child with honey and expect it to protect her against all the vinegar in the world. Like other aspects of personality, self-esteem is tied to the social context in which it is acquired. A child can feel good about herself at home and bad about herself elsewhere, or—like Cinderella in Chapter 4—vice versa. Parents can make a child think she's something special by favoring her over her siblings, but the boost to her ego doesn't travel well. Researchers found no tendency for college students who believed they were their parents' favorite child to have higher self-esteem in general. They had higher self-esteem only in one area of their lives: the area the researchers called "home–parent relationships."

Self-esteem in general—the kind that travels well—is a function of one's status in one's group. School-age children are aware of how they compare with their classmates and how they are regarded by them. Low status in the peer group, if it continues for long, leaves permanent marks on the personality. And it can sure wreck a kid's childhood.

Status in the peer group is a chancy thing. Groups typecast their members, often for trivial reasons—random events, superficial differences. The kid who wets his pants on the first day of first grade, the kid who uses a three-syllable word on the first day of first grade, may be pinned with labels that they will wear for years, maybe even forever. I know a middle-aged woman who is still called "Pudgy" by her old friends, though she lost her puppy fat somewhere around third grade.

Parents cannot prevent their children from being typecast in negative ways in the peer group. However, they can make it a little less likely to happen. They do have some control over the way their children look, and their goal should be to make them look as normal and as attractive as possible, because looks do count. "Normal" means dressing the child in the same kind of clothing the other kids are wearing. "Attractive" means things like dermatologists for the kid with bad skin and orthodontists for the one whose teeth came in crooked. And, if you can afford it or your health insurance will cover it, plastic surgery for any serious sort of facial anomaly.

Children don't want to be different, and for good reason: oddness is not considered a virtue in the peer group. Even giving a kid a weird or silly name can put him at a disadvantage. I heard of a father who thought it clever to name his son after his favorite poet. Unfortunately, his favorite poet was Homer.

Parent–Child Relationships

People sometimes ask me, "So you mean it doesn't matter how I treat my child?" They never ask, "So you mean it doesn't matter how I treat my husband or wife?" and yet the situation is similar. I don't expect that the way I act toward my husband is going to determine what kind of person he will be ten or twenty years from now. I do expect, however, that it will affect how happy he is to live with me and whether we will still be good friends in ten or twenty years.

You can learn things from the person you're married to. Marriage can change your opinions and influence your choice of a career or a religion. But it doesn't change your personality, except in temporary, context-dependent ways. A man might be tender with his wife and tough with his employees, or vice versa. A woman married to a man who constantly belittled her might look sad or angry whenever she was near him. If she stuck with him despite the belittling and wore a hang-dog expression even when he wasn't around, you couldn't be sure—could you?—whether her personality problems were the cause of her unhappy situation (the reason why she married this jerk and why she doesn't leave him) or an effect (the result of all the belittling). In fact, you might blame her depression and passivity on her mother, who got her used to being belittled when she was a child. You would be wrong, but you would be admitting that she had these problems before she married the jerk.

The researchers who study babies' attachment to their mothers like to speak of "working models": they believe that a baby's mind has a working model of his relationship with his mother and it tells him what to expect from her. Okay, I'll buy that. Only the attachment researchers wind up the model and think it will keep going forever: they think it tells the baby what to expect from *other* people too. If the baby expected the world to come running when he cries because his mother does, he would suffer no end of disappointments. But he doesn't expect that. He doesn't expect the mobile with red doodads to work the same as the mobile with blue doodads, so why should he expect the babysitter to work the same as Mommy?

I think the relationship department of the mind contains working models of all the important relationships in our lives. Only for unimportant relationships might we generalize—act the same way to all the people in the category *peers* or the category *employees*—and only as a default. As soon as we get to know someone better, we give them a working model of their own. A child does not act in the same way to his mother and his teacher, or to his brother and his friend. He does not act in the same way, once he gets to know them, to Jonathan, who is nice, and Brian, who's a bully.

A parent, too, can be a bully, and children are quick to learn that. It doesn't make them expect everyone to be that way, but it does mess up their relationship with the parent. If the bullying goes on long enough, their relationship might be messed up forever. If you don't think the moral imperative is a good enough reason to be nice to your kid, try this one: be nice to your kid when he's young so that he will be nice to you when you're old.

Children are keenly aware, not only of how their parents treat them, but of how they are treated relative to their brothers and sisters. If they feel that their siblings are getting a better deal than they are, the resulting resentments can poison their relationships with their parents (and with their siblings), sometimes for a lifetime. A researcher studied the adult relationships of people in Sweden who as children had considered themselves to be the least favored among their siblings—the one their parents loved the least or punished the most. She found that these people were less likely than other Swedes to have close and warm relationships with their aging parents.

I mention this study hesitantly, because there is a cause-or-effect problem here. Perhaps the parents had some reason for not liking this offspring: perhaps he or she was difficult as a child and continues to be difficult as an adult. It is possible. But it makes sense to me that people will feel closer in adulthood to parents who treated them well when they were little. I was not my parents' favorite child: they liked my brother better. My brother remained in the same town with our parents and watched over them in their declining years, while I lived on the other side of the continent and visited occasionally.

On the other hand, it's true that I was a difficult child. Maybe my parents were right: my brother is nicer.

Evolution and Child-Rearing

You have little power to determine how your children will behave when they're not with you, but you have a great deal of power to determine how they will behave at home. You have little power to determine how the world will treat them, but you have a great deal of power to determine how happy or unhappy they will be at home.

There are child-rearing manuals that might give you some pointers on how to make life at home a little more pleasant for you and your kids. Unfortunately, all of these books are based on what I consider a false premise, most take insufficient account of the fact that children are born different, and many are complete hogwash.

Let's say, for the sake of argument, that I have convinced you that the advice-givers are talking through their hats. What might my book tell you about raising kids?

Of course, I hope I've made you more aware of the importance of peers to your child's current life and future prospects. But I hope I've also made you more aware of the importance of the evolutionary history of our species. An understanding of what childhood was like for thousands of generations of our ancestors can shed light on why things sometimes go wrong in modern homes.

In Chapter 5, I talked about child-rearing in tribal and small village societies. I've also talked from time to time about hunter-gatherer societies, about which less is known because there are not many left in the world. The observations of traditional societies give us clues about how young humans were designed by evolution to be reared. In these societies, babies receive attentive care in their first two years. The baby goes with its mother wherever she goes, carried around during the day and sleeping with her at night. In most societies around the world, even today, babies sleep with their mothers.

The baby-care problem that brings the most complaints from American parents is sleep disturbances: the baby won't sleep. The baby keeps them up at night. The recommendation usually given is that the parents should get the baby accustomed to sleeping alone. But a baby in a wandering hunter-gatherer band was never left alone under normal circumstances. If he found himself alone, and if his first whimpers did not immediately fetch his mother, he was in serious trouble. Chances are that either his mother was dead or she had decided she couldn't take care of him. The group was moving on and they weren't taking him! He was a goner if he couldn't quickly persuade them to change their minds.

Screaming was the only persuader he had. He screamed because he was terrified and angry, and with good reason.

Babies are amazingly adaptable. Most American babies adapt quite well to sleeping alone. But some do not. Many parents—my younger daughter among them—are relieved when you tell them it's okay to let their baby sleep with them, that it's what nature intended. They *hate* letting the baby cry. It is going against nature to let a baby cry, and yet parents do it—though they suffer almost as much as the baby—because it's what the advice-givers recommend.

The advice-givers also tell you that you have to provide babies with the proper "stimulation" in order to make their little brains grow properly and encourage the right synapses to form. You're supposed to talk to them and read to them and give them interesting things to look at. This advice is based on two kinds of data, both poorly understood or misinterpreted. The first is the finding that severe sensory deprivation in young animals—rats, cats, and monkeys—can lead to permanent neurological deficits. The second is correlational: parents who read to their children and hang fancy mobiles over their cribs tend to have smarter children.

If the brain required poetry readings and fancy mobiles in order to hook up its synapses correctly, our ancestors would have been wandering around with defective brains. The experiences of babies in traditional societies give us clues about what sort of environment the developing human brain was programmed to expect. Babies in these societies are not read to; they are not even talked to very much. They have plenty to look at and listen to, but every baby does. Although these babies learn very little during their two years in their mothers' arms, that does not keep them from learning, when the time is ripe, all the things they need to know to become successful adults.

As for the correlations, I trust you know by now what to make of them. The reason why parents who read to their children have smarter children is that these are smarter parents. Their children are smarter because intelligence is partly inherited. If there were an environmental reason why parents who read to their children have smarter children, then we wouldn't find a zero correlation in IQ between two adoptive siblings reared by the same parents. There is no scientific basis for the belief that it is possible to make babies smarter by giving them fancy things to listen to or look at.

In a recent netnews posting on the Internet, a young mother who identifies herself as "a graduate student studying brain development" tells about her exceptionally bright and alert twenty-month-old son. Her parents attribute the boy's brightness to the fact that he has a bright mother and a bright father, but she finds this explanation "insulting to my parenting." "I have worked hard," the mother explains, "to create a well-attached, loving relationship and to provide plenty of appropriate stimulation."

She has worked hard; I give her full marks. But parenting is not supposed to feel like hard work, any more than sex is. Evolution provides carrots as well as sticks. Nature gets us to do what she wants us to do by making it pleasurable for us to do it. If parenting were hard work, do you think chimpanzees would bother? Parents are meant to *enjoy* parenting. If you are not enjoying it, maybe you're working too hard.

Parents As Pals

Evolution provides sticks as well as carrots. Nature makes big, strong creatures dominant over smaller, weaker members of their species. The big ones get to tell the smaller ones what to do and to punish them if they do not do it. No, it isn't fair, but what can I tell you? Nature doesn't give a tinker's damn about fairness. In chimpanzee groups, big males dominate smaller males and beat them up if they don't show the proper respect. Males beat up females for the same reason. Young animals do the same to younger ones.

This unpretty pattern is preserved intact in traditional societies. It is very old. Our current obsession with fairness and niceness is very new.

Parents are meant to be dominant over their children. They are meant to be in charge. But nowadays they are so hesitant about exerting their authority—a hesitancy imposed upon them by the advice-givers—that it is difficult for them to run the home in an effective manner.

I do not believe that children are better off today than they were before the nurture assumption made wimps out of parents. The experiences of previous generations show that it is possible to rear well-adjusted children without making them feel that they are the center of the universe or that a time-out is the worst thing that could happen to them if they disobey. Parents know better than their children and should not feel diffident about telling them what to do. Parents, too, have a right to a happy and peaceful home life.

In traditional societies, parents are not pals. They are not playmates. The idea that parents should have to entertain their children is bizarre to people in these societies. They would fall down laughing if you tried to tell them about "quality time."

Former Secretary of Labor Robert Reich resigned his Washington position and went home to Massachusetts, in part because he wanted to spend more time with his sons, ages twelve and sixteen. It didn't work out quite the way he envisioned.

> Forget what you've heard about "quality time." Teenage boys don't want it, can't use it, have better things to do. When I came home from Bill Clinton's Cabinet and suddenly had weekend time to spare, I waited for one of my boys to take me up on my offer of hours of quality time with them. "Sorry Dad. I'd really like to go to the game with you, but . . . well, see, David and Jim and I are gonna hang out in the Square." "That's a cool movie, Dad, but . . . well, to tell the truth, I'd rather see it with Diane."

The boys didn't shun him entirely. Sometimes they did ask his advice, which made him feel better. They really didn't want to hurt his feelings. They love him, but . . . well.

Younger kids are less likely than teenagers to play hard to get. But that may be only because they have less freedom to go places by themselves, so they have fewer options. If given a choice, even toddlers seem to prefer the company of other children, though they do like to have a parent hanging around in the background.

Siblings As Allies

In traditional societies, toddlers graduate from their mothers' arms into a play group that consists mainly of their relatives—brothers and sisters, half-brothers and -sisters, cousins. A common pattern in these societies is to put the next-older sibling in charge of the toddler and not to interfere. The older sibling is held responsible for any injury that befalls the younger one. The younger one, remember, is the very child who succeeded him in their mother's arms. The very one who monopolized their mother's attention for the last couple of years.

The older one is allowed—indeed, is expected—to dominate his younger sibling. It is natural for older kids to dominate younger ones, and in traditional societies no effort is made to prevent it, because there is not so much concern about equality and fairness.

In our society the concern about equality and fairness leads to trouble in sibling relationships. Parents' efforts to prevent the older child from dominating the younger one produce a great deal of ill will between them. The parents can prevent domination only by exerting their power on the younger child's behalf, and this makes the older one feel—correctly, in many cases—that the parents are favoring his little sister or brother.

I am not suggesting that you put your five-year-old in charge of your three-year-old—at least not abruptly. But if you understood what was going wrong between them, perhaps you would be more sympathetic to the older one's plight. He has been deprived, first of parental attention, because parents in every society give more attention to younger than to older kids; and second, of his natural right to boss the younger one around. In traditional societies, you lose one, you win one. In our society the score is 0 and 2.

I told you a story in an earlier chapter of a young boy in Africa who was badly injured when he ran after a big chimpanzee that had grabbed his baby brother. The boy saved his brother's life (the chimpanzee would have killed and eaten the baby) but almost lost his own. His mother had put him in charge of his baby brother, something most American mothers would never dream of doing. It was a responsibility the boy took seriously. In traditional societies, siblings are not rivals—they are allies.

Go Figure

You never know. One mother was lackadaisical about offering her child piano lessons and her child cannot play a note; another was equally lackadaisical but her child became an accomplished pianist. Some kids who have everything going for them get on the road to success and stay there, while others triumph over adversity and attain even greater success. Having a silly name or moving around a lot can be a disaster for a child, but kids with silly names or peripatetic parents sometimes go on to become presidents, poets, or famous biologists. Kids do well if they go to schools where all the kids are bright, but I did better in Arizona than in the snooty suburb because on my first day in the new school in Arizona I aced a biology test and got labeled a "brain." You never know.

If it makes you feel any better, neither do the advice-givers.

You've followed their advice and where has it got you? They've made you feel guilty if you don't love all your children equally, though it's not your fault if nature made some kids more lovable than others. They've made you feel guilty if you don't give them enough quality time, though your kids seem to prefer to spend their quality time with their friends. They've made you feel guilty if you don't give your kids two parents, one of each sex, though there is no unambiguous evidence that it matters in the long run. They've made you feel guilty if you hit your child, though big hominids have been hitting little ones for millions of years. Worst of all, they've made you feel guilty if anything goes wrong with your child. It's easy to blame parents for everything: they're sitting ducks. Fair game ever since Freud lit his first cigar.

Somehow the advice-givers always manage to take the joy and spontaneity out of child-rearing and turn it into hard work. A long time ago, John Watson inveighed against the dangers of "loving our children to death." He described, with barely contained revulsion, a car ride in which his warnings were ignored but his numerical skills put to good use:

> Not long ago, I went motoring with two boys, aged four and two, their mother, grandmother, and nurse. In the course of the two-hour ride, one of the children was kissed thirty-two times—four by his mother, eight by the nurse and twenty times by the grandmother. The other child was almost equally smothered in love.

The reason, I think, that the mother gave the fewest kisses was that she was Watson's wife. She was going against her husband's wishes in kissing her sons. Those were stolen kisses.

Today the advice-givers have gone in the other direction and made kissing your child a duty instead of a crime. If I were a kid, I'd rather have one stolen kiss a year than three a day given because the doctor prescribed them.

The Guilt Trip Stops Here

In this chapter I have been talking about what parents can do to influence their child's personality, behavior, attitudes, and knowledge. I haven't said anything about giving your child a healthy diet or seeing that he has his inoculations, because these are not the sort of things this book is about. Nor do I feel qualified to give advice about mental illnesses. There are things that can go wrong with children that are outside the scope of this book. If you see signs of them in your child, of course you should take him or her to a qualified professional.

As for what you can do to influence your child's personality, behavior, attitudes, and knowledge, I recognize that you might not be satisfied with my answer. Some people are not relieved to hear that they can stop blaming themselves for whatever they don't like about their children. Some people find the news upsetting, especially if their kids are young. They want to feel that they can make a difference as parents. They want to feel that there is still something they can do to improve their child's chances, some way they can change the things they don't like about their child. If they work at it hard enough, surely there is something they can do!

They have been sold a bill of goods. They have a right to feel cheated. Parenting does not match its widely publicized job description. This is a job in which sincerity and hard work do not guarantee success. Through no fault of their own, good parents sometimes have bad kids.

We have all kinds of marvelous technology. We have learned how to eliminate many of the diseases that used to kill or cripple so many children. We have been successful in dealing with many of the curve balls nature throws at us, and perhaps that is why we have the illusion we can deal with all of them.

The idea that we can make our children turn out any way we want is an illusion. Give it up. Children are not empty canvases on which parents can paint their dreams.

Don't worry about what the advice-givers tell you. Love your kids because kids are lovable, not because you think they need it. Enjoy them. Teach them what you can. Relax. How they turn out is not a reflection on the care you have given them. You can neither perfect them nor ruin them. They are not yours to perfect or ruin: they belong to tomorrow.

15 THE NURTURE ASSUMPTION ON TRIAL

They fuck you up, your mum and dad.
They may not mean to, but they do.
They fill you with the faults they had
And add some extra, just for you.
—Philip Larkin

Poor old Mum and Dad: publicly accused by their son, the poet, and never given a chance to reply to his charges. They shall have one now, if I may take the liberty of speaking for them.

How sharper than a serpent's tooth
To hear your child make such a fuss.
It isn't fair—it's not the truth—
He's fucked up, yes, but not by us.

Philip's mum and dad are not on trial here, however. The defendant is the nurture assumption itself, which their son summed up so succinctly in his four lines of doggerel. Ladies and gentlemen of the jury, I ask you to find the defendant guilty of fraud and grand larceny. The people have been robbed of the truth and the nurture assumption is the perpetrator.

Fooling the People All of the Time

Philip Larkin isn't the only one who blames his failings on his parents. Everybody does (even, in my weaker moments, me). It sure beats blaming yourself. But self-interest alone cannot account for the way the nurture assumption has worked itself so deeply into our culture. Nor is the explanation I gave in Chapter 1—that it's a product of the combined influence of psychoanalytic theory (Freud) and behaviorism (Watson and Skinner)—enough to account for its pervasiveness. What started out as part of academic psychology has long since spread beyond its ivory tower origins. Talk-show hosts and talk-show guests, poets and potato farmers, your accountant and your kids—they all blame their parents for their own failings and themselves for their kids'.

Parenting has been oversold. You have been led to believe that you have more of an influence on your child's personality than you really do. At the beginning of the book I quoted the science journalist who said we don't have to wait for the day when parents can choose their children's genes because parents already have a good deal of power to determine how their children will turn out. "Parents play the largest and most important role in shaping their children's sense of themselves," said another science journalist in the pages of the *New York Times.* You are expected to give them a positive sense of themselves by showering them with praise and physical affection. The professional advice-giver who calls herself "Dr. Mom" tells you to make sure to give your child "daily nonverbal messages of love and acceptance." All children need touching and hugging, she says, no matter how old they get. If you do the job right, your child will emerge happy and self-confident, according to Penelope Leach, another professional advice-giver. "His foundations are laid in his relationship with you and all that you have taught him." Physical punishment and verbal criticism are outlawed by the advice-givers. You don't tell the child that he was bad, you tell him that what he did was bad. No, better not go even that far: tell him that what he did made you feel bad.

Kids are not that fragile. They are tougher than you think. They have to be, because the world out there does not handle them with kid gloves. At home they might hear "What you did made me feel bad," but out on the playground it's "You shithead!"

The nurture assumption is the product of a culture that has as its motto, We can overcome. With our dazzling electronic devices, our magical biochemical elixirs, we can overcome nature. Sure, children are born different, but that's no problem. Just put them through this marvelous machine here—step right up, ladies and gentlemen!—and add our special patented mixture of love, limits, time-outs, and educational toys. Voilà! A happy, smart, well-adjusted, self-confident person!

Perhaps it's a fin-de-siècle phenomenon: the tendency to carry things to extremes, to push ideas beyond their logical limits. The nurture assumption has become so overblown, so oppressive in the demands it makes upon parents, that it looks to be past ripe and well on its way to rot.

First of All, Do No Harm

I wouldn't feel so strongly about it if I thought it was a harmless fantasy. After all, the nurture assumption might have some beneficial side effects. At least in theory, it should make parents kinder. If they think that any mistake they make could mark their kids for life, shouldn't that encourage them to be more careful? To bite back those mean words? To spare the rod? It's a nice thought, but there are no signs of a decline in parental abusiveness. Nor are there any signs that children are happier today than they were two or three generations ago.

There is no evidence that the nurture assumption has done any real good. But it has done some real harm. It has put a terrible burden of guilt on parents unfortunate enough to have a child whose pass through the marvelous machine has for some reason failed to produce a happy, smart, well-adjusted, self-confident person. Not only must these parents suffer the pain of having a child who is difficult to live with or who fails in some other way to live up to the community's standards: they must also bear the community's opprobrium. Sometimes it's more than mere opprobrium: sometimes they are held legally responsible. Fined. Threatened with jail sentences.

The nurture assumption has turned children into objects of anxiety. Parents are nervous about doing the wrong thing, fearful that a stray word or glance might ruin their child's chances forever. Not only have they become servants to their children: they have been declared unsatisfactory servants, because the standards set by the promulgators of the nurture assumption are so high that no one can meet them. Parents who lack the time to get a full night's sleep are being told that they're not giving their kids enough quality time. Parents are being made to feel that they've fallen short. They try to make it up to their kids by buying them lots of toys. The modern American child owns an astonishing quantity of toys.

The nurture assumption has introduced an element of phoniness into family life. It has made sincere expressions of love meaningless because they are drowned out by the obligatory, feigned expressions of love.

The nurture assumption has also held back the progress of scientific inquiry. The proliferation of meaningless research—one more dreary study showing a correlation between parents' sighs and children's yawns

—has been substituted for useful investigation. Here are some of the things that researchers *should* be looking at, a few of the questions they *should* be asking. How can we keep a classroom of children from splitting up into two dichotomous groups, pro-school and anti-school? How do some teachers, some schools, some cultures, manage to prevent this split and keep the kids united and motivated? How can we keep kids who start out with disadvantageous personality characteristics from getting worse? How can we step in and break the vicious cycle of aggressive kids becoming more aggressive because in childhood they are rejected by their peers and in adolescence they get together with others like themselves? Is there any way to influence the norms of children's groups for the better? Is there any way to keep the larger culture from having deleterious effects on the norms of teenagers' groups? How many does it take to make a group?

I have been unable, in this book, to give you answers to these questions because the research has not yet been done.

The Case for the Defense

According to the nurture assumption, parents have important effects on the way their children turn out. *Important* effects. We are not talking about an IQ point here or there, or one more "yes" on a questionnaire of a hundred items. We are talking about popular versus friendless, college graduate versus high school dropout, neurotic versus well-adjusted, virgin versus pregnant. We are talking about psychological characteristics that affect how you behave and how well you do in life—characteristics that are noticeable to you and to the people you live and work with. Characteristics that will remain with you for the rest of your days. That's what people think, isn't it? That parents have big effects on their children, lasting effects.

But if they do have effects, it must be a different effect for each child, because children raised by the same parents do not turn out alike, once you skim off the similarities due to shared genes. Two adopted children reared in the same home are no more similar in personality than two adopted children reared in separate homes. A pair of identical twins reared in the same home are no more alike than a pair reared in separate homes. Whatever the home is doing to the children who grow up in it, it is not making them more conscientious, or less sociable, or more aggressive, or less anxious, or more likely to have a happy marriage. At least, it's not doing any of these things to all of them.

The behavioral geneticists were the ones who made this discovery and it put them in a pickle, because most of them do believe in the importance of the home environment, just like everyone else. So they came up with the idea that what matters in the home are the things that *differ* for each child who lives there. The things that two siblings have in common had been shown not to matter—or at least to have no predictable effects—so the things the siblings *don't* have in common were left to bear the full weight of supporting the nurture assumption.

This is not quite as far-fetched as it sounds. After all, there's no reason to expect parents to treat their children all alike. Shouldn't good parents want each of their children to be unique, each to do whatever he or she does best? It's the Marxist view of parenting: from each according to his abilities, to each according to his needs.

And it's true, to a point. Yes, parents should want their children to be different, at least in some ways. If the first child is active and talkative, a quiet one would be a welcome change. If the first is a pianist, they might be perfectly happy to have the second one take up the tuba. But that doesn't mean they would be equally happy to have the second one become a prizefighter or a drug dealer. When our second child came along, my husband and I didn't say, "Well, we have one academic achiever, no point doing *that* again. Let's make the second one into something else." On the contrary, we could have put up quite nicely with the boredom of having two academic achievers. There are certain qualities parents would like to see in all their children—kindness, conscientiousness, intelligence—and other qualities they are willing to let vary within reasonable limits. But the findings for the universally desired qualities are the same as for the optional ones: no evidence of a long-term effect of the home environment.

Parents treat each of their children differently and the children *are* different—these two facts are indisputable. But for the behavioral geneticists to hold on to the nurture assumption, it is necessary for them to show that the differences in parental behavior are producing or contributing to the differences among the children and are not just a response to preexisting differences. That has not been shown. In fact, there is evidence that parental treatment is actually more uniform than the children themselves—that there is more variation in the way two siblings behave than in the way their parents treat them.

One thing that could have worked in favor of the nurture assumption, but didn't, is birth order. Parents treat firstborns and laterborns differently, and the differential treatment isn't just a response to characteristics the kids were born with. But researchers have been trying for more than half a century to find convincing proof that birth order leaves lasting marks on personality and their efforts have not panned out. Nor have efforts to show differences in personality between only children and children with siblings. If parents have major effects on their children, how come they don't mess up the personality of the only child?

These two disappointments—no birth order effects, no only-child effect—should knock the last remaining prop from under the nurture assumption.

Hmm, it hasn't fallen down yet; something still seems to be holding it up. Ah yes, I see it. It is the claim that the behavioral genetic evidence—the data showing that the overall home environment has no predictable effects—does not cover a wide enough range of home environments. The trouble is that all the subjects came from "good-enough" homes—homes within the normal range. Some theoreticians are now willing to admit in public that it doesn't matter which home a child grows up in, as long as it is within the range of normal, good-enough homes. But they still think it is possible that homes *outside* the normal range—that is, exceptionally bad homes—have an effect on the child.

What they're saying is that there's no relationship between the goodness of the home and the goodness of the offspring over the entire range of homes for which they have data, the range that begins at "excellent" and extends through "bad" but stops short of "terrible." The relationship holds only for the small proportion of homes for which they do not have data. All the evidence they've collected so far—and they've collected a great deal of it—either is irrelevant or indicates that the nurture assumption is wrong. But there is a little bit of evidence they haven't collected yet, and *that,* they believe, will prove that the nurture assumption is right.

It's an awfully thin prop. The idea is that ordinary, run-of-the-mill parents like you and me don't have any distinctive effects on our children: we are interchangeable, like factory workers. The only parents who do have distinctive effects are the super-bad ones who abuse their kids so severely they wind up in the hospital, or who leave them unattended in cold apartments stinking with unchanged diapers and rotted food. It's the nurture assumption's last slim hope: that a home environment can be bad enough to inflict permanent damage on the children who grow up in it.

I will leave the proponents of the nurture assumption clinging to their last slim hope, that their assumption might hold true for the small proportion of families that qualify as super-bad. It does not hold true for the vast majority of families. It does not hold for families like yours and mine. There is no justification for using it as a weapon against ordinary parents whose children are not turning out quite the way we hoped they would.

Five Wrong Ideas

How are children shaped by the experiences they have while they are growing up? That is the question the nurture assumption was designed to answer. The answer is wrong because it is based on a number of mistaken ideas about children.

The first mistake has to do with the child's environment. The natural environment of the child was assumed to be the nuclear family—the arrangement that was so popular during the first half of this century. Mother, father, two or three kids, living cozily together in a private house. But there is nothing particularly natural about this arrangement. The separateness of the nuclear family—its ability to carry on its activities free from the prying eyes of neighbors—is a modern invention, only a few hundred years old. The monogamous bond between one man and one woman is also something of a novelty. In 80 percent of the cultures known to anthropologists, men who could afford them have been allowed extra wives. Polygyny is ancient and widespread in our species. Children have often been required to share their fathers with the children of their father's other wives. Or they've had to grow up without a father or without a mother, because parental death was as prevalent in the past as parental divorce is today.

The second mistake has to do with the nature of socialization. A child's job is not to learn how to behave like all the other people in his or her society, because all the other people in the society do not behave alike. In every society, acceptable behavior depends on whether you're a child or an adult, a male or a female. Children have to learn how to behave like the other people in their own social category. In most cases they do this willingly. Socialization is not something that grownups do to kids—it is something kids do to themselves.

The third mistake has to do with the nature of learning. It was assumed that learned behavior is carried along like a backpack from one place to another—from the home to the schoolyard, for example—even though it has always been clear that people of every age behave differently in different social contexts. They behave differently because they have had different experiences—in one place they are praised, in another they are laughed at—and because different behaviors are called for. It was also assumed, also incorrectly, that if children behave one way at home and in a different way in the schoolyard, it must be the home behavior that matters most.

The fourth mistake has to do with the nature of nature—of heredity. The power of the genes has still not been given its due, even though everybody has heard the stories about the separated identical twins who meet in adulthood and find they are both wearing blue shirts with pockets on both sides and epaulets on the shoulders. Philip Larkin noticed that he shared many of his faults with his parents, but that did not give him the idea that he inherited them: he thought it was something his parents did to him after he was born.

The fifth mistake is to ignore our evolutionary history and the fact that, for millions of years, our ancestors lived in groups. It was the group that enabled those delicate creatures, unequipped with fangs or claws, to survive in an environment that did have fangs and claws. But animal predators were not their greatest threat: the most dangerous creatures in their world were the members of other groups. That is still true today.

The Alternative: Group Socialization Theory

The group is the natural environment of the child. Starting with that assumption takes us in a different direction. Think of childhood as the time when young humans turn themselves into accepted and valued members of their group, because that is what they needed to do in ancestral times.

During childhood, children learn to behave the way people of their age and sex are expected to behave in their society. Socialization is the process of adapting one's behavior to that of the other members of one's social category. In the novel *The Shipping News,* a father is counseled by his aunt to put off worrying about his young daughter's peculiarities:

> "Why don't you just wait, Nephew. See how it goes. She starts school in September. . . . I agree with you that she's different, you might say she is a bit strange sometimes, but you know, we're all different though we may pretend otherwise. We're all strange inside. We learn how to disguise our differentness as we grow up. Bunny doesn't do that yet."

We learn how to disguise our differentness; socialization makes us less strange. But the disguise tends to wear thin later in life. I see socialization as a sort of hourglass: you start out with a bunch of disparate individuals and as they are squeezed together the pressure of the group makes them more alike. Then in adulthood the pressure gradually lets up and individual differences reassert themselves. People get more peculiar as they grow older because they stop bothering to disguise their differentness. The penalties for being different are not so severe.

Children identify with a group of others like themselves and take on the norms of the group. They don't identify with their parents because parents are not people like themselves—parents are grownups. Children think of themselves as kids, or, if there are enough of them, as girls and boys, and these are the groups in which they are socialized. Most socialization occurs today in same-age, same-sex groups because developed societies make it possible for children to form such groups. In the past, when humans were spread thinly across the planet, children were socialized in mixed-age, mixed-sex groups.

There has always been a bond between parents and their children, but the intense, guilt-ridden form of parenting we see today is unprecedented. In societies that don't send their kids to school and have not yet been penetrated by the advice-givers, children learn most of what they need to know from other children. Although parenting styles differ drastically from one culture to another—too hard in some places, too soft in others—children's groups are pretty much the same around the world. That is why children get socialized in every society, even though their parents don't read Dr. Spock. Their brains develop normally in every society too, even though their parents don't read *Goodnight Moon.*

Modern children do learn things from their parents; they bring to the group what they learned at home. The language their parents taught them is retained if it turns out that the other kids speak the same language, and the same is true for other aspects of the culture. Since most children grow up in culturally homogeneous neighborhoods—their parents speak the same language and have the same culture as the parents of their peers—most children are able to retain a good deal of what they learned at home. This makes it look as though the parents are the conveyers of the culture, but they are not: the peer group is. If the peer group's culture differs from the parents', the peer group's always wins. The child of immigrant parents or deaf parents invariably learns the language of her peers and favors it over the language her parents taught her. It becomes her native language.

You can see it happening as early as nursery school, when three-year-olds start bringing home the accents of their peers. Perhaps it begins even earlier than that. Psychologists Susan Savage and Terry Kit-fong Au tell this story in a recent issue of the journal *Child Development:*

> A baby we know had to face a dilemma very early on. From the age of about 12 months on, she was quite successful in requesting a bottle by saying "Nai nai!" (the Chinese term for milk) to her parents. Meanwhile, she noticed that other babies at her day-care center got their bottles by saying "Ba ba!" and followed suit at age 15 months. The demands of leading a double life apparently were too great for her to bear. A day or two later, when her mother asked "Nai nai?" she shook her head vigorously and said emphatically "Ba ba!"

Even when their parents belong to the same culture as the parents of their peers, children cannot count on being able to export the behaviors they acquired at home. A boy can whine and complain with impunity at home; he can express anxiety and affection. But in the peer group he is expected to be tough and cool. It is the tough, cool persona that will become his public personality and that he will carry with him to adulthood. The personality acquired at home won't be lost completely, however: it will turn up at Christmas dinners like the Ghost of Christmas Past.

In the peer groups of childhood and adolescence, kids take on the behaviors and attitudes of their peers and contrast themselves with the members of other groups—groups that differ in sex or race or social class or in their propensities and interests. The differences between these groups widen because the members of each group like their own group best and are at pains to distinguish themselves from the others. Differences *within* the group widen, too, especially when the group is not actively engaged in competing with another group. At the same time that children are becoming more similar to their peers in some ways, they are becoming more distinctive in other ways. Children learn about themselves by comparing themselves to their groupmates. They vie for status within the group; they win or lose. They are typecast by their peers; they choose or are chosen for different niches. Identical twins do not end up with identical personalities, even if they are members of the same peer group, because they have different experiences within the group.

Experiences in childhood and adolescent peer groups modify children's personalities in ways they will carry with them to adulthood. Group socialization theory makes this prediction: that children would develop into the same sort of adults if we left their lives outside the home unchanged—left them in their schools and their neighborhoods—but switched all the parents around.

A Penny for Your Thoughts

Arguments based on scientific evidence are not enough to change your mind. Your belief in the nurture assumption is not based on cold science but on feelings, thoughts, and memories. If your parents weren't important in your personal history—if they didn't have a powerful influence on you—why is it that your memories of childhood, along with so many of the memories you've stored away since then, have your parents playing leading roles? Why are they so often in your thoughts?

In his book *How the Mind Works,* evolutionary psychologist Steven Pinker discusses the fact that the conscious mind has access to some kinds of information and not to others.

> I ask, "A penny for your thoughts?" You reply by telling me the content of your daydreams, your plans for the day, your aches and itches, and the colors, shapes, and sounds in front of you. But you cannot tell me about the enzymes secreted by your stomach, the current settings of your heart and breathing rate, the computations in your brain that recover 3-D shapes from the 2-D retinas, the rules of syntax that order the words as you speak, or the sequence of muscle contractions that allow you to pick up a glass.

It isn't that your daydreams and itches are more important than the computations in your brain that enable you to see three-dimensional objects or speak grammatical sentences or pick up a glass. It's just that some of these things are accessible to consciousness and some of them aren't.

The other thing about the way the mind works (as Pinker and his fellow evolutionary psychologists have pointed out) is that it's modular. The mind is composed of a number of specialized departments, each collecting its own data and issuing its own reports or orders. Just as the body is organized into physical organs that each do one specific job—the lungs oxygenate the blood, the heart pumps it around the body—so the mind is organized into mental organs or modules or departments. One department lets you see the world in three dimensions, another enables you to pick up the glass. Some of the mind's departments issue reports that are accessible to consciousness; others do not.

I believe the human mind has at least two different departments for dealing with social behavior. One has to do with personal relationships, the other has to do with groups.

The group department has a long history and is found in many species. Fish, for example, swim around in schools. They have to adapt their behavior to the group's but they don't have to recognize their schoolmates. Though they may distinguish between males and females, between bigger fish and smaller ones, between kin and nonkin, they don't remember individuals, not even their own children.

The social life of primates is more complex. Primates, too, have to adapt their behavior to the group's, but they also have to keep track of the individuals in their lives. They must learn which members of their community they can count on for support and which they'd better keep away from. It's a talent that has flowered in our species. Humans remember who did them a favor and who owes them one. They know—both from personal experience and by word of mouth—who can be trusted and who cannot be. They hold grudges, sometimes forever, against those who did them wrong, and they look for opportunities to take revenge. And those who did the wrongs had better not forget who their victims were. We have very good memories for people. Our brains have a special area devoted to recognizing faces.

The department of the brain that keeps track of relationships is accessible to the conscious mind. The department of the brain that adapts your behavior to that of your group is no less important but it is less accessible to consciousness. A lot of its work goes on at an automatic level, like the muscle movements that enable you to pick up the glass.

Much of the information we collect about the world is collected unconsciously. We don't know how we know many of the things we know; we just know them. Children learn that red fruits are sweeter than green ones and if you give them a choice they'll pick the red one, but they can't tell you why. The gathering of data, the construction of categories, and the averaging of data within categories was all carried out below the level of consciousness.

The processes that I have been talking about in this book generally go on below the level of consciousness. We identify with a group of people. We learn to speak and behave like these people; we take on their attitudes. We adapt our speech and behavior to different social contexts. We develop stereotypes of our own group and of other groups. These things can be brought into consciousness but that's not where they live. In this book I have been talking about things that children do without noticing them, without having to exert conscious effort. It leaves the tops of their minds free to do other things.

Groups and relationships: they are both important to us but in different ways. Our childhood experiences with peers and our experiences at home with our parents are important to us in different ways.

The bond between parent and child lasts a lifetime. We kiss our parents goodbye not once but many times; we do not lose track of them. Each visit home gives us opportunities to take out family memories and look at them again. Meanwhile, our childhood friends have scattered to the winds and we've forgotten what happened on the playground.

When you think about childhood you think about your parents. Blame it on the relationship department of your mind, which has usurped more than its rightful share of your thoughts and memories.

As for what's wrong with you: don't blame it on your parents.

APPENDIXES

APPENDIX 1: PERSONALITY AND BIRTH ORDER

Do firstborn children carry with them through life a sense of specialness? Are people who grew up with older siblings more likely to rebel? Such questions are of interest to anyone who has a sibling and of theoretical importance to social scientists. For the better part of a century, psychologists from Alfred Adler to Robert Zajonc have been constructing theories about birth order and searching for evidence to back them up—evidence that firstborns and laterborns differ from each other in personality, intelligence, creativity, rebelliousness, what have you. Such differences, when they are found, are called birth order effects.

Such differences are often found. But they usually turn out to be spurious or misleading. The evidence for birth order effects has been knocked down again and again, whenever careful reviewers—reviewers with no theory of their own to promote—have looked closely at the data.

The careful reviewers of the data, knowing that their conclusions were not in line with their readers' expectations, have peppered their papers with exclamation points and italics. Carmi Schooler's 1972 article in the *Psychological Bulletin* was entitled "Birth Order Effects: Not Here, Not Now!" Cécile Ernst and Jules Angst stated firmly in their 1983 book that *"Birth order and sibship size do not have a strong impact on personality.* . . . An environmental variable that is considered highly relevant is thus disaffirmed as a predictor for personality and behavior." Judy Dunn and Robert Plomin, in their 1990 book on sibling relationships, acknowledged that their conclusion "goes against many widely held and cherished beliefs" but asserted that "individual differences in personality and psychopathology in the general population . . . are *not* clearly linked to the birth order of the individuals."

These emphatic statements were ignored not just by the public but by many social scientists as well. The resilience of the faith in birth order effects—its ability to pop back up each time it is knocked down—was remarked upon by Albert Somit, Alan Arwine, and Steven Peterson in their 1996 book on birth order and political behavior. Somit and his colleagues spoke of the "inherent non-rational nature of deeply held beliefs" and mused that "permanently slaying a vampire"—the belief in birth order effects—may require something drastic. They suggested a stake through its heart at high noon.

What makes this vampire so difficult to slay? The answer is that it is protected by a potent amulet, a magic shield: the nurture assumption. Psychologists and nonpsychologists alike take it for granted that a child's personality, to the degree that it is shaped by the environment, receives that shaping primarily at home. Since it is perfectly clear that a child's experiences at home are affected by his or her position in the family—oldest, youngest, or in the middle—researchers take it for granted that birth order must leave permanent marks on the child's personality. They start out with that assumption, look for evidence that it is true, and refuse to take no for an answer. So the belief in birth order effects does not die: it just rests in its coffin until someone lifts the lid again.

The latest lifter of the lid is historian of science Frank Sulloway, whose theory of birth order effects is presented in his book *Born to Rebel.* Sulloway's theory is a sophisticated one; he uses concepts from evolutionary psychology to account for the behavioral genetic finding that children in the same family do not turn out alike. He points out that siblings compete with each other for parental attention and that it is adaptive for siblings to differentiate themselves from each other—for each to find a different specialty, a different niche in the family. The differences reflect the children's own strategies; they are not imposed upon them by the parents (not directly, at any rate). On all these points I agree with Sulloway. And he provides powerful support for his theory. *Born to Rebel* contains an impressive compilation of data from a wide variety of sources, painstakingly assembled.

We start out with similar premises but our paths soon deviate. Sulloway uses the idea of within-family niche-picking to account for variations in adult personality. He argues (see Chapter 3) that firstborns are stodgy stick-in-the-muds, laterborns are open to new experiences and new ideas. That firstborns are uptight, aggressive, status-hungry, and jealous; laterborns are easy-going and nicer. Sulloway, needless to say, is a laterborn. I am a firstborn: Born to Rebut.

Sulloway has assembled a mountain of data to support his theory. I looked closely at that mountain and came to different conclusions. The following critique is aimed, not at *Born to Rebel* in particular, but at social science in general, because the methods used and the mistakes that were made are common ones. My findings serve as a demonstration of what can go wrong when researchers first assume that something is true and then set out to prove it.

Sulloway's Reanalysis of Ernst & Angst's Survey

I was first alerted to the possibility that Sulloway's mountain might not be as solid as it looks by a review of *Born to Rebel* in the journal *Science.* The reviewer, historian John Modell, had much praise for the book but he also leveled some disturbing criticisms. Referring to Sulloway's reanalysis of data from a 1983 review of the birth order literature, Modell said:

> I was persuaded by Sulloway's reworking of these materials—until I tried to replicate it with the literature review in hand. I could not do so, try as I might, or even come near.

The literature review in question is the one I described in Chapter 3; it was carried out with great thoroughness by Swiss psychologists Cécile Ernst and Jules Angst and reported in a long chapter of their 1983 book. E & A searched the world literature for every study on birth order reported in the period 1949 through 1980. They concluded that most of the studies they found were of no value because they lacked the proper controls: the researchers had failed to control for sibship size (the number of children in the family) or for variations in socioeconomic status (SES). Because small sibships are relatively more prevalent at higher SES levels, and because firstborns are relatively more prevalent in small sibships, the failure to control for these variables leads to a confounding of demographic factors with birth order. Outstandingly successful people are more likely to be firstborns not because of their superior position in their family of origin but because their family of origin was more likely to be superior in education and income.

Once the variables are confounded there is no way to disentangle them; if the researchers who carried out a birth order study failed to record sibship size or SES, the study is useless. E & A therefore focused on the minority of studies that included one or both of these controls. On the basis of these studies they concluded that birth order has little or no effect on personality.

These same studies—the minority that controlled for sibship size and/or SES—provide the data upon which Sulloway rests his case for birth order effects on personality. In fact, they are virtually the only data he uses to support that case; most of the statistics in *Born to Rebel* do not pertain to personality directly but to the opinions and attitudes publicly expressed by historical figures. Although opinions and attitudes are no doubt related to personality, they are not the same. Personality generally does not change much in adulthood, whereas opinions can change at any time in the lifespan. Darwin's *Origin of Species* changed many people's opinions but it is unlikely to have changed their personalities.

Because Sulloway's case for birth order effects on personality rests so heavily upon his reanalysis of E & A's survey, the claim of the *Science* reviewer that he was unable to replicate that reanalysis had to be taken seriously. I decided to make a second attempt to replicate it.

"If we ignore all birth-order findings that lack controls for social class or sibship size," Sulloway says in *Born to Rebel,* "196 controlled studies remain in Ernst and Angst's survey, involving 120,800 subjects." Of these 196 studies, he reported, 72 provide support for his theory: firstborns were found to be more conforming, traditional, achievement-oriented, responsible, antagonistic, jealous, neurotic, or assertive than laterborns. Fourteen studies produced results contrary to his theory, and the remaining 110 found no significant difference based on birth order. These results were reported in Table 4 (page 73) of *Born to Rebel.* According to Sulloway's statistics, there was less than one chance in a billion billion that they had occurred by chance.

My first job was to comb through E & A's chapter on birth order and personality in search of the 196 controlled studies Sulloway found there. But in two passes through E & A's text and tables, I found only 179. I found about the same number of disconfirming studies (13) and about the same number of no-difference studies (109) as Sulloway had, but 20 fewer confirming studies. I also found five I couldn't categorize at all.

The mystery deepened when I entered the data I had extracted from E & A into a database and sorted by authors' names: it became clear that some of my 179 studies had appeared several times in E & A's survey. If a study yielded results that were relevant to several different questions about personality, it was mentioned by E & A several times. Eliminating redundant entries by lumping together all results reported by the same author(s) in the same publication, or that consisted of follow-ups of the same group of subjects, reduced the number of studies to 116.

It was then that I noticed the following statement in the note under Sulloway's Table 4: "Each reported finding constitutes a 'study.'" In correspondence, Frank Sulloway has taken me to task for not noticing that sentence sooner and not realizing at once what it meant, but the *Science* reviewer had been equally baffled. Sulloway has promised to make the point more clearly in the next edition of his book. The point is that a single study can produce several findings. More findings, in fact, than I found in my search through E & A's chapter.

Based on information Sulloway has sent me and on a statement he has added to an endnote in the paperback edition of *Born to Rebel* (Note 68, page 472), I now have a better understanding of how he carried out his reanalysis of E & A's survey.

First of all, he didn't take E & A's word on anything. Though the note under his table begins, "Data are tabulated from Ernst and Angst (1983:93–189)," what in fact he did in many cases was to go back to the original reports of the reviewed studies and make up his own mind about them. Often his opinion differed from E & A's about whether a study had or had not included the proper controls and whether it had or had not produced significant effects. His reassessments almost always resulted in an increase in the number of outcomes favorable to his theory and/or a decrease in the number of no-difference outcomes. Sulloway believes that E & A were biased against finding birth order effects.

Other studies were eliminated by Sulloway because the researchers hadn't been clear enough about the number of subjects tested or the number of tests given, or because they yielded results that didn't bear on his theory.

Sulloway calls his reanalysis of E & A's data a "meta-analysis." Correcting errors and eliminating poorly done studies are legitimate procedures in a meta-analysis. The next step, however, takes us off the beaten path. E & A had sometimes listed a single study two or more times in their chapter, if it produced results pertaining to more than one aspect of personality. However, they did no statistical analyses based on these multiple listings. By defining the word "study" as "finding," Sulloway carried the idea of multiple listings one step further. If a researcher gave a personality test to a group of subjects and found that firstborns among them were more conforming, responsible, antagonistic, anxious, and assertive than the laterborns, Sulloway's definition allowed him to count the results of that study as five favorable outcomes—five "studies."

As near as I can figure from the information he has provided, the actual number of research studies included in Sulloway's tally is no more than 115. The total number of subjects tested in those 116 studies was approximately 75,000. Sulloway's statement in his book that, if we ignore those that lack the proper controls, "196 controlled studies remain in Ernst and Angst's survey, involving 120,800 subjects," is misleading.

Granted, 75,000 is still a lot of subjects. But the statistical analysis that Sulloway carried out was based on the presumption that there were 120,800 subjects. The analysis requires each favorable outcome to be independent of all others, as it would be if you were flipping a coin. Multiple measures of a particular sample of subjects are not independent because any peculiarities of the sample—an unusually high proportion of neurotic firstborns, for example—can affect other measures made on the same sample. A sample that, by chance, produces one result significant at what statisticians call "the 5 percent level" has a greater than 5 percent likelihood of producing others.

A more serious problem is that Sulloway's calculations vastly understate the number of no-difference outcomes. His statistics are based on the assumption that if you flip a hundred coins 196 times, and on 72 of those trials you get significantly more than 50 percent heads, the overall outcome is highly unlikely to be a coincidence—something must be making those coins turn up heads. But what if you were really flipping the coins many more than 196 times and saying "That doesn't count" when the results don't turn out the way you hoped?

When researchers test a large number of subjects and find no significant results on their first pass through the data, they often resort to a method that I called, in Chapter 2, "divide-and-conquer": they split up the data in various ways in search of subgroups of subjects who yield significant effects. Such searches not only increase the chances of producing a publishable outcome: they also bias published results in favor of the researchers' preconceptions, because subgroup effects do not have to be reported if they do not match the researchers' preconceptions.

The slash marks of divide-and-conquer are clearly visible in many of the studies reviewed by E & A. Significant birth order effects were found for boys but not for girls, or vice versa. Or for middle-class subjects but not for working class, or vice versa. Or for people from small families but not from large ones, or for high school students but not college-age subjects. Researchers thought of some ingenious ways to divide up the data. Birth order effects were found in one study only if "firstborn" was defined as "firstborn of that sex." In another, birth order effects were found only for high-anxiety subjects. The examples in this paragraph all come from the 52 results I tallied as favorable to Sulloway's theory.

Technically, such findings are called "interactions." For an interaction to be meaningful, however, it has to be repeatable. An interaction that appears in only one study is not meaningful; it simply provides the researchers with another chance at finding the hoped-for outcome—another flip of the hundred coins that doesn't have to be reported if it doesn't yield a significant number of heads.

And dividing up the subjects is only the beginning. Once you have a bunch of people lined up you might as well give them a bunch of tests. Or give them one big test and separate their answers into various "factors," each of which can be looked at separately. The 52 results I tallied as favorable to Sulloway's theory included one that found firstborns yielded more often to group pressure but in only one of two conditions, one that found laterborns to be more interested in group activities but in only one of five factors, and one in which firstborns expressed high fear of more test items than laterborns but there was no significant birth order difference in the overall amount of fear expressed on the test. I know about these mixed outcomes only because the researchers reported them and E & A happened to mention them. I do not know about all the other tests that researchers gave and that were never reported because they produced uninteresting—that is, nonsignificant—results. Those hundred coins were not flipped just 196 times. We have no way of knowing how many times the coins had to be flipped in order to yield the 72 significant results Sulloway found in E & A.

The Trouble with Meta-Analyses

"The question we need to ask about any topic of research is whether significant results exceed 'chance' expectations," Sulloway states in *Born to Rebel. "Meta-analysis* allows us to answer this question. Meta-analysis involves pooling studies to gain statistical power."

True enough. But what Sulloway carried out was not a meta-analysis in the usual sense of the term. Ordinarily, a meta-analysis would take into account two important pieces of information that Sulloway did not consider: the size of each study—how many subjects were tested or observed—and the size of the effect. Large studies that produce large effects should count more than dinky ones that produce dinky ones. In a proper meta-analysis they do count more.

Birth order effects, when they are found, tend to be small. Small effects can be statistically significant, however, if the study is large enough—if there are enough subjects. Thus, if birth order effects were real but small, significant effects should be found more often in large studies than in small ones.

Just the opposite turns out to be true in the studies reviewed by E & A. I divided the 179 results I found in E & A into three nearly equal groups on the basis of the number of subjects who participated in the study, after eliminating the 16 results for which this information was not given. The table below shows the outcome. There is a trend opposite to what we would expect if birth order effects were real but small: significant results were found more often in smaller studies, infrequently in larger ones. Studies with more than 375 subjects yielded positive results only 10 times in 54 tries.

	Outcome			
Size of Study	*Favorable to Sulloway's Theory*	*Unfavorable to Sulloway's Theory*	*No Difference, Inconsistent, or Unclear*	*Total*
Small (31–140 Subjects)	22	4	29	55
Medium (141–371)	17	4	33	54
Large (384–7274)	10	4	40	54

These results do not indicate that small studies are more likely than large ones to yield significant effects. A more likely explanation is that small studies are less likely to be published if they do not yield significant effects. The researchers shrug their shoulders and go on to something else.

In the social sciences, the failure to publish no-difference results is an acknowledged problem but it is not considered life-threatening. The problem also exists in medical research, however, and here the consequences are more serious. A no-difference outcome is important if it indicates that patients' chances of getting better are not improved by an expensive new drug or a painful surgical procedure. And yet, even in medicine, no-difference outcomes are less likely to be published and when they are published they take longer to get into print.

Garbage in, garbage out—it's a saying from computer science but it holds for meta-analysis, too. Put together a lot of little studies and you've got a big one, but it is not necessarily a good one. In medical research, the little studies are less likely to be properly controlled. The patients are not chosen at random; perhaps the ones who got the new treatment were sicker—or not as sick—as the ones who got the old one. The study is not "double-blinded": the physician who administers the treatment is the same one who decides whether or not it worked, and the patients also know whether they're getting the old treatment or the new one.

Typically, a new medical treatment is first assessed by a lot of little, poorly controlled studies. But if it seems promising, someone eventually does a definitive study, the kind medical researchers refer to as the "gold standard." The gold standard study is large (at least a thousand patients), randomized, and double-blinded, and the researchers have no financial connection to the suppliers of the treatment or the drug. Such studies, alas, are never found in psychology. The psychological studies that occasionally find their way into medical journals (see Chapter 13) would never have made it if they had been judged by the same criteria used to accept or reject medical studies.

A recent article in the *New England Journal of Medicine* compared the results of gold standard medical studies with meta-analyses of the smaller studies that had preceded them. Here is the researchers' conclusion: "The outcomes of the 12 large randomized controlled trials that we studied were not predicted accurately 35 percent of the time by the meta-analyses published previously on the same topic." When there is a discrepancy, knowledgeable physicians rely on the results of the large, well-controlled study rather than the meta-analysis of a bunch of small ones.

The nearest thing to a gold standard in birth order research is the study that E & A themselves carried out. Its purpose was to confirm or disconfirm the results of their survey; it is reported in a later chapter of the same book. E & A's study is ironclad. They used all the proper controls, tested more subjects—7,582 young adults—than the most assiduous of the researchers whose work they reviewed, and measured (with a self-report questionnaire) twelve different aspects of personality, including openness. For sibships of two, E & A found no significant effects of birth order on any measured aspect of personality. For sibships of three or more they found one significant effect: the lastborn tested slightly lower on masculinity.

Unaccountably, Sulloway does not mention this study in *Born to Rebel.*

Birth Order After 1980

E & A's survey of the birth order literature stopped at 1980. So did Sulloway's. But birth order studies are still being done. I decided to search the literature for studies that were published after E & A's cutoff date of 1980. Nowadays it is not difficult to carry out such a search, even for someone who lacks access to a university library. My on-line service offers (for an additional fee) access to *Psychological Abstracts,* which can be searched for key terms and provides summaries of published articles.

I searched *Psychological Abstracts* from 1981 to the present for published articles associated with the terms "birth order" and "personality or social behavior"; the search yielded a total of 123 articles. After eliminating those that were not studies of birth order effects on personality or social behavior, and those that did not reveal the results in the summary, I ended up with 50 studies. I classified the outcome of each as favorable to Sulloway's theory, unfavorable, mixed, no difference, or unclear. The results are shown in the table below. I conclude, as E & A did, that birth order either has no effects on adult personality or has effects that are so small and unreliable that they are of no practical importance.

Outcome Relative to Sulloway's Theory	Number of Studies
Favorable	7
Unfavorable	6
Mixed	5
No Difference	20
Unclear	12

You Got It at Home

If birth order really doesn't have important effects on adult personality, how come everyone thinks it does? And how come popular views of firstborns and laterborns have been so consistent over the years? Sulloway's description of the younger brother agrees quite well with the popular stereotype of him: easy-going, cheerful, rebellious, and perhaps a tad immature. If this stereotype is inaccurate, where did you get it from?

You got it at home. It comes from parents observing the behavior of their children and children observing the behavior of their siblings. Observing the way they behave at home.

Among the studies reviewed by E & A were several in which parents were asked to describe their children's personalities or children were asked to describe their siblings. The results of such studies were generally right in line with Sulloway's theory and with popular stereotypes. Firstborns were described by their parents as serious, sensitive, responsible, worried, and adult-oriented. Laterborns were described as independent, cheerful, and rebellious. Secondborns said their elder siblings were bossy and aggressive.

The small group of studies that used assessments by parents or siblings must have contributed a disproportionate amount of data to Sulloway's meta-analysis: most of them yielded several findings and most of the findings were favorable to Sulloway's theory. In fact, of the results in E & A's survey based on ratings by family members, I counted 75 percent as favorable to his theory, compared with 22 percent of those based on self-report questionnaires.

E & A noticed the lack of agreement between the two kinds of measures and criticized the use of family members to assess personality. They pointed out, first of all, that parents' judgments of their children are of doubtful validity; as I have mentioned elsewhere in this book, such judgments agree poorly with those of people outside the family. Moreover, parents' descriptions of their children perforce involve comparisons between an older individual (the firstborn) and a younger one (the laterborn), and older children do tend to be, well, more mature.

Birth order effects are frequently found in ratings by parents and siblings; they are generally absent from measurements made outside the family context. E & A came up with several possible explanations for this discrepancy. One of their hypotheses was that personality is linked to social context. Firstborns behave like firstborns, and laterborns like laterborns, only when they're in the presence of their parents or their siblings. "The firstborn personality," said E & A, "may be *parent-specific.*" The evidence I presented in Chapter 4 agrees with this hypothesis. Children learn ways of behaving with their parents and siblings that they do not transfer to other people and other situations.

Birth order effects on personality do exist: they exist in the home. People leave them behind when they leave home. That is why most studies of adult subjects that do not involve ratings by family members show no birth order effects.

Innovation and Rebellion

The main focus of *Born to Rebel* is not on personality in general but on innovation and rebellion. Laterborns, according to Sulloway, are more likely to accept the radical or innovative ideas of others and to reject the outmoded ideas of their parents. To support these hypotheses, Sulloway presents data on the publicly expressed opinions and behavior of historical figures—people who were important enough to have had their opinions and behavior recorded for posterity.

In his review of *Born to Rebel*, historian John Modell noted the difficulties of assessing the historical data in the book: the author's "passionate advocacy has produced a text seemingly designed to overwhelm readers rather than to lay before them what they need in order to evaluate its ideas." I came to a similar conclusion. To test the claims made in the book, therefore, I must rely on evidence furnished by other investigators.

Sulloway's theory predicts that firstborns and laterborns should differ in their political opinions: firstborns should be more conservative, laterborns more liberal. Albert Somit, Alan Arwine, and Steven Peterson surveyed the literature in their 1996 book on birth order and political behavior and concluded:

> We have looked at all of the literature dealing with the relationship between birth order and political behavior which we have been able to identify. This research ranges across a wide behavioral spectrum—personal participation in politics, interest in politics, liberalism–conservatism, attitudes toward free speech, leadership preferences, political socialization, Machiavellianism and non-traditional behavior, as well as elective and appointive office holding. In many of these studies the data show no meaningful relationship with birth order; in those where such a linkage was reported, critical analysis raised grave doubts, to put it mildly, about the validity of the findings.

Sulloway alleges that laterborns are more rebellious and less willing to conform to parental standards. One way that children and adolescents commonly rebel is by refusing to do their schoolwork; in taking this path they leave behind a paper trail of easily collectable data. The data have been collected and they contradict popular beliefs: the tendency to perform below capacity in school is not related to birth order. According to psychologist Robert McCall, "Systematic research . . . fails to confirm that underachievement is more common in later-borns than in firstborns."

Sulloway claims that laterborns are more open to innovative ideas. Psychologist Mark Runco has studied "divergent thinking" in children—

thinking that is off the beaten track. Firstborns and only children outscored laterborns.

Research has shown that, on the whole, marriages work better if the husband and wife are similar in personality and attitudes. If birth order had important effects on personality and attitudes, marriages should be happier if firstborns married other firstborns and laterborns married laterborns. The only evidence I've seen on this question suggests the opposite: psychologist Walter Toman reported that couples of mismatched birth order were less likely to divorce.

Finally, Sulloway's theory predicts that social upheavals should be more likely to occur during periods when the population contains a high proportion of laterborns. Frederic Townsend has tested this prediction with twentieth-century data and has soundly disconfirmed it. The generation of American 20- to 25-year-olds involved in the youth rebellion of the 1960s actually contained a relatively low proportion of laterborns. The proportion of laterborns was considerably higher during the placid 1950s. It rose again during the 1970s, just as the youth rebellion was petering out.

Birth Order, Evolution, and Social Change

Sulloway's theory is based on the Darwinian concept of survival of the fittest—the nature-red-in-tooth-and-claw view of evolution. In Sulloway's view, siblings are engaged in a life-or-death struggle for family resources. His models for fraternal relationships are Cain and Abel and the blue-footed booby, a species of bird in which the first chick to hatch reduces competition in the nest by pecking to death one of its smaller siblings.

Siblicide, however, is found mainly in species in which litters of young are reared in parallel. Primates generally rear their young serially, one at a time. As I related in Chapter 6, chimpanzee brothers are playmates in childhood and are likely to be valued allies in adulthood. The same is true, in traditional societies, for human brothers. Cain and Abel notwithstanding, fratricide is one of the rarest forms of murder in most human societies, including our own.

But fratricide becomes more common under some circumstances. It is more common at times and in places where everything goes to the eldest brother—the kingdom, the title, the farm—and younger siblings are left wanting. The homicides that occur under such circumstances appear, on the surface, to be exactly the sort of sibling rivalry Sulloway describes: a struggle for parental favor, for family resources. I believe, however, that what motivates these murders is not the younger brother's desire to improve his status with his parents—killing their firstborn child would hardly accomplish that!—but to improve his status in the society in which he is destined to spend his adult life. Primogeniture makes older brothers dominant in their *group,* not just within their family. Within-group vying for dominance can lead to murder, and this is true in many species and in every human society.

Relationships between siblings depend on factors not only within the family but outside of it as well, and that is why birth order effects may occur under some conditions. When primogeniture was the rule in European countries, younger siblings grew up under the shadow of their oldest brother, not just within the family but wherever they went. In an era when the children of the rich were educated at home and those of the poor weren't educated at all, children spent most of their day with their siblings. A younger brother was dominated by his elder brother not just at home but in the play group as well. My theory predicts that low status in the group, especially if it persists for years, will leave permanent marks on a child's personality.

In Western societies today, primogeniture is dead and children spend time with their siblings mainly at home. Outside the home they are in the company of their agemates. A younger brother who is dominated by his older sibling at home can be a dominant member of his peer group. The patterns of behavior developed in sibling relationships are left behind —left at home—when the modern child goes out the door, just the way the parents' language is left behind by the child of immigrants.

Perhaps birth order effects were real in the days of primogeniture; that may be the explanation for the historical data in Sulloway's book. In recent studies birth order effects are not found or are found to be negligible. This is true even for intelligence, where older data provided fairly clear evidence of birth order effects. Recent tests have failed to substantiate the earlier findings of a slight firstborn advantage in intelligence.

I conclude that Carmi Schooler had the right idea when he titled his article "Birth Order Effects: Not Here, Not Now!"

APPENDIX 2: TESTING THEORIES OF CHILD DEVELOPMENT

A test of group socialization theory—the first to be identified as such—appeared in the *Journal of Personality and Social Psychology* in 1997. The researcher was behavioral geneticist John Loehlin of the University of Texas. Reanalyzing data from a twin study carried out some years earlier, Loehlin found that adolescent twins who said they had the same friends were more alike in personality than those whose parents said they treated the twins alike. He summed up his findings this way: "The results of the present study offer mild support for a couple of predictions from Harris's theory that peers shape personality more than parents do."

The effects were in the predicted direction but they were not strong. Why weren't they stronger? For one thing, according to group socialization theory it is the influence of the peer group, not the influence of friends, that has long-term effects on personality. The twins weren't asked about their peer group—they were asked whether they had the same friends. Although having the same friends can serve as an indicator of shared peer group membership (because kids who are friends are usually members of the same peer group), it is an imperfect indicator (because kids who are not friends can nonetheless belong to the same peer group).

Moreover, the twins were asked about shared friends in *adolescence*—there was no information about the twins' friendships at earlier ages. I believe the most important aspects of personality development occur in childhood, not in adolescence.

Finally, peer group influence doesn't necessarily lead to greater similarity between twins. Groups differentiate as well as assimilate, and differentiation would *decrease* the similarities between twins.

Twins Are a Special Case

The fact that identical twins reared in the same home are not very alike in personality—no more alike than identical twins reared in separate homes—demands an explanation. The explanation offered by believers in the nurture assumption is that twins reared together have different experiences within the home. They may be treated differently by their parents or typecast by the family in different ways.

Twins reared together do have different experiences within the home, but I attribute their personality differences to the different experiences they have *outside* the home. However, for twins (and for siblings so close in age that they are in the same grade in school) the line between the two social contexts is blurred. A twin is a peer as well as a sibling. Twins see each other at school and in the neighborhood play group, not just at home.

Children who grow up in the same family are likely to be typecast or labeled in different ways—often, in contrasting ways—by the members of their family. But most children leave their home labels behind when they go out. If they took their home labels with them, we would see sizable and persistent effects of birth order on personality, and we do not (see Appendix 1). Twins, however, may take their home labels along when they go out, because they go out together. They provide a social context for each other both within the home and outside of it. Any asymmetries in their relationship—and there are bound to be some—go with them wherever they go.

There is a pair of conjoined twins growing up in the midwestern United States: two happy, pretty girls who share a single body. Abigail controls the right arm and leg, Brittany the left. Their separate brains were constructed according to identical instructions, coded in identical DNA. Their environments are identical too—they go everywhere together, they have no choice. And yet they are not identical in personality. I saw them on the Oprah Winfrey show and it was very clear: they each have a distinct personality. One twin—I think it's Abigail—is dominant. Her attitude toward her sister is maternal, protective. The other appears less sure of herself, younger. Perhaps these differences had their origin in the fact that Abigail has been healthier than Brittany (who is prone to lung and ear infections). However the asymmetry in their relationship began, it became a self-perpetuating pattern of interaction.

Abigail and Brittany will bring this pattern with them to the peer group. Their peers will notice the differences between them, just as I did when I saw them on TV. Their friends will see them as distinct individuals (once you get to know a pair of twins, you'll have a different working model for each of them) and will typecast them in distinct ways. Maybe they will grant one higher status in the group than the other, and address their questions to the higher-status twin. As a result, the personality differences between the twins will be stamped in, perhaps widened. Because the differences are expressed in the peer group and not only at home, they will become permanent parts of the twins' personalities. It doesn't work that way for ordinary siblings because they're different ages, and nowadays most children spend their time outside the home mostly in same-age groups. A child who is dominated by her older brother at home can leave her second-class status behind her when she goes out.

In former times, kids spent most of their day, inside the home and out, in the company of their siblings. Childhood is still like that for twins. That is why identical twins reared together are not identical in personality, even though they have the same genes and are socialized in the same peer group. Thus, for the special case of twins, group socialization theory makes a prediction that, in most cases, is indistinguishable from the prediction made on the basis of within-family environmental differences. It makes a different prediction only in those cases—and they are probably rare—where identical twins are typecast in one way inside the family and in another way by their peers, or one has higher status inside the family and the other has higher status outside.

Parent-to-Child Effects or Child-to-Parent Effects?

Let us turn, then, to the far more common case of ordinary siblings reared in the same home—siblings who are identical neither genetically nor in age. They may be full siblings, half siblings, or biologically unrelated step- or adoptive siblings.

If these various kinds of siblings are included in a single study (along with identical and fraternal twins), behavioral geneticists can use the data to calculate the heritability of the behavioral or personality characteristics they are studying. They can calculate how much of the variation in those characteristics is due to the siblings' genes, how much to the fact that they live in the same home (their "shared environment," behavioral geneticists call it), and how much of the variation remains unexplained by genes and home environment. Such a study was recently carried out by behavioral geneticist David Reiss and his colleagues.

Behavioral geneticists attribute nongenetic differences between siblings to the fact that each sibling occupies a unique "microenvironment." Parents do not treat their children all alike; therefore, each child has different experiences within the home. The problem is that parent-to-child effects —the kind needed by proponents of the nurture assumption to account for the differences in the children's personalities—are very difficult to distinguish from child-to-parent effects. If the parents are simply reacting to preexisting differences between their children, that doesn't explain how the differences got there. The differences between the children are not all due to their genes—we know that. But theoretically, at least, the differences in parental treatment *could* be due entirely to the children's genes. The parents could be reacting to genetic differences between their children even if not all the differences between the children are genetic.

That is exactly what Reiss and his colleagues found. "The results," he admits, "were stunning." Genetic differences between the children could account for almost all of the differences in the way their parents treated them. For example, the researchers found a high correlation between a mother's negative behavior toward one of her adolescent children and antisocial behavior by that adolescent. The traditional explanation of this finding would be: the kid is acting up because his mother isn't nice to him—his mother doesn't like him as much as she likes his sibling. But the data analysis indicated that genetic influences could account for most of the correlation between the mother's behavior and the kids'. The mother was *reacting* to inborn differences between her children—she wasn't causing them. As Reiss puts it:

> A better interpretation of our data is that genetic differences between the adolescent and the sibling occasion differential treatment: the child with heritable behavioral difficulties is treated more harshly.

Unlike Loehlin, Reiss does not identify his study as a test of group socialization theory. But it is. Reiss has verified a prediction made by the theory: he has shown that differences in the way their parents treat them cannot explain why siblings in the same family behave differently. We already knew that *similarities* in the way their parents treat them cannot explain why siblings in the same family sometimes behave alike: they behave alike only because they share genes. If they don't share genes, they don't behave alike.

If parents' behavior toward their children cannot account for either the similarities or the differences in the children's behavior, then the nurture assumption must be wrong.

Explaining the Variation

If the parents' behavior doesn't account for the nongenetic variation in their children's personalities, what does account for it? "What could these nonshared environmental factors be?" David Reiss wonders. Chance experiences? Not a happy thought, since chance events are "unmeasurable and unknowable." Perhaps, he muses, there are "other less capricious variables" that he and his colleagues have not yet explored. But this, Reiss says, seems unlikely because they have already explored so many variables. They've looked at families, friends, teachers, and peers. Yes, they've even looked at peer group membership.

But they haven't looked at the different experiences kids have *within* their peer group. According to group socialization theory, the members of a peer group became more alike in some ways and less alike in others. The variations in personality and social behavior that researchers measure are probably influenced as much by within-group differentiation as by assimilation to the group. There are differences in the way children are treated by their peers: some kids are picked on, others are imitated or have questions and suggestions addressed to them. There are differences in the way kids are typecast by their peers, or typecast themselves by comparing themselves to their groupmates.

Children have different experiences within the group; they also have different experiences at home. My theory predicts that only the group experiences will have long-term consequences. But it is easy to mistake the effects of group experiences for the effects of home experiences because experiences in the two contexts tend to be correlated. For example, many of the characteristics that put children at risk of being abused by their parents—retarded development, unattractive appearance, difficult temperament—also put them at risk of being victimized by their peers. Some children, therefore, are at the receiving end of both kinds of abuse. The long-term effects usually attributed to parental abuse (if they're not due to genetically influenced characteristics) may actually be the results of abuse by peers.

It should be possible to separate the effects of these two kinds of abuse, because some children who are popular with their peers are unpopular with their parents, or vice versa. Group socialization theory predicts that abuse by peers, but not abuse by parents, will have deleterious long-term effects on the personality. Developmentalist David Perry of Florida Atlantic University is currently running a longitudinal study that will test this prediction.

There is another way to test the theory, but it will work only for boys. Among boys (but not girls), height can serve as a rough indicator of status in the peer group: taller boys tend to have higher status than shorter ones. If, as seems likely, parents are as nice to short sons as to tall ones, height can be used to distinguish the effects of peer-group status from the effects of parental treatment. I would like to see a study that looks for associations between boys' height and variations in personality characteristics. There is old evidence that early-maturing boys (who tend to be taller than their agemates in childhood) have higher self-esteem, and newer evidence (see Chapter 8) that very short children are more likely to suffer from a variety of emotional problems, but so far, to my knowledge, there have been no systematic attempts to link variations in personality to height in childhood.

Among girls, prettiness can serve as a rough indicator of status in the peer group. However, prettiness makes girls popular with their parents as well, so this characteristic cannot be used to distinguish the effects of peer-group status from the effects of parental treatment.

The Right Kind of Research

In order to test theories of child development, it is necessary to tease apart three possible influences on children's behavior and personality: their genes, their experiences at home, and their experiences outside the home.

Behavioral genetic studies are the most straightforward way to estimate genetic influences, which can then be skimmed off in order to study environmental influences. For example, David Rowe of the University of Arizona studied genetic and environmental influences on teenage smoking. He showed that genetic influences account for the tendency of parents who smoke to have children who smoke, but that the environment also has effects. Once the genetic influences were subtracted, it became possible to see that the environment's influence on teenage smoking was entirely through the peer group: a teenager who belonged to a peer group that approved of smoking was more likely to experiment with tobacco. It was heredity, however, that determined whether or not the teenager would become addicted to nicotine.

Not every researcher has the interest or the resources to do behavioral genetic research. Fortunately, there are other ways of getting around the fact that children differ from each other partly for genetic reasons. One method is to let them serve as their own controls. Thomas Kindermann of Portland State University has done this. He studied cliques of fourth- and fifth-graders—small groups of kids who hang around together—and found that kids in the same clique generally have the same attitude toward schoolwork, pro or con. By high school such groups have solidified into fairly fixed social categories with labels like "brains" and "burnouts," but in elementary school the categories still have permeable borders. Over the course of a school year, many children switch cliques. Kindermann found that when children switch, their attitude toward schoolwork tends to change to match that of their new group. The change in attitude can be attributed to peer group influences because neither genetically influenced characteristics nor the parents' attitudes are likely to change over the course of a single school year.

Separating the effects of peer group variables from those of parental variables is difficult because they are correlated on so many different levels. Within a given neighborhood, the norms of the children's groups are likely to be similar to the norms of the parents—more similar, at any rate, than those of children and parents from different neighborhoods. Because parents who live in the same neighborhood tend to have similar styles of child-rearing, the effects of the kids on each other can be mistaken for the effects of the parents' child-rearing style, particularly if the study mixes together kids from a variety of neighborhoods. Group socialization theory makes the following prediction: that two biologically unrelated children of about the same age who are reared in the same home will be no more alike in personality and behavior (measured outside the home) than two biologically unrelated children of about the same age who are reared in different homes but who live in the same neighborhood and go to the same school.

This prediction has already been verified, because two biologically unrelated children reared in the same home are not alike at all by the time they are in high school. For younger children, however, there are some modest similarities between unrelated siblings reared in the same home. My theory predicts that these similarities will be no greater than those between unrelated children reared in different homes but in the same neighborhood.

If you are a developmental psychologist, you probably think that there is already more than enough evidence—indeed, a mountain of evidence—to disprove group socialization theory. I have, in the course of the book, mentioned some of the reasons why I do not believe this evidence is valid. Let me summarize here why I think the existing evidence does not prove what, at first glance, it appears to prove.

1. Very few of the studies provide a way to distinguish genetic influences from environmental influences.

2. Hardly any of the studies provide a way to distinguish child-to-parent effects from parent-to-child effects.

3. The researchers have not distinguished between the child's behavior at home and the child's behavior outside the home: they've simply assumed that measuring one tells you something about the other. In some cases they haven't even bothered to mention where the measurements were made.

4. The researchers have failed to consider circumstances that might influence children's experiences outside the home. For example, in studying the effects of divorce, they have failed to take into account the effects of moving to a new residence. My colleague David G. Myers has pointed out that the necessity of moving can legitimately be considered one of the consequences of divorce, and he is right. However, the two are separable: not all children of divorced parents move, and not all children who move have divorced parents. If the effects usually attributed to children's experiences within the family are really due to their experiences outside of it, people are being given the wrong kind of advice and their children are receiving the wrong kind of therapy.

5. Demographic factors have not been adequately controlled. When children from different ethnic groups, social classes, or neighborhoods are mixed together in the same study, misleading correlations between parents and children are likely to be found. The correlations reflect the fact that parents and children belong to the same ethnic group and social class and live in the same neighborhood. The children bear a closer resemblance to their own parents (and to the parents of their peers) than they do to parents from other ethnic groups, social classes, or neighborhoods.

6. Many methodological errors have been made. For example, in many studies the same informants are asked to make statements about (a) their parents' child-rearing methods and (b) their own behavior or psychological well-being, or (a) their own child-rearing methods and (b) their children's behavior. The correlations found between (a) and (b) are then taken as proof that (a) causes (b).

7. More generally, the research was not carried out in the impartial manner that is demanded in other scientific disciplines. The data collection was not "double blinded." Many variables were measured and then the data were scrutinized for significant correlations in any subset of the variables. Studies that didn't produce the desired results didn't get published. Studies that produced any results at all were overinterpreted.

Studying Context-Specific Behavior

Group socialization theory predicts that children will behave differently in different social contexts because learned behavior is specific to the context in which it is learned. Thus, any similarity between how children behave in different contexts (except in the case of twins, where the social contexts may not really be so different) must be due to genetic factors. Inherited characteristics, including physical appearance, affect the child's behavior in every context.

This prediction has been upheld. For example, Kimberly Saudino of Boston University recently reported that children who are shy and quiet at home also tend to be shy and quiet outside the home, and that this uniformity of behavior can be attributed almost entirely to built-in aspects of their temperament:

> This finding of cross-situational genetic effects for shyness in the lab and the home means that genetic factors contribute to the stability of shyness across the two situations. Indeed, the observed correlation between the two measures of shyness was almost entirely due to overlapping genetic effects. In contrast, environmental factors contributed to differences between shyness in the lab and in the home.

But genetic factors do not always produce similarities in behavior across contexts: *differences* in behavior can also be due to inherited characteristics. Saudino presents evidence that inherited characteristics sometimes have different effects in different contexts. This finding, too, is consistent with group socialization theory. Cinderella found that her beauty was a liability in her home but an asset outside of it.

Group socialization theory makes a strong prediction about learned behavior: that behavior acquired outside the home may leak into the home, but that the reverse will not occur. That, when the two contexts overlap, the behavior acquired outside the home will take precedence. To test this prediction, researchers can look at situations where children's contexts overlap. When a child brings friends home from school and is playing with them at home, whose rules of behavior does he follow, his parents' or his peers'? When a father takes his teenage daughter to a restaurant and invites a few of her friends to come along, does he see her behaving in a way that is unfamiliar to him? When parents visit the school, their children are pleased and/or embarrassed, but do they revert to behaving the way they behave at home? What does a boy do when he skins his knee in the presence of both his mother and his peers: does he cry, as he would with his mother, or act tough, as he would with his peers?

Studying Language and Accents

Group socialization theory can illuminate areas of development even when it doesn't make specific predictions about them. Consider, for example, the acquisition of a second language. When children are moved from one country to another at an early age, they can acquire the new language and speak it like a native. Though the parents will always have a foreign accent, the kids will speak their new language without an accent—if they were young enough when they made the move. But there are questions about second language acquisition that haven't been answered. How young do children have to be to acquire a new language and speak it without an accent? Why is this ability lost during development, somewhere around the early teens? And why do some individuals lose it earlier than others?

The answer to the last question will surely have to include genetic differences in language aptitude. Some people are apparently born with an ear for languages. A small fraction can continue to pick up new languages, and learn to speak them like a native, even when they are well advanced in age. They are natural mimics.

It is the converse finding—the finding that some people end up with "foreign" accents even if they immigrated at age four or five—that group socialization theory can help explain. Psycholinguists have been baffled by the variability in acquisition of a second language, especially by the fact that some people never lose their accent, even though they were very young when they immigrated. The psycholinguists haven't considered the variability in the outside-the-home environment of the immigrant child.

When people immigrate to a new country, most end up living in areas where there are others who came from the same homeland. It makes the transition easier for the adult generation. The new arrivals can continue to speak their native language. They can ask the previous immigrants for advice. The local stores stock familiar foods and label them with familiar names.

In such areas, children grow up with both languages. They may always speak their second language with an accent, even if they arrived in their new country as babies—even if they were born there—because that is how they hear it spoken. That is how their peers speak it.

In order to do an adequate study of second language acquisition, it is necessary for researchers to distinguish between two kinds of subjects: those who grow up in places such as Chinatowns or Mexican-American neighborhoods, and those like Joseph (see Chapter 11) who grow up in areas where no one other than their parents speaks the language of their homeland. The researchers should ask: Do the children use their first language or their second when they're with their peers? Do their peers speak the second language with an accent? I predict that when researchers control for these differences in outside-the-home language environment, much of the apparent variability in second language acquisition will disappear. They will find that children can acquire a second language without an accent (or, if the move is from one part of the country to another, that children can acquire a new accent) at least up to the age of eleven or twelve.

But most people eventually do lose this ability. Why is it lost? Because the brain loses its plasticity when the body matures, or because the social consequences of mispronouncing a word are less severe for older adolescents and adults? Both theories have their advocates. Group socialization theory doesn't take sides but provides a helpful tip on deciding between them: researchers should be careful to control for differences in their subjects' outside-the-home language environments. Once that is done, it may be possible to decide between the alternative hypotheses with an interesting test: look for sex differences. Physical maturation is completed at an earlier age in females than in males, so if the loss of language plasticity is due to a loss of brain plasticity, females should lose it at an earlier age. If researchers find that a boy of 13 can acquire a second language as readily as as a girl of 12, that would be evidence in favor of the physical maturation theory. (Steven Pinker of the Massachusetts Institute of Technology independently thought of the idea of looking for sex differences; he plans to put it to use in future research.)

One more question about second language acquisition; this is another one for which group socialization theory makes no prediction. Psycholinguists often say that babies lose their ability to hear the difference between sounds that are not distinguished in their language. The evidence is that babies stop responding to the differences. But if they had really lost their ability to distinguish between these sounds—if, for example, the part of the brain needed to distinguish between them had been taken away and reassigned to some other purpose—children would never be able to learn to speak a second language without an accent. So the loss of the ability to distinguish the sounds must not be a sensory loss: it must be more like learning not to pay attention to something. The question is: How do the children of immigrants learn to pay attention to sound distinctions that earlier they had learned to ignore? As far as I know, this has not been studied. To study it, it will again be necessary to take into account differences in children's outside-the-home language environments. The ideal subjects would be children like Joseph—children whose outside-the-home environment contains no one who speaks the language of their homeland. Or children from foreign countries who are adopted into homes where no one—neither their new parents nor their new peers—speaks the language of their homeland.

Proving It

My colleague David Lykken—once a clinical psychologist, now a member of the University of Minnesota team studying twins reared apart—disagrees with me about the efficacy of parents. He believes that parents *can* make a difference, at least at the extremes. That outstandingly good parents can succeed with a kid whom others might find unmanageable, and that outstandingly bad parents can turn a kid who might have been okay into a ne'er-do-well or criminal.

Perhaps our personal histories are relevant here. David and his wife have reared three successful, well-adjusted sons and I suspect it is difficult for him to give up the idea that he and his wife were at least partly responsible for this happy outcome. I, on the other hand, don't feel my husband and I deserve any credit at all for the way our two daughters turned out. The paths they took to adulthood were so diverse—so circuitous, in the case of the younger one—that it's hard to believe we had any influence on either of them. I am proud of both my daughters, but I don't believe that my parenting skills, or lack thereof, had anything to do with the way they turned out.

Although David Lykken and I don't agree on everything, there is much we do agree on. Today I received in the mail a chapter of a book he is writing. Here is a sentence from it:

> I think Harris makes a very powerful case, one that cannot be refuted on the basis of evidence collected under existing paradigms.

I believe Lykken is right: my theory of development cannot be refuted on the basis of existing evidence.

Nor can his. There is still a slim hope for the nurture assumption: the possibility that very, very bad parents may cause irremediable harm to their children.

The evidence indicates that differences between one home and another, between one set of parents and another, do not have long-term effects on the children who grow up in those homes. But all the evidence has come from "good-enough" homes—normal homes. The evidence covers a wide range of homes but it does not include the very, very bad ones where parents are viciously cruel or criminally negligent.

No one can deny that there are circumstances under which a child cannot possibly turn into a normal adult, even though he or she may survive childhood. The case of Genie is an example. Genie was kept alone in a room for thirteen years, tied to a potty chair. When she was discovered, she was unable to speak or walk, and she never did learn to speak grammatical English. Her social behavior remains highly abnormal; she lives in an institution. But Genie had no peers.

Group socialization theory predicts that no matter how bad their home environment might be, children will turn into normal adults if the following conditions are met: they do not inherit any pathological characteristics from their parents (so it will be necessary to use adopted children or stepchildren to test this prediction), their brains are not damaged by the neglect or abuse, and they have normal relationships with their peers. We can call this the Cinderella experiment.

Cinderella, by the way, turned out fine.

NOTES

Preface

page

xv Do parents have important long-term effects? Harris, 1995.

xvi Concentrates the mind wonderfully: Samuel Johnson, 1777. Some of it got published: Harris, Shaw, & Bates, 1979; Harris, Shaw, & Altom, 1985.

xvii The effectiveness of an effort: Waring, 1996, p. 76.

Chapter 1. "Nurture" Is Not the Same as "Environment"

1 Genotype choice: Morton, 1998, p. 48.

3 Parents who care: Clinton, 1996.

4 Nature and nurture: *The Tempest* (1611–1612), Act IV, Scene 1. An earlier use: Quoted in Gray, 1994, p. 49.

5 You can't marry your mother: Spock, 1968, p. 375 (first edition published in 1946).

6 A dozen healthy infants: Watson 1924, p. 104.

6 Recommendations to parents: Watson, 1928.

6 Reinforcing responses: Skinner, 1938.

6 The negative consequences of rewards: Deci, 1971; Lepper, Greene, & Nisbett, 1973. Rewards without negative consequences: Eisenberger & Cameron, 1996.

7 The effects of a "bad environment": Goodenough, 1945, p. 656 (first edition published in 1934).

7 Experiment in primate-rearing: Kellogg & Kellogg, 1933.

8 Your four-year-old: Gesell, 1940; Gesell & Ilg, 1943.

8 Disappointing results: Maccoby, 1992, p. 1008. (Maccoby's paragraph contained some references in parentheses; I have left them out.)

10 Neither seen nor heard: Glyn, 1970, p. 128.

11 Julia and the eggs: Fraiberg, 1959, p. 135.

11 Proper behavior for Polynesian children: Martini, 1994.

Chapter 2. The Nature (and Nurture) of the Evidence

17 The trouble with epidemiology: Taubes, 1995.

18 Obnoxiousness with parents versus obnoxiousness with peers: Dishion, Duncan, Eddy, Fagot, & Fetrow, 1994.

19 Mothers' total expressiveness: Parke, Cassidy, Burkes, Carson, & Boyum, 1992, p. 114. (Italics in the original.)

20 Long and thorough review of socialization research: Maccoby & Martin, 1983, p. 82.

21 Dog-rearing experiments: Freedman, 1958; Scott & Fuller, 1965.

21 The kind of parents they are: Pérusse, Neale, Heath, & Eaves, 1994.

21 Gauging the effects of heredity and environment: For a good introduction to the methods of behavioral genetics, see Plomin, 1990.

23 Fifty percent heredity: Bouchard, 1994; Plomin & Daniels, 1987; Tellegen, Lykken, Bouchard, Wilcox, Segal, & Rich, 1988. More precisely, heredity accounts for about 50 percent of the *reliable* variation in measured personality traits. Reliable variation is what's left after measurement error (which is around .20 for personality tests) is deducted. Estimated heritability of personality traits is often closer to .40 than .50 because in behavioral genetic analyses all the variation due to measurement error gets put on the other side—lumped into the estimate of environmental influences. Measurement error is lower for IQ tests (about .10), and that is one reason why calculated heritability of IQ is higher than that of personality traits.

24 The need to control for the effects of heredity: Plomin & Daniels, 1987; Scarr, 1993.

27 Autism: Bettelheim, 1959, 1967. A mother attacks Bettelheim: Gold, 1997. Born that way: Plomin, Owen, & McGuffin, 1994.

28 Descriptions of parental treatment by identical and fraternal twins: Rowe, 1981.

29 Descriptions of parental treatment by twins reared apart: Plomin, McClearn, Pedersen, Nesselroade, & Bergeman, 1988; Hur & Bouchard, 1995.

29 Cute and homely babies: Langlois, Ritter, Casey, & Sawin, 1995, p. 464.

30 Timidity is partly genetic: Stavish, 1994.

30 Hypersensitive nervous systems: Kagan, 1989; Fox, 1989.

31 The relationship is two-way: Bugental & Goodnow, 1998.

Chapter 3. Nature, Nurture, and None of the Above

33 Identical twins reared apart: Bajak, 1986; Lykken, McGue, Tellegen, & Bouchard, 1992; Wright, 1995.

34 Identical twins reared together: Plomin & Daniels, 1987; Tellegen, Lykken, Bouchard, Wilcox, Segal, & Rich, 1988.

34 The Giggle Twins: Lykken et al., 1992.

35 Unexpected results: Loehlin & Nichols, 1976.

36 Making them less alike: Plomin & Daniels, 1987.

36 Predictable effects on children: Smetana, 1994, p. 21.

37 Reared-together siblings: Plomin & Daniels, 1987. IQ correlations: Plomin, Chipuer, & Neiderhiser, 1994; Plomin, Fulker, Corley, & DeFries, 1997.

37 Toxic parents: Forward, 1989. Their presumed effects: Myers, 1998, p. 112.

38 Review of socialization research: Maccoby & Martin, 1983, p. 82.

41 The firstborn is "dethroned": Adler, 1927.

41 First-, second-, and thirdborns: Bradshaw, 1988, pp. 33–35.

42 Birth order can't be very significant: Dunn & Plomin, 1990, p. 85.

43 Is there a "firstborn personality"? Ernst & Angst, 1983, p. x.

43 Family size and socioeconomic status (SES) can bias the results: For example, firstborns more often achieve eminence not because they were born first but because they are more likely to come from small, upper-class families. Small families produce a dearth of laterborns, so a laterborn picked at random from the population is more likely to have come from a large family, which means that a firstborn is more likely to have come from a small one. Small families are, on the average, higher in SES than large families, so a firstborn is more likely than a laterborn to have come from a family of high SES.

43 Dynamic psychology will have to be revised: Ernst & Angst, 1983, p. 284.

44 Birth order and heterodox views: Sulloway, 1996.

45 Parental favoritism: Sulloway, 1996, p. 90; Dunn & Plomin, 1990, pp. 63, 74–75; McHale, Crouter, McGuire, & Updegraff, 1995.

45 A sheer waste of time: Ernst & Angst, 1983, p. xi.

47 Including my own: Harris & Liebert, 1991, pp. 322–325. At least I didn't swallow it whole—I did point out some of the problems with this kind of research.

47 Styles of parenting: Baumrind, 1967; Baumrind & Black, 1967.

47 Different for girls and for boys: Baumrind, 1989. Found only for white kids: Darling & Steinberg, 1993.

49 Chinese-American child-rearing style: Chao, 1994. Asian-American child-rearing styles and children's characteristics: Dornbusch, Ritter, Leiderman, Roberts, & Fraleigh, 1987; Steinberg, Dornbusch, & Brown, 1992. African-American child-rearing styles and children's characteristics: Deater-Deckard, Dodge, Bates, & Pettit, 1996.

50 Effects of maternal employment: Hoffman, 1989, p. 289.

50 Effects of nonmaternal care: Scarr, 1997b, p. 145. See also Andersson, 1992; Roggman, Langlois, Hubbs-Tait, & Rieser-Danner, 1994.

50 Unconventional families: Weisner, 1986.

51 Children of lesbian and gay parents: Flaks, Ficher, Masterpasqua, & Joseph, 1995; Gottman, 1990; Patterson, 1992, 1994.

51 Genes play a role in sexual orientation: Bailey & Pillard, 1991; Bailey, Pillard, Neale, & Agyei, 1993; Friedman & Downey, 1994.

51 Unintended pregnancies: Gottlieb, 1995. Children conceived with modern reproductive technologies: Golombok, Cook, Bish, & Murray, 1995.

51 Conceived through donor insemination: Chan, Raboy, & Patterson, 1998.

52 Only children versus children with one or two siblings: Chen & Goldsmith, 1991; Falbo & Polit, 1986; Falbo & Poston, 1993; Meredith, Abbott, & Ming, 1993; Veenhoven & Verkuyten, 1989; Yang, Ollendick, Dong, Xia, & Lin, 1995. I am restricting the comparison to small families because there are other differences, including socioeconomic ones, between small families and larger ones.

52 Parents aren't the be-all and end-all of the child's life: Rowe, 1994.

53 It remains a mystery: Bouchard, 1994, p. 1701.

Chapter 4. Separate Worlds

54 The story of Cinderella: Gruenberg, 1942, p. 181.

54 Families broken by death: Coontz, 1992.

55 Three faces of Eve: Thigpen & Cleckley, 1954.

56 Multiple selves: James, 1890, p. 294. (Italics in the original.)

56 Arguing about personality: Carson, 1989.

56 Blooming, buzzing confusion: James, 1890, p. 488.

57 No evidence for transfer: Detterman, 1993.

57 The learning ability of young babies: These experiments are summarized in Rovee-Collier, 1993.

58 Crybabies avoided: Kopp, 1989.

58 Playing different roles in games of pretend: Garvey, 1990.

59 Pretense appears early: Piaget, 1962.

59 Research on children's play: Fein & Fryer, 1995a, p. 367. Chuck 'em or change 'em: Fein & Fryer, 1995b, pp. 401, 402.

60 Babies of depressed mothers: Pelaez-Nogueras, Field, Cigales, Gonzalez, & Clasky, 1994, p. 358. See also Zimmerman & McDonald, 1995.

60 Parents' descriptions do not agree with those of other caregivers: Fagot, 1995; Goldsmith, 1996, p. 230.

60 No evidence of carryover: Abramovitch, Corter, Pepler, & Stanhope, 1986, p. 228.

60 Few significant associations: Stocker & Dunn, 1990, p. 239.

61 Equal and unequal relationships: Bugental & Goodnow, 1998. Sibling interactions more likely to involve conflict: Volling, Youngblade, & Belsky, 1997.

61 Born to be rivals: Sulloway, 1996.

61 Judgments by parents and siblings: Ernst & Angst, 1983, pp. 167–171.

62 Picky eaters: Rydell, Dahl, & Sundelin, 1995.

62 Obnoxious behavior in two settings: Dishion, Duncan, Eddy, Fagot, & Fetrow, 1994.

62 Heritability of disagreeableness and aggressiveness: Bouchard, 1994; Plomin & Daniels, 1987; van den Oord, Boomsma, & Verhulst, 1994.

62 What they are born with they cannot leave behind: Saudino, 1997.

63 Prettiness is an asset: Burns & Farina, 1992.

63 Trouble both directly and indirectly: Caspi, Elder, & Bem, 1987.

63 Language is acquired but innate: Pinker, 1994.

64 English-speaking child in Montreal: Baron, 1992, p. 183. Swedish-speaking child in Finland: P. Pollesello (1996, March 5), What is a native language? (Netnews posting in alt.usage.english.sci.lang).

65 The son of Polish immigrants: Winitz, Gillespie, & Starcev, 1995.

65 Code-switching in adults: Kolers, 1975, pp. 195, 190 (originally published in 1968).

66 The girl who said "shoot": A. Fletcher (1996, December 31), A word misspoken (Netnews posting in rec.humor.funny).

66 Language and social context: Levin & Garrett, 1990; Levin & Novak, 1991.

67 Multiple personality disorder: Eich, Macaulay, Loewenstein, & Dihle, 1997; Putnam, 1989.

67 Portnoy's kitchen utensil: Roth, 1967, p. 107.

68 Parents speak Korean: Lee, 1995, p. 167. Parents speak Yiddish: Meyerhoff, 1978, p. 43.

68 Parents speak Cantonese: Mar, 1995, p. 50.

69 Parents speak Bengali: Sastry, 1996, p. AA5.

69 Language learning is the child's job: Snow, 1991. Mothers don't speak to prelinguistic children: Pinker, 1994, p. 40. Two-year-olds retarded in language development: Kagan, 1978.

70 The king's experiment: Herodotus, Book 2.

70 Children of deaf become fluent speakers of English: Lenneberg, 1972. They find the question offensive: Preston, 1994.

71 Playing House: Garvey, 1990, pp. 88, 91. Girl whose mother was a physician: Maccoby & Jacklin, 1974, p. 364.

72 Mutilation is commonplace: Opie & Opie, 1969, p. 305.

72 Dad in his poodle hat: Barry, 1996.

73 Cheating in various contexts: Hartshorne & May, 1928.
74 Context effects: Council, 1993, p. 31.
76 If the child is observed outside the home: Such discrepancies are sometimes buried deep within the data. Scrutinize, for example, a report by two prominent researchers (Hetherington & Clingempeel, 1992) on the effects of parents' divorce and remarriage on children. Almost all the ill effects were either reported by parents or stepparents or by the children themselves, in interviews carried out in their homes. When teachers were asked to report on the children's behavior in school, on three occasions they reported no differences between children whose parents had divorced and remarried and those whose parents had never divorced (p. 60). Those whose parents had divorced and not remarried were, according to a single report by their teachers, showing more signs of behavioral problems. However, another teachers' report failed to turn up a difference, and the third teachers' report was missing from the data (p. 58).
76 Style-of-parenting research: The questions about how their parents treat them may even be embedded in the same questionnaire in which adolescents are asked to describe their own behavior. See, for example, Steinberg, Dornbusch, & Brown, 1992, p. 725.
76 Parents' reports versus teenagers' reports: Patterson & Yoerger, 1991.
76 The effects of parental favoritism: Brody & Stoneman, 1994; Stocker, Dunn, & Plomin, 1989. They may last a lifetime: Bedford, 1992.

Chapter 5. Other Times, Other Places

78 In his fate: Minturn & Hitchcock, 1963, p. 288. Not objects of anxiety: p. 317.
79 Parental guilt is a modern invention: Dencik, 1989, pp. 155–156.
79 Characteristics of the modern family: Hareven, 1985, p. 20.
79 Displays of emotion frowned upon: Dencik, 1989; Fine, 1981.
80 Private life distinguished from public life: Jacobs & Davies, 1991.
80 The home of the Brun family: Rybczynski, 1986.
81 (footnote) What the baby might put in its mouth: Eibl-Eibesfeldt, 1989, p. 600.
81 Babies sleep with their mothers: Anders & Taylor, 1994.
81 Reactions of Mayan mothers: Morelli, Rogoff, Oppenheim, & Goldsmith, 1992, p. 608.
82 Children were not "cared for": Schor, 1992, p. 92.
82 The home became private: Jacobs & Davies, 1991. More children surviving: Hareven, 1985.
83 Prussian common law: Schütze, 1987.
83 Puritan minister: Quoted in Moran & Vinovskis, 1985, p. 26.
84 Spartan meals for British children: Glyn, 1970. Holt's book: Cited in Hulbert, 1996. Benjamin was skeletally thin: Hulbert, 1996, p. 84.
84 Our posture was straight: Lewald, 1871, quoted in Schütze, 1987, p. 51.
84 Mother love: Schütze, 1987, p. 52. Tyrant of the house: Müller, 1922, quoted in Schütze, p. 52.
85 A sensible way of treating children: Watson, 1928, pp. 81–82.
85 Watson was the first: Schütze, 1987, p. 56. The mother can drop from exhaustion: p. 61.
85 The mother's fault: Ambert, 1994b; Sommerfeld, 1989.
85 (footnote) Female advice-givers: Hetzer, 1937, quoted in Schütze, 1987, p. 58.
86 Daily messages of love and acceptance: Neifert, 1991, p. 77. (Italics in the original.)
86 Mother-infant bonding: Klaus & Kennell, 1976.

87 Not permitted to bond: Schütze, 1987, p. 73.
87 Warnings of a British pediatrician: Jolly, 1978, quoted in Eyer, 1992, pp. 42–43.
87 Research on bonding: Eyer, 1992, pp. 3–4.
88 Maternal behavior in deer: Klopfer, 1971.
88 Deaths in pregnancy or childbirth: Crossette, 1996.
88 Childbirth is not a solitary activity: Trevathan, 1993.
89 Childbirth among the Efe: Morelli, Winn, & Tronick, 1987, p. 16.
89 Experts' advice to mothers: Sommerfeld, 1989.
90 Force feeding: LeVine & LeVine, 1963, p. 141.
90 No longer force fed: LeVine & LeVine, 1988.
90 The decision to keep the baby: Eibl-Eibesfeldt, 1989, p. 194; Pinker, 1997, pp. 443–444.
91 Sibling rivalry in traditional cultures: Eibl-Eibesfeldt, 1989; LeVine & LeVine, 1963; Whiting & Edwards, 1988.
92 Parents seem to lose interest: Youniss, 1992.
92 Children are raised in the play group: Eibl-Eibesfeldt, 1989, pp. 600–601.
92 His friends were waiting: Maretzki & Maretzki, 1963. Younger boys tag along: LeVine & LeVine, 1963.
93 Mothers prefer girls as babysitters: Whiting & Edwards, 1988. Boy who rescued his baby brother: Goodall, 1986, p. 282.
93 Domination by older children considered natural: Whiting & Edwards, 1988.
94 Futile attempts to prevent domination: Edwards, 1992.
95 Learning Zinacanteco weaving: Turok, 1972, quoted in Greenfield & Childs, 1991, p. 150.
95 Learned by imitation: Rogoff, Mistry, Göncü, & Mosier, 1993.

Chapter 6. Human Nature

98 (footnote) The swimming abilities of babies: McGraw, 1939.
99 Gua and Donald: Kellogg & Kellogg, 1933, pp. 69, 149.
99 Gua was more fun: Kellogg & Kellogg, 1933.
100 Humans imitate apes: de Waal, 1989, p. 36.
100 Donald was the imitator: Kellogg & Kellogg, 1933, p. 141.
100 Fifty words: Fenson, Dale, Reznick, Bates, Thal, & Pethick, 1994.
100 (footnote) Donald graduated from med school: L. T. Benjamin, personal communication, September 13, 1996.
101 Theory of mind in human children: Astington, 1993; Leslie, 1994; Perner, 1991; Wellman, 1990. It was Premack & Woodruff, 1978, who invented this term. They used it to raise some interesting questions about *chimpanzee* cognition.
102 The baby watches the mother's face: Klinnert, 1984; Sorce, Emde, Campos, & Klinnert, 1985. Reaction to a stranger: Eibl-Eibesfeldt, 1995.
102 Pointing in humans: Baron-Cohen, Campbell, Karmiloff-Smith, Grant, & Walker, 1995. In apes: Tomasello, 1995.
102 An ape's reaction to an object: Terrace, 1985, p. 1022. Terrace concluded that chimpanzees can learn to use sign language *words* but they cannot produce genuine sign language *sentences.*
103 Mindreading: Baron-Cohen et al., 1995. Mindblindness: Baron-Cohen, 1995.
103 Williams syndrome: Karmiloff-Smith, Klima, Bellugi, Grant, & Baron-Cohen, 1995.
104 A touchy-feely bunch: Goodall, 1986.

104 The victor grants forgiveness: de Waal, 1989. Males may attempt to monopolize a female: Wrangham & Peterson, 1996. Males take turns: Goodall, 1986, p. 443.

105 Defending McGregor: Goodall, 1988, p. 222. Goodall says: "Hugo and I went to stand in front of the cripple. To our relief, the displaying male turned aside." (Hugo van Lawick was the photographer who took the magnificent photos reproduced in Goodall's book.)

105 "Us" versus "them" in chimpanzees: Russell, 1993. Not complete strangers: Goodall, 1986, p. 331.

106 The attack on Godi: Goodall, 1986, p. 506.

106 Rahab the harlot was spared: Joshua 6:22–25.

106 No instinct to make war: Montagu, 1976, p. 59. He cites Julian Huxley. The word *instinct:* Pinker, 1994.

106 Preadaptations of the chimpanzee: Goodall, 1986, p. 531.

107 Man is a social being: Darwin, 1871, p. 480.

107 Paleontological evidence of war: Keeley, 1996. Our prehuman heritage: Diamond, 1992b, p. 297.

107 Attacking one's neighbors: Wrangham & Peterson, 1996.

108 Human hallmarks: Diamond, 1992b, p. 294.

108 The noble savage: Darwin, 1871, p. 481.

108 Giving up his life to save his relatives: According to kinship theory, it would make sense for the man to give up his life if he could thereby save more than two of his children or siblings (with whom he shares 50 percent of his genes) or more than eight of his cousins (with whom he shares $12\frac{1}{2}$ percent). See Pinker, 1997, pp. 398–402.

108 Born selfish: Dawkins, 1976, p. 3.

109 The evolution of cheater-detectors: Cosmides & Tooby, 1992; Pinker, 1997, pp. 403–405.

109 Early warfare: Goodall, 1986, p. 531.

109 Six million years: The timespans given here are rough estimates, based on my reading of the paleoanthropological literature. When I say "six million years," for example, I mean "six million years, give or take a couple of million." The theory of hominid evolution recounted here is the one that, in my opinion, best fits the available data.

109 DNA differences: Diamond, 1992b.

110 Furry faces: Holden, 1995.

111 The great leap forward: Diamond, 1992b, p. 32. Cultural takeoff: M. Harris, 1989, p. 64.

111 A village's mortal enemy: Quoted in de Waal, 1989, p. 247.

111 Roughing up the kings: Joshua 10:24–26.

112 Hunter-gatherer groups: Eibl-Eibesfeldt, 1989, p. 323.

112 Childhood xenophobia: Eibl-Eibesfeldt, 1995, p. 256.

112 Speciation: Gould, 1980.

113 Grasshoppers that don't interbreed: Parker, 1996.

113 Groups demarcate themselves from others: Eibl-Eibesfeldt, 1995, p. 260–261.

114 Genetic diversity in New Guinea: Diamond, 1992b, p. 229.

115 Neanderthals couldn't sew: Diamond, 1992b, p. 43.

116 The defeat of Ai: Joshua 8:1–29.

117 Larger groups are a cognitive achievement: Dunbar, 1993.

117 A stranger at Jericho: Joshua 5:13.

117 Figan keeps a secret: Goodall, 1986, p. 579.

117 Chimpanzees after a fight: de Waal, 1989, p. 43.
118 Whether chimpanzees have a theory of mind: Povinelli & Eddy, 1996.
118 We treat machines like humans: Caporael, 1986.
119 Children of deaf couples: Preston, 1994.
119 Useful innovations: Rowe, 1994.
120 Loss of parents among the Yanomamö: Chagnon, 1992, p. 177.
121 Resisting parental manipulation: Trivers, 1985, p. 159.
121 The blue-footed booby: Sulloway, 1996, p. 61.
121 Chimpanzee brothers: Goodall, 1986, pp. 176–177.

Chapter 7. Us and Them

123 Stranded on an island: Golding, 1954.
123 Little boys seek out older ones: Whiting & Edwards, 1988.
124 Antiwar views: Montagu, 1976. A pack of British boys: Golding, 1954, p. 242.
125 The tribes inhabiting adjacent districts: Darwin, 1871, pp. 480–481. (Italics added.)
125 A study of group relations: Sherif, Harvey, White, Hood, & Sherif, 1961.
126 They wanted to run them off: Sherif et al., 1961, p. 78.
128 The mere fact of division into groups: Tajfel, 1970, p. 96.
128 Us south-siders: Sherif et al., 1961, p. 76.
129 A cap with a silver badge: Golding, 1954, p. 18.
129 Class distinctions among British schoolboys: Glyn, 1970; Hibbert, 1987.
129 Cheering the losers: Sherif et al., 1961, p. 104.
129 Naming is classifying: Hayakawa, 1964, p. 216.
130 The advantages of categorization: Pinker, 1997. The dangers of categorization: Hayakawa, 1964, p. 220.
130 Categories are not arbitrary: Pinker, 1994; Rosch, 1978.
130 The categorization skills of pigeons: Roitblat & von Fersen, 1992; Wasserman, 1993.
130 Three-month-old babies can categorize: Eimas & Quinn, 1994. Babies can form concepts: Mandler, 1992. An underestimator of babies: Piaget, 1952.
131 The categorization skills of babies: Eimas & Quinn, 1994; Mandler & McDonough, 1993; Levy & Haaf, 1994; Leinbach & Fagot, 1993. The facial differences between grownups and children: Bigelow, MacLean, Wood, & Smith, 1990; Brooks & Lewis, 1976.
131 Three ways we categorize people: Fiske, 1992.
131 Dislike of being categorized: Hayakawa, 1964, p. 217.
132 The consequences of categorization: Krueger, 1992; Krueger & Clement, 1994.
132 Contrast effects in groups that were alike to begin with: Wilder, 1986.
132 Boys talk dirty: Fine, 1986.
132 The Eagles opt for prayer: Sherif et al., 1961, p. 106.
133 The nail that sticks up: WuDunn, 1996. Teenagers aren't pushed to conform: Lightfoot, 1992.
134 Experiments on group conformity: Asch, 1987, pp. 462, 464 (originally published in 1952).
134 The gang seizes on idiosyncrasies: Stone & Church, 1957, p. 207.
135 The group clown: Sherif et al., 1961, p. 78. The nickname "Nudie": p. 92.
135 Reluctant to fight: Chagnon, 1988, p. 988.
136 Family members stuck together: Diamond, 1992a, p. 107.

137 The birds all died: "Effort to reintroduce thick-billed parrots in Arizona is dropped," 1995.
138 Self-categorization: Turner, 1987.
138 Self-esteem as a motive: Turner, 1987.
139 The gene is preserved: Dawkins, 1976.
139 Kin recognition: Pfennig & Sherman, 1995.
139 Why we don't marry our siblings: Bem, 1996.
140 Married couples resemble each other: Diamond, 1992b, p. 102; O'Leary & Smith, 1991.
140 People who are similar to you: Segal, 1993.
140 The crippled chimpanzee: Goodall, 1988.
141 A unique individual: Turner, 1987.
142 Salience: Turner, 1987, hasn't completely solved the problem, because his answer doesn't explain why we divide up people into the particular social categories that are salient to us. Why not people with freckles versus people without them? People with long names versus people with short ones? Theoretically, there is no end to the ways we could categorize people and ourselves. Pinker (1994, pp. 416–417) has discussed this problem with regard to "similarity" and has come to the conclusion that our sense of similarity must be innate. The same must be true of social categories: we are predisposed to categorize people in certain ways, especially by age and sex.
142 Reference group: Shibutani, 1955. Psychological group: Turner, 1987, pp. 1–2.
142 Simultaneously friends and rivals: de Waal, 1989, p. 1.
143 The need for a common enemy: de Waal, 1989, p. 267.
143 A pair of !Kung San brothers: Eibl-Eibesfeldt, 1989, p. 596.
143 The *Little House* books: Wilder, 1971 (originally published in 1935).
144 The psychologically significant group: Turner, 1987, pp. 1–2.
145 Unification and simplification: Einstein, 1991, p. 40 (originally published in 1950).

Chapter 8. In the Company of Children

148 Primates seek playmates: Edwards, 1992; Fagen, 1993; Goodall, 1986; Kellogg & Kellogg, 1933; Napier & Napier, 1985.
148 A pair of unacquainted babies: Eckerman & Didow, 1988.
148 Secure base: Ainsworth, 1977, p. 59.
149 Primate mother–offspring relationships: Goodall, 1986, p. 275.
149 The evolution of love: Eibl-Eibesfeldt, 1995.
150 Letting go of mother, in primates: Goodall, 1986, p. 166. In humans: Leach, 1972; McGrew, 1972.
150 Attachment theory: Ainsworth, 1977; Ainsworth, Blehar, Waters, & Wall, 1978. For a recent summary of attachment research, see Rubin, Bukowski, & Parker, 1998.
150 Attachment in abused children: Egeland & Sroufe, 1981.
151 (footnote) Stepping on the duckling's feet: Hess, 1970.
151 Testing security of attachment: Ainsworth et al., 1978; Belsky, Rovine, & Taylor, 1984; Sroufe, 1985.
152 Working model: Bowlby, 1969, 1973. See also Bretherton, 1985; Main, Kaplan, & Cassidy, 1985.
152 Security of attachment and peer relations: Erickson, Sroufe, & Egeland, 1985; LaFreniere & Sroufe, 1985; Pastor, 1981. And problem solving: Matas, Arend, &

Sroufe, 1978. Contrary results: Howes, Matheson, & Hamilton, 1994; Youngblade, Park, & Belsky, 1993.

152 Little empirical support: Lamb & Nash, 1989, p. 240.

152 Independence of attachment relationships: Fox, Kimmerly, & Schafer, 1991; Main & Weston, 1981; Goossens & van IJzendoorn, 1990.

152 The effects of the child's characteristics: Ge et al., 1996; Jacobson & Wille, 1986; Scarr & McCartney, 1983.

153 Brain growth: Tanner, 1978. Development of the visual system: Mitchell, 1980.

153 Motherless monkeys: Harlow & Harlow, 1962. Reared with peers: Harlow & Harlow, 1962; Suomi & Harlow, 1975. Reared without peers: Harlow & Harlow, 1962, p. 146. According to Suomi, 1997, there are some subtle behavioral deficiencies in the monkeys reared with peers and without mothers—that is, there are some statistical differences between the behavior of these monkeys and that of normally reared monkeys. The important point, however, is that the behavior of these monkeys falls within the normal range of monkey behavior.

154 Concentration camp children: Freud & Dann, 1967, pp. 497–500 (originally published in 1951).

155 Effective lives: Hartup, 1983, pp. 157–158.

155 Romanian orphans: Kaler & Freeman, 1994, p. 778. See also Dontas, Maratos, Fafoutis, & Karangelis, 1985.

155 Unable to care deeply: Holden, 1996; Rutter, 1979.

156 Eritrean orphans: Wolff, Tesfai, Egasso, & Aradom, 1995, p. 633.

156 Childhood in an institution: Maunders, 1994, pp. 393, 399.

156 Children reared on isolated farms: Parker, Rubin, Price, & DeRosier, 1995. Children with chronic physical disorders: Ireys, Werthamer-Larsson, Kolodner, & Gross, 1994, p. 205; Pless & Nolan, 1991.

157 The ones who are off the charts: Winner, 1997.

157 The story of William James Sidis: Montour, 1977, p. 271; Primus IV, 1998, p. 80.

158 For the story of Victor, see Lane, 1976; for the story of Genie, see Rymer, 1993.

158 Isolated twins: Koluchová, 1972, 1976. No pathological symptoms: 1976, p. 182.

159 Baby imitates baby: Eckerman & Didow, 1996; Eckerman, Davis, & Didow, 1989. Baby imitates chimpanzee: Kellogg & Kellogg, 1933.

159 Development of play, at two and a half: Eckerman & Didow, 1996. At three: Göncü & Kessel, 1988; Howes, 1985.

159 Children's preferences for certain peers: Howes, 1987; Strayer & Santos, 1996; Rubin et al., 1998. For same-age peers: Bailey, McWilliam, Ware, & Burchinal, 1993. For same-sex peers: Maccoby & Jacklin, 1987; Strayer & Santos, 1996.

160 Rare for children to have agemates: Edwards, 1992; Konner, 1972; Smith, 1988. Older ones form their own groups: Edwards, 1992.

160 Older ones teach the younger: Eibl-Eibesfeldt, 1989. Teasing and ridicule: Martini, 1994; Nydegger & Nydegger, 1963. Serious aggression is uncommon: Edwards, 1992; Konner, 1972; Martini, 1994. Children less aggressive when playing by themselves: Lore & Schultz, 1993; Opie & Opie, 1969.

160 Three-year-olds just beginning to talk: Kagan, 1978; Zukow, 1989. Conversational partners: McDonald, Sigman, Espinosa, & Neumann, 1994; Rogoff, Mistry, Göncü, & Mosier, 1993.

161 Any adult can admonish a child: Maretzki & Maretzki, 1963; Youniss, 1992.

161 Socialization within the play group: Eibl-Eibesfeldt, 1989, p. 600.

161 Okinawan boy: Maretzki & Maretzki, 1963. Chewong children: Howell, 1988, pp. 160, 162.

162 Features found in young animals: Archer, 1992b, p. 77.

162 Social behavior in two chimpanzee troops: Mitani, Hasegawa, Gros-Louis, Marler, & Byrne, 1992. In two Mexican villages: Fry, 1988, p. 1016. "La Paz" and "San Andres" are not the real names of these villages.

163 Little boys pretending to shave: Harris & Liebert, 1991, p. 95.

163 Rules of behavior for Polynesian children: Martini, 1994.

163 Trophy wives: Chagnon, 1992.

164 Selective imitation: Jacklin, 1989; Perry & Bussey, 1984. Child's refusal to speak German: T. A. Kindermann, personal communication, August 9, 1995.

164 Role-playing games are rare in some societies: LeVine & LeVine, 1963; Martini, 1994; Pan, 1994. In every society, little girls make mud pies and pretend they are food. Playing House involves more than that—it means taking on a different persona, talking in a different voice, playing a role in a cooperative fantasy. Mud pies are universal; playing House is not.

164 Girls found in wolves' den: Maclean, 1977.

165 Donald imitated Gua: Kellogg & Kellogg, 1933. Children imitate older siblings: Brody, Stoneman, MacKinnon, & MacKinnon, 1985; Edwards, 1992; Zukow, 1989.

165 Children can learn by imitation: Rogoff et al., 1993. Organisms have to be rewarded: Skinner, 1938. Children can learn by observing: Bandura & Walters, 1963.

166 How to get a preschooler to like broccoli: Birch, 1987.

166 The daughter of a British psycholinguist: Baron, 1992, p. 181.

166 Groupness: Tajfel, 1970. Some of its characteristics: Turner, 1987.

167 Subsystems of the visual system: Farah, 1992; Pinker, 1997; Rao, Rainer, & Miller, 1997.

168 Most behavior *is* social behavior: Scott, 1987.

169 A three-year-old knows she's a girl: Ruble & Martin, 1998. Race doesn't matter: Stevenson & Stevenson, 1960.

169 Group socialization theory: Harris, 1995. "Socialization" implies something that is done to children: Corsaro, 1997.

169 Children take on the attitudes, behaviors, and styles of dress of their peers: Adler, Kless, & Adler, 1992; Readdick, Grise, Heitmeyer, & Furst, 1996.

169 "Warshington": Reich, 1986, p. 306.

170 Laughter as a weapon: Eibl-Eibesfeldt, 1989.

170 Group as a bunch of people versus group as a social category: Merten, 1996b, p. 40. Psychological group: Turner, 1987, p. 1.

170 American boy in a Tibetan monastery: "Daja Meston '96," 1995, p. 5.

171 Having a friend versus acceptance or rejection by the group: Bagwell, Newcomb, & Bukowski, 1998, p. 150. Having a friend in fifth grade had "unique predictive implications only for positive relations with family members" (p. 150). The two factors appeared to operate independently of each other, as group socialization theory would predict.

171 Friendship not the same as status in the peer group: Bukowski, Pizzamiglio, Newcomb, & Hoza, 1996; Parker & Asher, 1993. Friends are usually members of the same group: Hallinan, 1992.

171 Sex segregation: Edwards, 1992; Maccoby & Jacklin, 1987; Strayer & Santos, 1996.

172 Self-categorization and preference for own gender: Alexander & Hines, 1994; Powlishta, 1995a.

172 Little girl won't wear "boy shoes": T. A. Kindermann, personal communication, January 22, 1997.

172 Little girl thinks only boys can play with guns: S. M. Bellovin (1989, November 18), Toys and sexual stereotypes (Netnews posting in misc.kids).

172 Mothers don't play hopscotch: Maccoby & Jacklin, 1974, p. 363.

172 How to act toward the opposite sex: Sroufe, Bennett, Englund, & Urban, 1993; Thorne, 1993. Eleven-year-old girl explains the penalties: Maccoby & Jacklin, 1987, p. 245.

173 Other distinctions become increasingly important: Hallinan & Teixeira, 1987; Hartup, 1983.

173 Juan's science lab: Schofield, 1981, p. 63.

173 Emotionality is frowned upon: Dencik, 1989; Eisenberg, Fabes, Bernzweig, Karbon, Poulin, & Hanish, 1993; Hubbard & Coie, 1994.

173 Swedish study: Kerr, Lambert, Stattin, & Klackenberg-Larsson, 1994.

174 Coeducation leads to mutual dislike: Hayden-Thomson, Rubin, & Hymel, 1987. They disliked all the girls in their class: Bigler, 1995, p. 1083.

175 Romantic relationships in prepubescent children: Smart & Smart, 1978, pp. 198–200; Smith, Snow, Ironsmith, & Poteat, 1993.

175 Resistance to adult rules in the nursery school: Corsaro, 1993, p. 360.

176 Cognitive advances around the age of seven: Piaget & Inhelder, 1969. Leaving home at about that age: Rybczynski, 1986; Schor, 1992.

177 Hovering between "us" and "me": Turner, 1987. Finding a few ways to be different: Tesser, 1988. People in Western cultures—cultures called "individualistic" (Triandis, 1994)—tend to remain closer to the "me" end of the continuum than people in more traditional cultures.

177 "Better" means different things for boys and girls: Adler, Kless, & Adler, 1992; Maccoby & Jacklin, 1987; Maccoby, 1990; Tannen, 1990.

177 Brown was too bossy: Sherif et al., 1961, p. 77.

178 What makes a leader: Bennett & Derevensky, 1995; Masten, 1986; Hartup, 1983. Aggressive children are unpopular: Hayes, Gershman, & Halteman, 1996; Newcomb, Bukowski, & Pattee, 1993; Parker et al., 1995. Aggressive children aren't always unpopular: Bierman, Smoot, & Aumiller, 1993; Farmer & Rodkin, 1996. Those who flare up and lash out: Caspi, Elder, & Bem, 1987.

178 Attention structure: Chance & Larsen, 1976; Hold, 1977.

178 The change to summer clothes: Eckert, cited in Tannen, 1990, p. 218.

178 More mature children have higher status: Savin-Williams, 1979; Weisfeld & Billings, 1988. This is true especially for boys. Girls who mature early do not always have high status among their agemates. The reason, I believe, is that early-maturing girls are more likely to be overweight (Frisch, 1988) and our culture tends to assign low status to overweight people. If researchers look only at girls who are not overweight, I predict that they will find the same correlation between maturity and status that they find for boys.

179 Young male chimpanzees seek out older ones: Goodall, 1986. Little boys seek out older ones: Whiting & Edwards, 1988.

179 Older children have higher status: Edwards, 1992. Lower-status children have younger friends: Ladd, 1983.

179 Unpopular children have low self-esteem: Bennett & Derevensky, 1995; Parker et al., 1995.

179 Rejection as a cause or an effect: Hartup, 1983; Parker & Asher, 1987.

180 Short children more likely to have psychological problems: Brooks-Gunn & Warren, 1988; Jones & Bayley, 1950; Richman, Gordon, Tegtmeyer, Crouthamel, & Post, 1986; Stabler, Clopper, Siegel, Stoppani, Compton, & Underwood, 1994; Young-Hyman, 1986.

180 Slow maturers in adulthood: Jones, 1957. See also Dean, McTaggart, Fish, &

Friesen, 1986; Mitchell, Libber, Johanson, Plotnick, Joyce, Migeon, & Blizzard, 1986.

180 Remaining at the top or at the bottom: Coie & Cillessen, 1993; Parker et al., 1995.

180 School-age children compare themselves to their peers, younger children overestimate themselves: Harter, 1983; Newman & Ruble, 1988; Perry & Bussey, 1984; Stipek, 1992.

181 Comparisons are made to others in the same social category: Stipek, 1992. The term "social comparison": Festinger, 1954.

181 Dislike of strangeness, in chimpanzees: Goodall, 1988. In children: Diamond, LeFurgy, & Blass, 1993; Hayes et al., 1996.

182 Older children split up into more homogeneous groups: Hallinan & Teixeira, 1987; Hartup, 1983. They form cliques: Parker et al., 1995. Members of cliques become more alike: Cairns, Neckerman, & Cairns, 1989; Kindermann, 1995.

182 Attitudes are influenced by group affiliation: Kindermann, 1993.

182 To him who has: Matthew 13:12.

Chapter 9. The Transmission of Culture

183 Definition of culture: Mead, 1959, p. vii.

184 La Paz versus San Andres: Fry, 1988.

184 Moulding the Arapesh baby: Mead, 1963, p. 56 (originally published in 1935). A group of cannibals: p. 164.

185 The Arapesh engage in warfare: Daly & Wilson, 1988. Warlike peoples are nice to their babies: Eibl-Eibesfeldt, 1989. The Yanomamö: Chagnon, 1992.

185 The heritability of aggressiveness: Ghodsian-Carpey & Baker, 1987; Gottesman, Goldsmith, & Carey, 1997; van den Oord, Boomsma, & Verhulst, 1994.

185 Twice as many children: Chagnon, 1988. Systematically breeding warriors: Cairns, Gariépy, & Hood, 1990, reported that it is possible to produce strains of mice that differ markedly in aggressiveness in only four or five generations of selective breeding.

186 A deucedly uncomfortable fashion: Chagnon, 1992, p. 86.

186 Hit 'em back: Eibl-Eibesfeldt, 1989. Discouraged from play fighting: Fry, 1988.

189 My own children are foreigners: Parks, 1995, pp. 15, 175.

190 The Hutterites: Reader, 1988, pp. 215, 214.

190 Bicultural children: LaFromboise, Coleman, & Gerton, 1993.

191 Chinese father laments: Ungar, 1995, p. 49.

192 Daughter of Portuguese immigrants: Ferreira, 1996.

192 Speaks Japanese haltingly: Hayakawa, 1964, p. 217.

192 Mesquakie Indian boys: Polgar, 1960, cited in LaFromboise et al., 1993.

193 Language is a membership card: Schaller, 1991, p. 90.

193 Deaf jokes: Schaller, 1991, p. 90.

194 For a positive view of Deaf culture, see Padden & Humphries, 1988. For a negative view, see Bertling, 1994.

194 More prestigious language: Umbel, Pearson, Fernández, & Oller, 1992, p. 1013.

194 Beaten for using sign language: See, for example, Sidransky, 1990, p. 63.

195 The laundry woman: Schaller, 1991, p. 191.

195 Institutions for the retarded: Sacks, 1989.

195 For an explanation of the miracle, see Pinker, 1994.

195 Nicaraguan deaf children: A. Senghas, 1995; Kegl, Senghas, & Coppola, 1999.

196 Birth of a language: A. Senghas, 1995, p. 502–503.

196 Hawaiian children created a language: Bickerton, 1983.
196 Tower of Babel: Genesis 11:1–9.
197 The common language of their peers: Bickerton, 1983, p. 119.
197 Deaf culture in Nicaragua: R. Senghas & Kegl, 1994.
198 Prisoners and guards: This analogy was inspired by the classic experiment of Zimbardo, 1993 (originally published in 1972).
199 The prisoners' culture: Goffman, 1961, Chapter 1; Minton, 1971, pp. 31–32. Outwitting the guards: Goffman, 1961, pp. 54–60.
200 Evading rules in the nursery school: Corsaro, 1997, pp. 42, 140.
200 Waving at the garbage man: Corsaro, 1985.
201 Nyansongo dirty words: LeVine & LeVine, 1963.
201 Games children play: Opie & Opie, 1969, pp. 7, 1, 5–6.
202 Robbers Cave study: Sherif et al., 1961. See Chapter 7.
202 Storyknifing: deMarrais, Nelson, & Baker, 1994.
202 Innovative monkeys: Napier & Napier, 1985.
203 The true Brit: Glyn, 1970, pp. 128, 129, 135, 150.
204 Language as a membership card: Schaller, 1991, p. 90.
204 Piggy had the wrong accent: Golding, 1954.
205 Can't stand children: Glyn, 1970, p. 142. Yanomamö worries: Chagnon, 1992; Eibl-Eibesfeldt, 1989.
205 The fiscal father: Parks, 1995, pp. 63–64, 175.
206 Fear of crookedness: Lewald, 1871, quoted in Schütze, 1987, p. 51.
206 Female circumcision: Council on Scientific Affairs, 1995.
207 Two visits to the Gusii: LeVine & LeVine, 1963; LeVine & LeVine, 1988, p. 32.
207 Maladaptive bottle-feeding: Howrigan, 1988, p. 48.
208 Breast-feeding among the well-off: Bee, Baranowski, Rassin, Richardson, & Mikrut, 1991. Among the economically disadvantaged: Jones, 1993, p. AA5.
208 Maternal support networks: Melson, Ladd, & Hsu, 1993; Salzinger, 1990.
208 Fathers' networks: Riley, 1990.
208 Support networks and child abuse: Salzinger, 1990.
208 Parental punishment in San Andres: Fry, 1988, p. 1010.
208 Physical punishment in economically disadvantaged and ethnic minority groups: Coulton, Korbin, Su, & Chow, 1995; Deater-Deckard, Dodge, Bates, & Pettit, 1996; Dodge, Pettit, & Bates, 1994b; Kelley & Tseng, 1992; Knight, Virdin, & Roosa, 1994.
210 Portnoy's kitchen utensil: Roth, 1967, p. 107. See Chapter 4.
212 Association with delinquent peers: See, for example, Keenan, Loeber, Zhang, Stouthamer-Loeber, & van Kammen, 1995. This study found no relationship between the parents' child-rearing practices and the offspring's delinquency, once the influence of delinquent peers was taken into account.
212 Larry's story: Friend, 1995.
212 Delinquent London boys: Farrington, 1995; Rutter & Giller, 1983.
212 Neighborhood effects: Blyth & Leffert, 1995; Brooks-Gunn, Duncan, Klebanov, & Sealand, 1993.
212 Neighbors' socioeconomic status: Brooks-Gunn et al., 1993; Duncan, Brooks-Gunn, & Klebanov, 1994; see also Fletcher, Darling, Dornbusch, & Steinberg, 1995.
213 African-American youths: Peeples & Loeber, 1994, p. 141.
213 Protective effect of middle-class neighborhood: Kupersmidt, Griesler, DeRosier, Patterson, & Davis, 1995, pp. 366, 360.
214 Dr. Snyder: Kolata, 1993, p. C8.

215 The same version of creole: Bickerton, 1983.
215 Across cultural groups or within them: See, for example, Deater-Deckard et al., 1996.
216 Study of character: Hartshorne & May, 1928, 1971 (originally published in 1930).
216 Character education and the group: Hartshorne & May, 1971, p. 197 (originally published in 1930).

Chapter 10. Gender Rules

218 Changing dolls' diapers and playing with trucks: Bussey & Bandura, 1992, p. 1247.
218 As sexist as ever: Bussey & Bandura, 1992, p. 1248; Serbin, Powlishta, & Gulko, 1993, p. 1.
219 Sex differences or gender differences? There is a movement afoot to use "gender" for social categories and "sex" for biological ones, but the distinction is easier to make in theory than in practice. See Ruble & Martin, 1998.
219 Boys go to Jupiter: I thank Katherine Rappoport for providing the words to this song.
219 A similar (though not identical) conclusion: Archer, 1992a; Edwards, 1992; Maccoby, 1990; Maccoby & Jacklin, 1987; Martin, 1993; Powlishta, 1995b; Serbin et al., 1993; Tannen, 1990.
219 Minor differences in treatment of sons and daughters: Lytton & Romney, 1991.
220 Masculinity and femininity unrelated to same-sex parent: Maccoby & Jacklin, 1974, pp. 292–293. Fatherless boys: Serbin et al., 1993; Stevenson & Black, 1988. Daughters of lesbians: Patterson, 1992.
220 Timid boys get bolder: Kerr, Lambert, Stattin, & Klackenberg-Larsson, 1994.
221 Dana and David: The original study was by Condry & Condry, 1976; the one that used films of several babies was by Burnham & Harris, 1992. The Condry & Condry study gave rise to a number of similar ones and not all yielded the same results. In fact, a review of such studies concluded that labeling a baby male or female has inconsistent effects on the judgments of observers who don't know the baby's real sex; significant effects are "found only occasionally" (Stern & Karraker, 1989, p. 518).
221 Botched circumcision: Money & Erhardt, 1972. The baby was circumcised because he had a condition called phimosis, in which the foreskin cannot be retracted because it is too tight. Electrocautery was used to perform the surgery. The current was too high and the entire organ was burned beyond salvation.
222 Loves to have her hair set: Money & Ehrhardt, 1972, pp. 119–120. Some minor problems: p. 122.
222 The opposite-sex identical twins: Harris & Liebert, 1984, pp. 302–303; 1987, pp. 294–295; 1991, pp. 336–337.
223 Became a boy again: M. Diamond & Sigmundson, 1997, p. 300.
223 Genetic males who look like females at birth: J. Diamond, 1992c; Thigpen, Davis, Gautier, Imperato-McGinley, & Russell, 1992.
223 The Tchambuli: Mead, 1963 (originally published in 1935). The real Tchambuli: Brown, 1991, p. 20.
224 The same pattern everywhere: Brown, 1991. Males in positions of influence: Eibl-Eibesfeldt, 1989. Males are the warriors: Wrangham & Peterson, 1996. Girls preferred as babysitters, vie with each other to hold a baby: Maretzki & Maretzki, 1963; Whiting & Edwards, 1988. Doll abuse in Israeli homes: Goshen-Gottstein, 1981, p. 1261.
224 Male and female stereotypes: Williams & Best, 1986.

224 Stereotypes are generalizations: Williams & Best, 1986, p. 244; Hilton & von Hippel, 1996.
224 Detectors of statistical differences: Hilton & von Hippel, 1996; Pinker, 1997.
225 The accuracy of stereotypes: Swim, 1994. See also Halpern, 1997; Jussim, 1993.
225 The category we're not in: Hilton & von Hippel, 1996.
226 Girls have to be nurses: Maccoby & Jacklin, 1974, p. 364.
226 The sexes split up: Fabes, 1994; Leaper, 1994a, 1994b; Maccoby, 1994; Martin, 1994; Serbin, Moller, Gulko, Powlishta, & Colburne, 1994.
227 Hanging around with the guys: M. Diamond & Sigmundson, 1997, p. 299.
227 Born into the wrong body: Morris, 1974, p. 3.
227 Feminine boys: Green, 1987; Zuger, 1988.
227 Everybody has a penis: Bem, 1989, p. 662.
228 Children compare themselves to other boys and girls: M. Diamond, 1997, p. 205.
228 A white body that houses a Tibetan: "Daja Meston '96," 1995, p. 5.
228 Girls stood on the sidelines: Maccoby, 1990, p. 514.
228 Boys don't listen to girls: Fagot, 1994; Maccoby 1990; Serbin, Sprafkin, Elman, & Doyle, 1984.
229 The causes of mutual avoidance: Leaper, 1994a; Maccoby, 1994. Boys don't listen to girls: Fagot, 1994; Maccoby, 1990. Different behavioral styles: Archer, 1992a; Fabes, 1994; Serbin et al., 1994. Categorization into two groups: Archer, 1992a; Powlishta, 1995b; Martin, 1993; Serbin et al., 1993.
229 Older children form sex-segregated groups: Edwards, 1992; Schlegel & Barry, 1991; Whiting & Edwards, 1988.
229 Behavioral differences between boys and girls: Maccoby, 1995, p. 351. (Maccoby's paragraph contained some references in parentheses; I have left them out.)
230 Girls who are good at sports: Thorne, 1993.
230 Street games: Opie & Opie, 1969. Tomboys in their youth: Thorne, 1993, p. 113–114.
230 Borderwork: Thorne, 1993; Sroufe, Bennett, Englund, & Urban, 1993. Kissing as a weapon: Thorne, 1993, p. 71.
230 Adult influence increases interaction: Edwards, 1992; Maccoby, 1990; Thorne, 1993.
231 Underground friendships: Gottman, 1994; Thorne, 1993.
231 Groupness versus personal relationships: For example, Gilligan, 1982; Tiger, 1969; Wrangham & Peterson, 1996.
231 Groupness stronger in males: Bugental & Goodnow, 1998.
231 Boys run faster, throw harder: Thomas & French, 1985. Males launch attacks on other groups: Wrangham & Peterson, 1996. All wars are boyish: Melville, 1866 (cited in *Bartlett's Familiar Quotations,* 1992).
232 Friends became enemies: Sherif et al., 1961, pp. 9–10.
232 Girls show hostility indirectly: Björkqvist, Lagerspetz, & Kaukiainen, 1992; Crick & Grotpeter, 1995.
232 Different cultures: Maccoby, 1990; Tannen, 1990. See also Adler, Kless, & Adler, 1992; Archer, 1992a.
233 Unfairly punished: Thorne, 1993, p. 56. Thorne has other objections to the "two cultures" idea: sex differences in behavior (like the mutual avoidance of the sexes) are more or less visible depending on the social context, and not all girls and boys fit neatly into the stereotypes for their gender.
234 Dodgeball players: Weisfeld, Weisfeld, & Callaghan, 1982.
235 The most traumatic experience: Glyn, 1970, p. 129.

235 Girls dominate less aggressively: McCloskey, 1996; Whiting & Edwards, 1988. Inhibition of aggression in females: Bjorklund & Kipp, 1996.

235 Life among the Efe: Morelli, 1997, p. 209.

236 !Kung hunter-gatherers versus farmers: Draper, 1997; Draper & Cashdan, 1988.

237 Greater aggressiveness of males: Eibl-Eibesfeldt, 1989; Maccoby & Jacklin, 1974; Wrangham & Peterson, 1996.

237 Girls with congenital adrenal hyperplasia: Collaer & Hines, 1995; Money & Ehrhardt, 1972. In most cases, the genital abnormalities are rectified through surgery. However, some women who had this surgery in infancy are protesting that it left them "maimed" and incapable of having orgasms (Angier, 1997). M. Diamond, 1997, recommends that surgery be put off until the individual is old enough to participate in the decision.

237 Girls avoid boys sooner: Maccoby, 1994.

237 Males talk more, females laugh more: Maccoby, 1990; Provine, 1993; Tanner, 1990; Weinstein, 1991.

237 Girls' self-esteem plummets: American Association of University Women, 1991; Daley, 1991; but see Sommers, 1994. A smaller effect than you were led to believe: Block & Robins, 1993.

238 The importance of being pretty: Leaper, 1994b; Granleese & Joseph, 1994. Granleese & Joseph found that for girls who attended a coed high school, self-esteem was closely related to their physical attractiveness. For girls who attended an all-girls school, physical attractiveness was less important. According to Buss, 1994, men all over the world place high value on a woman's physical attractiveness. Beautiful women are sought after as mates and have higher social status.

238 Low status leads to low self-esteem: Leary, Tambor, Terdal, & Downs, 1995. Depression more common in females: Culbertson, 1997; Weissman & Olfson, 1995. Link between depression and self-esteem: King, Naylor, Segal, Evans, & Shain, 1993; Myers, 1992.

238 No sex differences for bipolar disorder: Culbertson, 1997.

238 Inhibitory control higher in females: Bjorklund & Kipp, 1996; Kochanska, Murray, & Coy, 1997.

239 Females are battered: Wrangham & Peterson, 1996.

239 Things are better for women today: This didn't happen spontaneously—we have a long line of courageous women to thank for it. I would like to thank my dear friend Naomi Weisstein (1971, 1977) for the part she played in making our culture less sexist.

Chapter 11. Schools of Children

240 Playing the fine line: Carere, 1987, pp. 125, 127, 129–130.

242 The Rattlers and the Eagles: Sherif et al., 1961. See Chapter 7.

242 Ability grouping: Dornbusch, Glasgow, & Lin, 1996; Winner, 1997.

242 Anti-school attitudes: Neckerman, 1996, pp. 140–141. Things that might have made them smarter: see Ceci & Williams, 1997.

243 Cliques in elementary school: Kindermann, 1993.

243 Self-esteem of African Americans: In some studies it is higher than that of European Americans. See Steele, 1997. People compare themselves to the members of their own group: McFarland & Buehler, 1995.

243 An apple for Miss A: Harris & Liebert, 1991, pp. 404–405; E. Pedersen, Faucher, & Eaton, 1978.

245 Wishful thinking: E. Pedersen et al., 1978, p. 19.
245 Jaime Escalante: Mathews, 1988, p. 217. Jocelyn Rodriguez: Pogrebin, 1996, p. B7.
246 Japanese classrooms: Kristof, 1997. Bullying on Japanese playgrounds: Kristof, 1995. Asian kids are ahead: Vogel, 1996.
247 Heritability of IQ in older adults: N. Pedersen, Plomin, Nesselroade, & McClearn, 1992.
248 Some of the same points: Herrnstein & Murray, 1994; Seligman, 1992.
248 The X factor: Seligman, 1992, p. 160.
249 Children learn more in smaller classes: Mosteller, 1995.
249 Black–white relations at Wexler: Schofield, 1981, pp. 74–76, 78, 83. (Ellipses in the original.)
250 High expections of minority parents: Galper, Wigfield, & Seefeldt, 1997. Greater emphasis: Stevenson, Chen, & Uttal, 1990.
250 Caught reading a book: Herbert, 1997.
250 Haitian achievers: Kosof, 1996, p. 60. Jamaican achievers: Roberts, 1995.
251 The children of American servicemen: Eyferth, Brandt, & Wolfgang, 1960, cited in Hilgard, Atkinson, & Atkinson, 1979.
251 Daja Meston: See Chapter 8.
251 Other people's expectations: Jussim, McCauley, & Lee, 1995; Jussim & Fleming, 1996. Although teachers' expectations can, under some conditions, weakly influence their students' performance, the students' race, ethnicity, sex, or social class appear to play little or no role in these effects. Teachers' expectations are generally based on the characteristics of the individual student, take into account previous academic performance, and tend to be accurate. For that reason they are likely to be verified. See Madon, Jussim, & Eccles, 1997, pp. 804–805.
251 Stereotype threat: Steele, 1997.
252 Fear of success: Horner, 1969.
252 Women's colleges: Alper, 1993; Sadker & Sadker, 1994.
252 Long-term effects of enrichment programs: Mann, 1997 (the supporter); Scarr, 1997a (the critic).
253 Effects on parents' behavior: Olds et al., 1997. Lack of effects on the children: White, Taylor, & Moss, 1992.
253 Home-based versus group interventions: Barnett, 1995; St. Pierre, Layzer, & Barnes, 1995.
253 Reducing aggressive behavior: Grossman et al., 1997.
254 Joseph learns English: Winitz, Gillespie, & Starcev, 1995.
254 Talking to Tony: Winitz et al., 1995, p. 133.
255 They start acting deaf: Evans, 1987, p. 170. (Ellipses in the original.)
256 A dismal failure: Ravitch, 1997, p. A35.
256 Russian immigrants: Kosof, 1996, pp. 26, 54.
257 Like a fish out of water: I am assuming that Joseph followed the typical pattern for the children of immigrants. See Chapter 4.
258 La Paz and San Andres: Fry, 1988. See Chapters 8 and 9.
258 Bullies: Marano, 1995.
258 He will be bicultural: Like the Mesquakie Indian boys described in Chapter 9 (LaFromboise et al., 1993).
259 The effects of group size: Brewer, 1991.
259 Neighborhoods and aggressive behavior: Kupersmidt, Griesler, DeRosier, Patterson, & Davis, 1995, p. 366; see also Peeples & Loeber, 1994.

260 All the homes are full of books: Dornbusch, Glasgow, & Lin, 1996, pp. 412–413.
260 A dictionary and a computer: Vogel, 1996. The effects of the neighborhood: Duncan, Brooks-Gunn, & Klebanov, 1994.
261 Adoption can raise a child's IQ: Personal communication, T. A. Kindermann, October 22, 1997.
261 The French adoption study: Capron & Duyme, 1989. See also Locurto, 1990.
261 IQ correlations between adoptive siblings: Plomin, Chipuer, & Neiderhiser, 1994. The IQ correlation between adopted children and their adoptive parents also declines to zero in adolescence; see Plomin, Fulker, Corley, & DeFries, 1997.
262 As they get older: Scarr & McCartney, 1983.
262 The long-term effects of adoption: Personal communication, M. McGue, October 23, 1997.
262 A dozen healthy infants: Watson, 1924. See Chapter 1.
262 High schools tend to be larger: Eccles et al., 1993, reported that school performance of "marginal" students tends to decline when they move from elementary school to junior high or from a small school to a larger one.
263 All-girl schools: Alper, 1993; Sadker & Sadker, 1994. Traditionally black colleges: Steen, 1987.
263 Socks filled with stones, the camp's water system: Sherif et al., 1961. See Chapter 7.

Chapter 12. Growing Up

264 Normal part of teen life: Moffitt, 1993, p. 675. Power and privilege: p. 686.
265 Parents count zilch: Harris, 1995. See the Preface.
266 A well-dressed man: Chagnon, 1992, p. 85.
267 Being left back: Yamamoto, Soliman, Parsons, & Davies, 1987.
267 Maturity and status: See the note to p. 178 on this topic.
269 A girl of consequence: Valero, 1970, pp. 82–84.
270 Puberty rites: Benedict, 1959, pp. 69–70, 103 (originally published in 1934); Delaney, 1995.
270 Why so harsh? Eibes-Eibesfeldt, 1989, p. 604.
271 Growth during childhood: Weisfeld & Billings, 1988.
272 The ones who resemble us: Smith, 1987. The death of an eight-year-old: Wright, 1994, pp. 174–175. The one-year-old gets the kisses: Dunn & Plomin, 1990, pp. 74–75; McHale, Crouter, McGuire, & Updegraff, 1995.
273 Eagles and Rattlers: See Chapter 7.
274 Little Leaguers: Fine, 1986, p. 63.
274 Their parents are not racists: In a recent poll, only one out of eight white teenagers said they had heard their parents say something negative about another race (Farley, 1997).
274 Adolescent rebellion isn't found: Schlegel & Barry, 1991.
275 Socrates: Quoted in Rogers, 1977, p. 6. Aristotle: Quoted in Cole, 1992, p. 778.
275 (footnote) The appalling manners of the young: Martin, 1995.
276 Cohort differences: Baltes, Cornelius, & Nesselroade, 1979.
276 Kids who hang around together: Kindermann, 1993.
277 Social categories in high school: Brown, Mounts, Lamborn, & Steinberg, 1993; Eckert, 1989. Male homosexuality rare in rural areas: Laumann, Gagnon, Michael, & Michaels, 1994.
277 Braininess no asset: Brown et al., 1993; Juvonen & Murdock, 1993.
277 Mels: Merten, 1996a, pp. 11, 20.

278 Sensation seekers: Arnett & Balle-Jensen, 1993; Zuckerman, 1984. Rejected by their peers: Parker, Rubin, Price, & DeRosier, 1995; Coie & Cillessen, 1993. Similar to begin with: Rowe, Woulbroun, & Gulley, 1994. The brains get brainier: Social psychologists call it "group polarization"; see Myers, 1982.
279 The wrong child-rearing style: Brown et al., 1993; Mounts & Steinberg, 1995.
280 Peer pressure: Lightfoot, 1992, pp. 240, 235. See also Berndt, 1992.
281 Best predictor of smoking: Stanton & Silva, 1992. Teenagers who smoke: Collins et al., 1987; Eckert, 1989; "Study Cites Teen Smoking Risks," 1995.
282 Genes, environment, and smoking: Rowe, 1994.
282 Arguments for smoking: Barry, 1995.
283 Making it harder to buy cigarettes: Rigotti, DiFranza, Chang, Tisdale, Kemp, & Singer, 1997.
284 Temporary delinquents: Moffitt, 1993, p. 674.
284 San Andres: See Chapter 9.
284 Fusiwe broke her arm: Valero, 1970, pp. 167–168.
285 Seeking similar others: Caspi, 1998; Rowe et al., 1994.
285 The best fit: Dobkin, Tremblay, Mâsse, & Vitaro, 1995; Rowe et al., 1994.
285 Trying to cure delinquency: Lab & Whitehead, 1988; Mann, 1994; Tate, Reppucci, & Mulvey, 1995.
286 Conformity: Asch, 1987, pp. 481–482 (originally published in 1952).
286 They say they listen more to their parents: For example, Berndt, 1979.
286 Tender with his children: James, 1890, p. 294.
287 Stability of adult personality: Caspi, 1998; McCrae & Costa, 1994. Set like plaster: James, 1890, p. 121.
287 Kissinger's accent: Pinker, 1994, p. 281.

Chapter 13. Dysfunctional Families and Problem Kids

289 A child murderer: "Maternal impressions," 1996, p. 1466 (originally published in 1896).
291 Cathy's weight problem: Guisewite, 1994.
292 Behaviors the child will copy: Pitts, 1997, p. 23.
292 Heredity, environment, and obesity: The data have been summarized by Grilo & Pogue-Geile, 1991.
293 Spooky similarities: Lykken, McGue, Tellegen, & Bouchard, 1992.
293 Amy and Beth (not their real names): Lykken et al., 1992; Wright, 1995.
295 Making Oliver into a thief: Dickens, 1990 (originally published in 1838).
295 Children who are difficult to socialize: Lykken, 1995; Mealey, 1995; Rutter, 1997.
296 How things go wrong: Patterson & Bank, 1989.
296 Obnoxious behavior: Dishion, Duncan, Eddy, Fagot, & Fetrow, 1994.
296 Children who break rules: Hartshorne & May, 1928.
297 My description of Oliver's personality is based on the book; I've never seen the show or any of the movies. Dickens said Oliver was "a child of a noble nature and a warm heart" (1990, p. 314). He described the boy "shaking from head to foot at the mere recollection of Mr. Bumble's voice" (p. 35).
297 Danish adoption study: Mednick, Gabrielli, & Hutchings, 1987.
298 Only in Copenhagen: Gottfredson & Hirschi, 1990.
298 Criminal behavior in siblings and twins: Rowe, Rodgers, & Meseck-Bushey, 1992; Rowe & Waldman, 1993; Rutter, 1997.
299 Teenage delinquency: Moffitt, 1993.
299 Push him off the ice: Murphy, 1976, cited in Lykken, 1995.

300 Among the Ache: Buss, 1994, pp. 49–50.

301 Worse off without a father: McLanahan & Sandefur, 1994, p. 1. (Italics in the original.) A decision to live apart: p. 3.

302 Things that don't matter: McLanahan & Sandefur, 1994. They controlled for racial and social class differences. Frequent father contact: p. 98. (Italics in the original.)

302 Widowed mothers: McLanahan, 1994, p. 51; Krantz, 1989.

303 Single mothers in poverty: Crossette, 1996; McLanahan & Booth, 1989. Kids' standing among their peers: Adler, Kless, & Adler, 1992. If the financial deprivation were severe enough to make it difficult for the children to get enough to eat, it could jeopardize their growth, vitality, and even intelligence. However, this degree of deprivation doesn't appear to be common in the United States, judging from the statistics on teenage pregnancies. Malnutrition delays sexual maturation and lowers fertility.

304 Poverty and neighborhood: Ambert, 1997, pp. 97–98.

304 Boys from an inner city: Zimmerman, Salem, & Maton, 1995, p. 1607.

304 Kids who live with both parents are no better off than those who live with one: The same result has been found by Chan, Raboy, & Patterson, 1998, within an economically *advantaged* group.

305 Changes of residence: McLanahan & Sandefur, 1994.

305 The consequences of moving: Rejection by peers, Vernberg, 1990. Behavioral problems, Wood, Halfon, Scarlata, Newacheck, & Nessim, 1993. Academic problems, Eckenrode, Rowe, Laird, & Brathwaite, 1995.

305 The children of divorce: Wallerstein & Kelly, 1980; Wallerstein & Blakeslee, 1989. An eight-year-old boy: Santrock & Tracy, 1978.

306 British study of children of divorce: Chase-Lansdale, Cherlin, & Kiernan, 1995, pp. 1618–1619.

307 A decision to live apart: McLanahan & Sandefur, 1994, p. 3.

308 Fail in their own marriages: McGue & Lykken, 1992.

308 Twin study of divorce: McGue & Lykken, 1992. The subjects in this study ranged in age from thirty-four to fifty-three years.

309 Personality and divorce: Jockin, McGue, & Lykken, 1996, concluded, "Thus, personality predicts divorce risk, and more specifically, it does so largely because of the genetic rather than the environmental influences they share" (p. 296).

309 Antisocial personality disorder can be inherited: Caspi, 1998; Gottesman, Goldsmith, & Carey, 1997. The Danish adoption study of criminal behavior (Mednick et al., 1987) indicated that men with antisocial tendencies are more likely to father children they are unwilling or unable to rear. For genetic reasons, the offspring of such men are more likely to have antisocial tendencies. Taken together, these observations can explain why fatherless boys are more likely to commit crimes (see Popenoe, 1996).

309 Troublesome behavior precedes the divorce: Block, Block, & Gjerde, 1986. Divorce, antisocial personality disorder, and conduct disorder: Lahey, Hartdagen, Frick, McBurnett, Connor, & Hynd, 1988.

309 Less divorce in families with sons: Glick, 1988.

310 Divorce *is* bad for kids: D. G. Myers, personal communication, February 2, 1998.

310 Child abuse by stepparents: Daly & Wilson, 1996.

310 Strong emotions: Pinker, 1997.

311 Parents' reports: Kagan, 1994. See my comments on Hetherington & Clingempeel, 1992, in the note to page 76.

311 The majority of American homes: Straus, Sugarman, & Giles-Sims, 1997.

312 Time-outs for white people: Gilbert, 1997.

313 Physical punishment in minority or low-income groups: Coulton, Korbin, Su, & Chow, 1995; Deater-Deckard, Dodge, Bates, & Pettit, 1996; Dodge, Pettit, & Bates, 1994b; Kelley & Tseng, 1992; Knight, Virdin, & Roosa, 1994.

313 Asian Americans do use physical punishment: Chao, 1994, Table 1.

313 Use of corporal punishment: Straus et al., 1997, p. 761.

313 Abstract in *JAMA:* November 12, 1997, vol. 278, p. 1470. Picked up by the AP: Coleman, 1997. Claims seem unfounded: Gunnoe & Mariner, 1997, p. 768.

314 Multiple personality disorder: Eich, Macaulay, Loewenstein, & Dihle, 1997.

315 Abused kids more aggressive: Dodge, Bates, & Pettit, 1990; Malinowsky-Rummell & Hansen, 1993. Trouble with friendship: Dodge, Pettit, & Bates, 1994a. Trouble with schoolwork: Perez & Widom, 1994. Abuse their own children: Wolfe, 1985.

315 They hardly ever do: An exception is Rothbaum & Weisz, 1994, who discuss both genetic effects and child-to-parent effects in their review of parental child-rearing methods.

315 Mental illness runs in families: Plomin, Owen, & McGuffin, 1994.

316 Victimized again: Vasta, 1982.

316 Abused children and their peers: Ladd, 1992.

316 Peer abuse: Ambert, 1994a, p. 121; 1997, p. 99. The percentages given are from the most recent batch of autobiographies analyzed, collected in 1989. See also Kochenderfer & Ladd, 1996.

317 Abused children moved more often: Eckenrode et al., 1995.

317 Reasonable control: Smolowe, 1996. Chained to the radiator: Gibbs, 1991.

318 Exaggerated by mutual influence: Myers, 1982.

319 Study of preschoolers: Baumrind, 1967.

319 Shared method variance: See, for example, Wagner, 1997, p. 291.

320 What parents themselves say: Smetana, 1995. No significant advantage of just-right parenting: Weiss & Schwarz, 1996. These researchers defined six parenting types; offspring of "Authoritative" parents did not have significantly better personalities or fewer problems. Offspring of "Unengaged" and "Authoritarian-Directive" parents did score significantly lower but the differences were small.

321 Multiple intelligences: Gardner, 1983. Scores on different tests are correlated: D. Seligman, 1992.

321 Everything related: Cohen, 1994, p. 1000. Fifteen items and 105 correlations: Because many of the items did not yield numerical responses, the researchers actually used chi-square tests. The work was done by Meehl & Lykken and reported in Cohen, 1994.

322 Parental bond: Foreman, 1997. Parent-family connectedness: Resnick et al., 1997.

323 Skeptical essayist: Carlson, 1997.

323 The New Zealand study: Caspi et al., 1997.

324 The parents' fault: Bradshaw, 1988; Forward, 1989.

325 I'll leave it to other writers: For example, Dawes, 1994; M. Seligman, 1994.

327 Happiness and unhappiness: Myers, 1992. Depression and memory: Dawes, 1994, pp. 211–216. Memories of identical twins: Hur & Bouchard, 1995. Genetic influences on happiness: Lykken & Tellegen, 1996.

Chapter 14. What Parents Can Do

328 Reared-apart twins: Lykken, 1995, p. 82.

329 Piano-teacher mother: Lykken, 1995, p. 82.

330 How to run a home: There is evidence from one twin study (Waller & Shaver, 1994)

that children may learn at home their attitudes toward romantic love. However, a twin study of divorce (McGue & Lykken, 1992, discussed in Chapter 13) yielded contradictory results: twins' experience of their parents' marriage does not appear to affect the success or failure of their own marriages. It is too soon to come to any conclusions on this issue.

330 Boys have not become more permissive: Serbin, Powlishta, & Gulko, 1993.

331 The whole family was executed: Heckathorn, 1992.

332 Reduces competition between siblings: Sulloway, 1996. Parents occupy family niches: Tesser, 1988.

333 It started as a joke: Thornton, 1995, pp. 3–4. They can play with each other: p. 43.

334 A brave corps: Mathews, 1988, p. 217.

334 Women of accomplishment: Thornton, 1995; Moore, 1996.

335 Not the parents who are pushing: Gottfried, Gottfried, Bathurst, & Guerin, 1994; Winner, 1996.

335 Psychological problems in prodigies: Winner, 1996, 1997.

335 Determine their child's peers: Ladd, Profilet, & Hart, 1992.

336 A higher proportion of smart kids: Rutter, 1983. Less likely to get into trouble, more likely to be rejected: Kupersmidt, Griesler, DeRosier, Patterson, & Davis, 1995.

336 Wilson's childhood: Quoted in Norman, 1995, p. 66.

337 Midwood High: Hartocollis, 1998.

339 Self-esteem, pro: Brody, 1997, p. F7; Clark, 1995, p. 1970.

339 Self-esteem, con: Dawes, 1994, pp. 9–10. (Italics in the original.) See also M. Seligman, 1995, pp. 31–33.

339 Self-esteem and violence: Baumeister, Smart, & Boden, 1996, p. 5. Self-esteem and risky behavior: Smith, Gerrard, & Gibbons, 1997.

340 Self-esteem and parental favoritism: Zervas & Sherman, 1994.

342 The mobile with red doodads: Rovee-Collier, 1993.

342 Poison the relationship between siblings: Brody & Stoneman, 1994. Least favored children in adulthood: Bedford, 1992.

343 Babies sleep with their mothers: Anders & Taylor, 1994.

344 Poorly understood data: Bruer, 1997.

344 IQ correlations in adopted children: Plomin, Fulker, Corley, & DeFries, 1997. No scientific basis: Bruer, 1997.

344 A bright baby: L. J. Miller (1997, September 10), Einstein and IQ (Netnews posting in sci.psychology.misc).

345 Parents are not playmates: Rogoff, Mistry, Göncü, & Mosier, 1993.

346 Quality time: Reich, 1997, pp. 10–11.

346 No effort is made to prevent it: Edwards, 1992.

347 Saved his brother's life: Goodall, 1986, p. 282.

348 Loving our children to death: Watson, 1928, pp. 69, 70.

Chapter 15. The Nurture Assumption on Trial

350 Your mum and dad: Larkin, "This Be the Verse," 1989, p. 140 (originally published in 1974).

351 Parents already have power: Morton, 1998. Children's sense of themselves: Brody, 1997, p. F7. Daily messages of love and acceptance: Neifert, 1991, p. 77. His foundations: Leach, 1995, p. 468 (originally published in 1989).

352 No signs of a decline in parental abusiveness: There has been an *increase* in reports

of child abuse (Lung & Daro, 1996), but it is not clear whether this is due to an actual increase in parental abusiveness or just to an increased willingness to report it. No signs that children are happier today: The increase in adolescent suicide and in rates of depression over the past thirty years (Myers, 1992) suggests that, if anything, children are less happy today.

354 Parental treatment is more uniform: O'Connor, Hetherington, Reiss, & Plomin, 1995.

355 Good-enough homes: Lykken, 1997; Rowe, 1997; Scarr, 1992.

356 Polygyny practiced in 80 percent of cultures: Pinker, 1997.

357 Disguise our differentness: Proulx, 1993, p. 134.

359 A baby's dilemma: Savage & Au, 1996.

360 A penny for your thoughts: Pinker, 1997, p. 135.

360 Mental modules: Buss, 1991; Cosmides & Tooby, 1992; Pinker, 1997.

361 Fish don't remember their schoolmates: Eibl-Eibesfeldt, 1995.

361 Unconscious data gathering: Lewicki, Hill, & Czyzewska, 1992.

Appendix 1: Personality and Birth Order

365 A to Z: Adler, 1927; Zajonc, 1983.

365 Exclamation points and italics: Schooler, 1972; Ernst & Angst, 1983, p. 284; Dunn & Plomin, 1990, p. 85.

366 A stake through its heart: Somit, Arwine, & Peterson, 1996, p. vi.

367 Persuaded until he tried to replicate: Modell, 1997, p. 624.

368 Virtually the only data: Sulloway also discusses the work of Koch, who published ten articles on her study of a single group of 384 five- and six-year-old children from sibships of two. This work is included in E & A's survey so it does not provide additional evidence.

368 Opinions versus personality: Sulloway uses the change of an opinion in adulthood —for example, the acceptance of Darwin's theory of evolution—as a measure of an enduring personality characteristic, openness. However, a single change (or non-change) of opinion isn't the same as a standardized personality questionnaire that has been tested and validated with a large number of subjects. It is more like a single item from a personality questionnaire—an item of unknown validity. Whether the change of opinion is correlated with other measures of personality has not been established.

368 The studies I found in E & A: I counted a study as no-difference if one subgroup of subjects—for example, boys—produced results favorable to Sulloway's theory and the other subgroup, girls, produced results in the opposite direction. I counted a study as confirming if one subgroup of subjects produced favorable results and the other produced no-difference results. An example of a study I couldn't categorize was summarized by E & A as follows: "Middle children appeared at the same time more excitable and more phlegmatic, less fearful and more mature than first- and lastborns" (E & A, 1983, p. 167).

369 E & A were biased: Sulloway, unpublished manuscript, January 25, 1998.

371 Interactions: In his unpublished manuscript (January 25, 1998), Sulloway writes that he has taken account of the additional no-difference outcomes produced by studies that yielded interactions. In the case of a two-way interaction—where, for example, sex interacts with birth order in such a way that favorable results are found for boys but not for girls—he reports that he has counted the results as one favorable outcome and one no-difference outcome; in the case of a three-way interaction,

he reports that he counted all four possible outcomes. Since there were many interactions in the studies reviewed by E & A, this procedure would noticeably increase the number of findings per study. Thus, in order to end up with 196 findings, Sulloway's analysis must have included considerably fewer than the 116 studies I found in E & A (hence, fewer than 75,000 subjects). In his unpublished manuscript, Sulloway reports that he eliminated from his analysis, for a variety of reasons, a number of studies that E & A had considered acceptable (most of which were included in my tally). However, it also appears that he included in his analysis other studies that E & A had written off as unacceptable or inconclusive. I have been unable to determine the precise number of studies included in Sulloway's analysis.

371 Meta-analysis: Sulloway, 1996, p. 72. (Italics in the original.)

371 A proper meta-analysis: Hunt, 1997.

372 No-difference outcomes less likely to get published: Hunt, 1997. Take longer to get into print: Ioannidis, 1998.

373 Meta-analyses versus gold standard studies: LeLorier, Grégoire, Benhaddad, Lapierre, & Derderian, 1997, p. 536.

374 Outcomes of post-1980 studies: Unclear results were those that did not relate in any obvious way to Sulloway's theory and those that were not specified with adequate clarity in the abstract. The search was performed on August 20, 1997; the most recent item retrieved was published in March 1997.

375 Descriptions by parents: Ernst & Angst, p. 167. By siblings, p. 97.

375 The firstborn personality may be parent-specific: Ernst & Angst, p. 171. (Italics in the original.)

376 Outmoded ideas of parents: Note that the parents' ideas are more likely to be outmoded by the time the laterborn comes along. If firstborns are more likely to share their parents' attitudes, it may be because the age gap between firstborns and their parents is not as wide as the age gap between laterborns and their parents. When families were larger and childbearing was spread over a period of twenty years or more, this difference could have been important, especially during periods of social change.

376 Passionate advocacy: Modell, 1997, p. 624.

376 Grave doubts about the validity: Somit, Arwine, & Peterson, 1997, pp. 17–18.

376 Underachievement is not more common in laterborns: McCall, 1992, p. 17.

376 Divergent thinking: Runco, 1991 (originally published in 1987).

377 Rebellious youth: Townsend, 1997.

377 Marriages work better if the couple is similar: O'Leary & Smith, 1991. Mismatched birth orders less likely to divorce: Toman, 1971.

377 Rear their young serially: More precisely, humans rear their young in overlapping fashion. See Harris, Shaw, & Altom, 1985, p. 186, note 1.

377 Fratricide is rare: Daly & Wilson, 1988.

378 No birth order effects on intelligence: Retherford & Sewell, 1991.

Appendix 2: Testing Theories of Child Development

379 Mild support for Harris's theory: Loehlin, 1997, p. 1201.

380 Conjoined twins: Wallis, 1996.

382 The results were stunning; a better interpretation of our data: Reiss, 1997, p. 102.

383 What could these factors be? Reiss, 1997, p. 103.

384 Teenage smoking: Rowe, 1994.

385 Fifth-grade cliques: Kindermann, 1993.
386 The consequences of divorce: D. G. Myers, personal communication, April 30, 1998.
387 Cross-situational genetic effects: Saudino, 1997, p. 88.
389 Joseph: My pseudonym for the subject of Winitz, Gillespie, & Starcev, 1995.
390 Cannot be refuted: Lykken, 1999.
391 The story of Genie: Rymer, 1993.

REFERENCES

Abramovitch, R., Corter, C., Pepler, D. J., & Stanhope, L. (1986). Sibling and peer interaction: A final follow-up and a comparison. *Child Development,* 57, 217–229.

Adler, A. (1927). *Understanding human nature.* New York: Greenberg.

Adler, P. A., Kless, S. J., & Adler, P. (1992). Socialization to gender roles: Popularity among elementary school boys and girls. *Sociology of Education,* 65, 169–187.

Ainsworth, M. D. S. (1977). Attachment theory and its utility in cross-cultural research. In P. H. Leiderman, S. R. Tulkin, & A. Rosenfield (Eds.), *Culture and infancy: Variation in the human experience.* New York: Academic Press.

Ainsworth, M. D. S., Blehar, M. C., Waters, E., & Wall, S. (1978). *Patterns of attachment: A psychological study of the Strange Situation.* Hillsdale, NJ: Erlbaum.

Alexander, G. M., & Hines, M. (1994). Gender labels and play styles: Their relative contribution to children's selection of playmates. *Child Development,* 65, 869–879.

Alper, J. (1993, April 16). The pipeline is leaking women all the way along. *Science,* 260, 409–411.

Ambert, A.-M. (1994a). A quantitative study of peer abuse and its effects: Theoretical and empirical implications. *Journal of Marriage and the Family,* 56, 119–130.

Ambert, A.-M. (1994b). An international perspective on parenting: Social change and social constructs. *Journal of Marriage and the Family,* 56, 529–543.

Ambert, A.-M. (1997). *Parents, children, and adolescents: Interactive relationships and development in context.* New York: Haworth Press.

American Association of University Women (1991). *Shortchanging girls, shortchanging America: A call to action.* Washington, DC: AAUW.

Anders, T. F., & Taylor, T. R. (1994). Babies and their sleep environments. *Children's Environments,* 11, 123–134.

Andersson, B.-E. (1992). Effects of day-care on cognitive and socioemotional competence of thirteen-year-old Swedish schoolchildren. *Child Development,* 63, 20–36.

Angier, N. (1997, May 17). New debate over surgery on genitals. *New York Times,* pp. C1, C5.

Archer, J. (1992a). Childhood gender roles: Social context and organization. In H. McGurk (Ed.), *Childhood social development: Contemporary perspectives* (pp. 31–61). Hove, UK: Erlbaum.

Archer, J. (1992b). *Ethology and human development.* Savage, MD: Barnes & Noble Books.

Arnett, J., & Balle-Jensen, L. (1993). Cultural bases of risk behavior: Danish adolescents. *Child Development,* 64, 1842–1855.

Asch, S. E. (1987). *Social Psychology.* Oxford, UK: Oxford University Press. (Originally published in 1952.)

Astington, J. W. (1993). *The child's discovery of the mind.* Cambridge, MA: Harvard University Press.

Bagwell, C. L., Newcomb, A. F., & Bukowski, W. M. (1998). Preadolescent friendship and peer rejection as predictors of adult adjustment. *Child Development,* 69, 140–153.

Bailey, D. B., Jr., McWilliam, R. A., Ware, W. B., & Burchinal, M. A. (1993). Social interactions of toddlers and preschoolers in same-age and mixed-age play groups. *Journal of Applied Developmental Psychology,* 14, 261–275.

Bailey, J. M., & Pillard, R. C. (1991). A genetic study of male sexual orientation. *Archives of General Psychiatry,* 48, 1089–1096.

Bailey, J. M., Pillard, R. C., Neale, M. C., & Agyei, Y. (1993). Heritable factors influence sexual orientation in women. *Archives of General Psychiatry,* 50, 217–223.

Bajak, F. (1986, June 19). Firemen are twins, too. *The New Jersey Register,* p. 8A.

Baltes, P. B., Cornelius, S. W., & Nesselroade, J. R. (1979). Cohort effects in developmental psychology. In J. R. Nesselroade & P. B. Baltes (Eds.), *Longitudinal research in the study of behavior and development* (pp. 61–87). New York: Academic Press.

Bandura, A., & Walters, R. H. (1963). *Social learning and personality development.* New York: Holt.

Barnett, W. S. (1995, Winter). Long-term effects of early childhood programs on cognitive and school outcomes. *The Future of Children,* 5(3), 25–50.

Baron, N. S. (1992). *Growing up with language: How children learn to talk.* Reading, MA: Addison-Wesley.

Baron-Cohen, S. (1995). *Mindblindness: An essay on autism and theory of mind.* Cambridge, MA: MIT Press.

Baron-Cohen, S., Campbell, R., Karmiloff-Smith, A., Grant, J., & Walker, J. (1995). Are children with autism blind to the mentalistic significance of the eyes? *British Journal of Developmental Psychology,* 13, 379–398.

Barry, D. (1995, September 17). Teen smokers, too, get cool, toxic, waste-blackened lungs. *Asbury Park (N.J.) Press,* p. D3.

Barry, D. (1996, August 11). That awful sound is a parent singing within earshot of a teen. *Asbury Park (N.J.) Press,* p. D3.

Baumeister, R. F., Smart, L., & Boden, J. M. (1996). Relation of threatened egotism to violence and aggression: The dark side of high self-esteem. *Psychological Review,* 103, 5–33.

Baumrind, D. (1967). Child care practices anteceding three patterns of preschool behavior. *Genetic Psychology Monographs,* 75, 43–88.

Baumrind, D. (1989). Rearing competent children. In W. Damon (Ed.), *Child development today and tomorrow* (pp. 349–378). San Francisco: Jossey-Bass.

Baumrind, D., & Black, A. E. (1967). Socialization practices associated with dimensions of competence in preschool boys and girls. *Child Development,* 38, 291–327.

Bedford, V. H. (1992). Memories of parental favoritism and the quality of parent-child ties in adulthood. *Journal of Gerontology: Social Sciences,* 47, S149–S155.

Bee, D. E., Baranowski, T., Rassin, D. K., Richardson, J., & Mikrut, W. (1991). Breastfeeding initiation in a triethnic population. *American Journal of Diseases of Children,* 145, 306–309.

Belsky, J., Rovine, M., & Taylor, D. G. (1984). The Pennsylvania Infant and Family

Development Project: III. The origins of individual differences in infant-mother attachment: Maternal and infant contributions. *Child Development,* 55, 718–728.
Bem, D. J. (1996). Exotic becomes erotic: A developmental theory of sexual orientation. *Psychological Review,* 103, 320–335.
Bem, S. L. (1989). Genital knowledge and gender constancy in preschool children. *Child Development,* 60, 649–662.
Benedict, R. (1959). *Patterns of culture.* New York: Houghton Mifflin. (Originally published in 1934.)
Bennett, A., & Derevensky, J. (1995). The medieval kingdom topology: Peer relations in kindergarten children. *Psychology in the Schools,* 32, 130–141.
Berndt, T. J. (1979). Developmental changes in conformity to peers and parents. *Developmental Psychology,* 15, 606–616.
Berndt, T. J. (1992). Friendship and friends' influence in adolescence. *Current Directions in Psychological Science,* 1, 156–159.
Bertling, T. (1994). *A child sacrificed to the deaf culture.* Wilsonville, OH: Kodiak Media Group.
Bettelheim, B. (1959, March). Joey: A "mechanical boy." *Scientific American,* 200, 116–127.
Bettelheim, B. (1967). *The empty fortress.* New York: Free Press.
Bickerton, D. (1983, July). Creole languages. *Scientific American,* 249, 116–122.
Bierman, K. L., Smoot, D. L., & Aumiller, K. (1993). Characteristics of aggressive-rejected, aggressive (nonrejected), and rejected (nonaggressive) boys. *Child Development,* 64, 139–151.
Bigelow, A., MacLean, J., Wood, C., & Smith, J. (1990). Infants' responses to child and adult strangers: An investigation of height and facial configuration variables. *Infant Behavior and Development,* 13, 21–32.
Bigler, R. S. (1995). The role of classification skill in moderating environmental influences on children's gender stereotyping: A study of the functional use of gender in the classroom. *Child Development,* 66, 1072–1087.
Birch, L. L. (1987). Children's food preferences: Developmental patterns and environmental influences. *Annals of Child Development,* 4, 171–208.
Bjorklund, D. F., & Kipp, K. (1996). Parental investment theory and gender differences in the evolution of inhibitory mechanisms. *Psychological Bulletin,* 120, 163–188.
Björkqvist, K., Lagerspetz, K. M. J., & Kaukiainen, A. (1992). Do girls manipulate and boys fight? Developmental trends in regard to direct and indirect aggression. *Aggressive Behavior,* 18, 117–127.
Block, J., & Robins, R. W. (1993). A longitudinal study of consistency and change in self-esteem from early adolescence to early adulthood. *Child Development,* 64, 909–923.
Block, J. H., Block, J., & Gjerde, P. F. (1986). The personality of children prior to divorce: A prospective study. *Child Development,* 57, 827–840.
Blyth, D. A., & Leffert, N. (1995). Communities as contexts for adolescent development: An empirical analysis. *Journal of Adolescent Research,* 10, 64–87.
Bouchard, T. J., Jr. (1994, June 17). Genes, environment, and personality. *Science,* 264, 1700–1701.
Bowlby, J. (1969). *Attachment and loss: Vol. 1. Attachment.* New York: Basic Books.
Bowlby, J. (1973). *Attachment and loss: Vol. 2. Separation.* New York: Basic Books.
Bradshaw, J. (1988). *Bradshaw on the family: A revolutionary way of self-discovery.* Deerfield Beach, FL: Health Communications.
Bretherton, I. (1985). Attachment theory: Retrospect and prospect. In I. Bretherton &

E. Waters (Eds.), Growing points of attachment theory and research (pp. 3–35). *Monographs of the Society for Research in Child Development,* 50 (1–2, Serial No. 209).

Brewer, M. B. (1991). The social self: On being the same and different at the same time. *Personality and Social Psychology Bulletin,* 17, 475–482.

Brody, G. H., & Stoneman, Z. (1994). Sibling relationships and their association with parental differential treatment. In E. M. Hetherington, D. Reiss, & R. Plomin (Eds.), *Separate social worlds of siblings: The impact of nonshared environment on development* (pp. 129–142). Hillsdale, NJ: Erlbaum.

Brody, G. H., Stoneman, Z., MacKinnon, C. E., & MacKinnon, R. (1985). Role relationships and behavior between preschool-aged and school-aged sibling pairs. *Developmental Psychology,* 21, 124–129.

Brody, J. E. (1997, November 11). Parents can bolster girls' fragile self-esteem. *New York Times,* p. F7.

Brooks, J., & Lewis, M. (1976). Infants' responses to strangers: Midget, adult, and child. *Child Development,* 47, 323–332.

Brooks-Gunn, J., Duncan, G. J., Klebanov, P. K., & Sealand, N. (1993). Do neighborhoods influence child and adolescent development? *American Journal of Sociology,* 99, 353–395.

Brooks-Gunn, J., & Warren, M. P. (1988). The psychological significance of secondary sexual characteristics in nine- to eleven-year-old girls. *Child Development,* 59, 1061–1069.

Brown, B. B., Mounts, N., Lamborn, S. D., & Steinberg, L. (1993). Parenting practices and peer group affiliation in adolescents. *Child Development,* 64, 467–482.

Brown, D. E. (1991). *Human universals.* Philadelphia: Temple University Press.

Bruer, J. T. (1997). Education and the brain: A bridge too far. *Educational Researcher,* 26, 4–16.

Bugental, D. B., & Goodnow, J. J. (1998). Socialization processes. In W. Damon (Series Ed.) & N. Eisenberg (Vol. Ed.), *Handbook of Child Psychology: Vol. 3. Social, emotional, and personality development* (5th ed., pp. 389–462). New York: Wiley.

Bukowski, W. M., Pizzamiglio, M. T., Newcomb, A. F., & Hoza, B. (1996). Popularity as an affordance for friendship: The link between group and dyadic experience. *Social Development,* 5, 189–202.

Burnham, D. K., & Harris, M. B. (1992). Effects of real gender and labeled gender on adults' perceptions of infants. *Journal of Genetic Psychology,* 153, 165–183.

Burns, G. L., & Farina, A. (1992). The role of physical attractiveness in adjustment. *Genetic, Social, and General Psychology Monographs,* 118, 157–194.

Buss, D. M. (1991). Evolutionary personality psychology. *Annual Review of Psychology,* 42, 459–491.

Buss, D. M. (1994). *The evolution of desire: Strategies of human mating.* New York: Basic Books.

Bussey, K., & Bandura, A. (1992). Self-regulatory mechanisms governing gender development. *Child Development,* 63, 1236–1250.

Cairns, R. B., Gariépy, J.-L., & Hood, K. E. (1990). Development, microevolution, and social behavior. *Psychological Review,* 97, 49–65.

Cairns, R. B., Neckerman, H. J., & Cairns, B. D. (1989). Social networks and the shadow of synchrony. In G. R. Adams, R. Montemayor, & T. Gullotta (Eds.), *Advances in adolescent development: Vol. 1. Biology of adolescent behavior and development* (pp. 275–305). Newbury Park, CA: Sage.

Caporael, L. R. (1986). Anthropomorphism and mechanomorphism: Two faces of the human machine. *Computers in Human Behavior,* 2, 215–234.

Capron, C., & Duyme, M. (1989). Assessment of the effects of socio-economic status on IQ in a full cross-fostering study. *Nature,* 340, 552–554.

Carere, S. (1987). Lifeworld of restricted behavior. In P. A. Adler, P. Adler, & N. Mandell (Eds.), *Sociological studies of child development: Vol. 2* (pp. 105–138). Greenwich, CT: JAI Press.

Carlson, M. (1997, September 22). Here's a precious moment, kid. *Time,* p. 101.

Carson, R. C. (1989). Personality. *Annual Review of Psychology,* 40, 227–248.

Caspi, A. (1998). Personality development across the life course. In W. Damon (Series Ed.) & N. Eisenberg (Vol. Ed.), *Handbook of Child Psychology: Vol 3. Social, emotional, and personality development* (5th Ed., pp. 311–388). New York: Wiley.

Caspi, A., Begg, D., Dickson, N., Harrington, H., Langley, J., Moffitt, T. E., & Silva, P. A. (1997). Personality differences predict health-risk behaviors in young adulthood: Evidence from a longitudinal study. *Journal of Personality and Social Psychology,* 73, 1052–1063.

Caspi, A., Elder, G. H., Jr., & Bem, D. J. (1987). Moving against the world: Life-course patterns of explosive children. *Developmental Psychology,* 23, 308–313.

Ceci, S. J., & Williams, W. M. (1997). Schooling, intelligence, and income. *American Psychologist,* 52, 1051–1058.

Chagnon, N. A. (1988, February 26). Life histories, blood revenge, and warfare in a tribal population. *Science,* 239, 985–992.

Chagnon, N. A. (1992). *Yanomamö: The last days of Eden.* San Diego: Harcourt Brace Jovanovich.

Chan, R. W., Raboy, B., & Patterson, C. J. (1998). Psychosocial adjustment among children conceived via donor insemination by lesbian and heterosexual mothers. *Child Development,* 69, 443–457.

Chance, M. R. A., & Larsen, R. R. (Eds.) (1976). *The social structure of attention.* London: Wiley.

Chao, R. K. (1994). Beyond parental control and authoritarian parenting style: Understanding Chinese parenting through the cultural notion of training. *Child Development,* 65, 1111–1119.

Chase-Lansdale, P. L., Cherlin, A. J., & Kiernan, K. E. (1995). The long-term effects of parental divorce on the mental health of young adults: A developmental perspective. *Child Development* 66, 1614–1634.

Chen, J.-Q., & Goldsmith, L. T. (1991). Social and behavioral characteristics of Chinese only children: A review of research. *Journal of Research in Childhood Education,* 5, 127–139.

Clark, L. R. (1995, June 28). Teen sex blues. *Journal of the American Medical Association,* 273, 1969–1970.

Clinton, H. R. (1996). *It takes a village, and other lessons children teach us.* New York: Simon & Schuster.

Cohen, J. (1994). The earth is round ($p < .05$). *American Psychologist,* 49, 997–1003.

Coie, J. D., & Cillessen, A. H. N. (1993). Peer rejection: Origins and effects on children's development. *Current Directions in Psychological Science,* 2, 89–92.

Cole, M. (1992). Culture in development. In M. H. Bornstein & M. E. Lamb (Eds.), *Developmental psychology: An advanced textbook* (3rd ed., pp. 731–789). Hillsdale, NJ: Erlbaum.

Coleman, B. C. (1997, August 14). Study: Spanking causes misbehavior. Associated Press (on-line).

Collaer, M. L., & Hines, M. (1995). Human behavioral sex differences: A role for gonadal hormones during early development? *Psychological Bulletin,* 118, 55–107.

Collins, L. M., Sussman, S., Rauch, J. M., Dent, C. W., Johnson, C. A., Hansen, W. B., & Flay, B. R. (1987). Psychosocial predictors of young adolescent cigarette smoking: A sixteen-month, three-wave longitudinal study. *Journal of Applied Social Psychology,* 17, 554–573.

Condry, J., & Condry, S. (1976). Sex differences: A study of the eye of the beholder. *Child Development,* 47, 812–819.

Coontz, S. (1992). *The way we never were: American families and the nostalgia trap.* New York: Basic Books.

Corsaro, W. A. (1985). *Friendship and peer culture in the early years.* Norwood, NJ: Ablex.

Corsaro, W. A. (1993). Interpretive reproduction in the "scuola materna." *European Journal of Psychology of Education,* 8, 357–374.

Corsaro, W. A. (1997). *The sociology of childhood.* Thousand Oaks, CA: Pine Forge Press.

Cosmides, L., & Tooby, J. (1992). Cognitive adaptations for social exchange. In J. Barkow, L. Cosmides, & J. Tooby (Eds.), *The adapted mind: Evolutionary psychology and the generation of culture* (pp. 163-228). New York: Oxford University Press.

Coulton, C. J., Korbin, J. E., Su, M., & Chow, J. (1995). Community level factors and child maltreatment rates. *Child Development,* 66, 1262–1276.

Council, J. R. (1993). Contextual effects in personality research. *Current Directions in Psychological Science,* 2, 31–34.

Council on Scientific Affairs, American Medical Association (1995, December 6). Female genital mutilation. *Journal of the American Medical Association,* 274, 1714–1716.

Crick, N. R., & Grotpeter, J. K. (1995). Relational aggression, gender, and social-psychological adjustment. *Child Development,* 66, 710–722.

Crossette, B. (1996, June 11). New tally of world tragedy: Women who die giving life. *New York Times,* pp. A1, A12.

Culbertson, F. M. (1997). Depression and gender: An international review. *American Psychologist,* 52, 25–31.

Daja Meston '96: West meets East meets West (1995). *Brandeis Review,* 15(2), 4–5.

Daley, S. (1991, January 8). Girls' self-esteem is lost on way to adolescence, new study finds. *New York Times,* pp. B1, B6.

Daly, M., & Wilson, M. I. (1988). *Homicide.* New York: Aldine de Gruyter.

Daly, M., & Wilson, M. I. (1996). Violence against stepchildren. *Current Directions in Psychological Science,* 5, 77–81.

Darling, N., & Steinberg, L. (1993). Parenting style as context: An integrative model. *Psychological Bulletin,* 113, 487–496.

Darwin, C. (1871). *The descent of man.* Reprinted in *The origin of species and The descent of man.* New York: Modern Library. No publication date given.

Dawes, R. M. (1994). *House of cards: Psychology and psychotherapy built on myth.* New York: Free Press.

Dawkins, R. (1976). *The selfish gene.* New York: Oxford University Press.

Dean, H. J., McTaggart, T. L., Fish, D. G., & Friesen, H. G. (1986). Long-term social follow-up of growth hormone deficient adults treated with growth hormone during childhood. In B. Stabler & L. E. Underwood (Eds.), *Slow grows the child: Psychosocial aspects of growth delay* (pp. 73–82). Hillsdale, NJ: Erlbaum.

Deater-Deckard, K., Dodge, K. A., Bates, J. E., & Pettit, G. S. (1996). Physical discipline among African American and European American mothers: Links to children's externalizing behaviors. *Developmental Psychology,* 32, 1065–1072.

Deci, E. L. (1971). Effects of externally mediated rewards on intrinsic motivation. *Journal of Personality and Social Psychology,* 18, 105–115.

Delaney, C. H. (1995). Rites of passage in adolescence. *Adolescence,* 30, 891–897.

deMarrais, K. B., Nelson, P. A., & Baker, J. H. (1994). Meaning in mud: Yup'ik Eskimo girls at play. In J. L. Roopnarine, J. E. Johnson, & F. H. Hooper (Eds.), *Children's play in diverse cultures* (pp. 179–209). Albany: State University of New York Press.

Dencik, L. (1989). Growing up in the post-modern age: On the child's situation in the modern family, and on the position of the family in the modern welfare state. *Acta Sociologica,* 32, 155–180.

Detterman, D. K. (1993). The case for the prosecution: Transfer as an epiphenomenon. In D. K. Detterman & R. J. Sternberg (Eds.), *Transfer on trial: Intelligence, cognition, and instruction* (pp. 1–24). Norwood, NJ: Ablex.

de Waal, F. (1989). *Peacemaking among primates.* Cambridge: Harvard University Press.

Diamond, J. (1992a, March). Living through the Donner Party. *Discover,* 13, 100–107.

Diamond, J. (1992b). *The third chimpanzee.* New York: HarperCollins.

Diamond, J. (1992c, June). Turning a man. *Discover,* 13, 70–77.

Diamond, K., LeFurgy, W., & Blass, S. (1993). Attitudes of preschool children toward their peers with disabilities: A year-long investigation in integrated classrooms. *Journal of Genetic Psychology,* 154, 215–221.

Diamond, M. (1997). Sexual identity and sexual orientation in children with traumatized or ambiguous genitalia. *Journal of Sex Research,* 34, 199–211.

Diamond, M., & Sigmundson, H. K. (1997). Sex reassignment at birth: Long-term review and clinical implications. *Archives of Pediatric and Adolescent Medicine,* 151, 298–304.

Dickens, C. (1990). *Oliver Twist.* New York: Bantam Books. (Originally published in 1838.)

Dishion, T. J., Duncan, T. E., Eddy, J. M., Fagot, B. I., & Fetrow, R. (1994). The world of parents and peers: Coercive exchanges and children's social adaptation. *Social Development,* 3, 255–268.

Dobkin, P. L., Tremblay, R. E., Mâsse, L. C., & Vitaro, F. (1995). Individual and peer characteristics in predicting boys' early onset of substance abuse: A seven-year longitudinal study. *Child Development,* 66, 1198–1214.

Dodge, K. A., Bates, J. E., & Pettit, G. S. (1990, December 21). Mechanisms in the cycle of violence. *Science,* 250, 1678–1683.

Dodge, K. A., Pettit, G. S., & Bates, J. E. (1994a). Effects of physical maltreatment on the development of peer relations. *Development and Psychopathology,* 6, 43–55.

Dodge, K. A., Pettit, G. S., & Bates, J. E. (1994b). Socialization mediators of the relation between socioeconomic status and child conduct problems. *Child Development,* 65, 649–665.

Dontas, C., Maratos, O., Fafoutis, M., & Karangelis, A. (1985). Early social development in institutionally reared Greek infants: Attachment and peer interaction. In I. Bretherton & E. Waters (Eds.), Growing points of attachment theory and research (pp. 136–146). *Monographs of the Society for Research in Child Development,* 50(1–2, Serial No. 209).

Dornbusch, S. M., Glasgow, K. L., & Lin, I.-C. (1996). The social structure of schooling. *Annual Review of Psychology,* 47, 401–429.

Dornbusch, S. M., Ritter, P. L., Leiderman, P. H., Roberts, D. F., & Fraleigh, M. J. (1987). The relation of parenting style to adolescent school performance. *Child Development,* 58, 1244–1257.

Draper, P. (1997). Institutional, evolutionary, and demographic contexts of gender roles: A case study of !Kung bushmen. In M. E. Morbeck, A. Galloway, & A. L. Zihlman (Eds.), *The evolving female: A life-history perspective* (pp. 220–232). Princeton, NJ: Princeton University Press.

Draper, P., & Cashdan, E. (1988). Technological change and child behavior among the !Kung. *Ethnology,* 27, 339–365.

Dunbar, R. I. M. (1993). Co-evolution of neocortex size, group size, and language in humans. *Behavioral and Brain Sciences,* 16, 681–735.

Duncan, G. J., Brooks-Gunn, J., & Klebanov, P. K. (1994). Economic deprivation and early childhood development. *Child Development,* 65, 296–318.

Dunn, J., & Plomin, R. (1990). *Separate lives: Why siblings are so different.* New York: Basic Books.

Eccles, J. S., Midgley, C., Wigfield, A., Buchanan, C. M., Reuman, D., Flanagan, C., & MacIver, D. (1993). Development during adolescence: The impact of stage-environment fit on young adolescents' experiences in schools and in families. *American Psychologist,* 48, 90–101.

Eckenrode, J., Rowe, E., Laird, M., & Brathwaite, J. (1995). Mobility as a mediator of the effects of child maltreatment on academic performance. *Child Development,* 66, 1130–1142.

Eckerman, C. O., Davis, C. C., & Didow, S. M. (1989). Toddlers' emerging ways of achieving social coordination with a peer. *Child Development,* 60, 440–453.

Eckerman, C. O., & Didow, S. M. (1988). Lessons drawn from observing young peers together. *Acta Paediatrica Scandinavica,* 77, 55–70.

Eckerman, C. O., & Didow, S. M. (1996). Nonverbal imitation and toddlers' mastery of verbal means of achieving coordinated action. *Developmental Psychology,* 32, 141–152.

Eckert, P. (1989). *Jocks and burnouts: Social categories and identity in the high school.* New York: Teachers College Press.

Edwards, C. P. (1992). Cross-cultural perspectives on family–peer relations. In R. D. Parke & G. W. Ladd (Eds.), *Family–peer relationships: Modes of linkage* (pp. 285–316). Hillsdale, NJ: Erlbaum.

Effort to reintroduce thick-billed parrots in Arizona is dropped (1995, May 30). *New York Times,* p. C4.

Egeland, B., & Sroufe, L. A. (1981). Attachment and early maltreatment. *Child Development,* 52, 44–52.

Eibl-Eibesfeldt, I. (1989). *Human ethology.* Hawthorne, NY: Aldine de Gruyter.

Eibl-Eibesfeldt, I. (1995). The evolution of familiality and its consequences. *Futura,* 10(4), 253–264.

Eich, E., Macaulay, D., Loewenstein, R. J., & Dihle, P. H. (1997). Memory, amnesia, and dissociative identity disorder. *Psychological Science,* 8, 417–422.

Eimas, P. D., & Quinn, P. C. (1994). Studies on the formation of perceptually based basic-level categories in young infants. *Child Development,* 65, 903–907.

Einstein, A. (1991). On the generalized theory of gravitation. *Scientific American,* Special Issue on "Science in the 20th Century," pp. 40–45. (Originally published in 1950.)

Eisenberg, N., Fabes, R. A., Bernzweig, J., Karbon, M., Poulin, R., & Hanish, L. (1993). The relations of emotionality and regulation to preschoolers' social skills and sociometric status. *Child Development,* 64, 1418–1438.

Eisenberger, R. & Cameron, J. (1996). Detrimental effects of reward: Reality or myth? *American Psychologist,* 51, 1153–1166.

Erickson, M. F., Sroufe, L. A., & Egeland, B. (1985). The relationship between quality of attachment and behavior problems in preschool in a high-risk sample. In I. Bretherton & E. Waters (Eds.), Growing points of attachment theory and research (pp. 147–166). *Monographs of the Society for Research in Child Development,* 50(1–2, Serial No. 209).

Ernst, C., & Angst, J. (1983). *Birth order: Its influence on personality.* Berlin, Germany: Springer-Verlag.

Evans, A. D. (1987). Institutionally developed identities: An ethnographic account of reality construction in a residential school for the deaf. In P. A. Adler, P. Adler, & N.

Mandell (Eds.), *Sociological studies of child development: Vol. 2* (pp. 159–182). Greenwich, CT: JAI Press.

Eyer, D. E. (1992). *Mother-infant bonding: A scientific fiction.* New Haven: Yale University Press.

Fabes, R. A. (1994). Physiological, emotional, and behavioral correlates of gender segregation. In Leaper, C. (Ed.), *Childhood gender segregation: Causes and consequences* (pp. 19–34). New Directions for Child Development, No. 65. San Francisco: Jossey-Bass.

Fagen, R. (1993). Primate juveniles and primate play. In M. E. Pereira & L. A. Fairbanks (Eds.), *Juvenile primates* (pp. 182–192). New York: Oxford University Press.

Fagot, B. I. (1994). Peer relations and the development of competence in girls and boys. In Leaper, C. (Ed.), *Childhood gender segregation: Causes and consequences* (pp. 53–65). New Directions for Child Development, No. 65. San Francisco: Jossey-Bass.

Fagot, B. I. (1995). Classification of problem behaviors in young children: A comparison of four systems. *Journal of Applied Developmental Psychology,* 16, 95–106.

Falbo, T., & Polit, D. F. (1986). Quantitative research of the only child literature: Research evidence and theory development. *Psychological Bulletin,* 100, 176–189.

Falbo, T., & Poston, D. L., Jr. (1993). The academic, personality, and physical outcomes of only children in China. *Child Development,* 64, 18–35.

Farah, M. (1992). Is an object an object an object? Cognitive and neuropsychological investigations of domain specificity in visual object recognition. *Current Directions in Psychological Science,* 1, 164–169.

Farley, C. J. (1997, November 24). Kids and race. *Time,* pp. 88–91.

Farmer, T. W., & Rodkin, P. C. (1996). Antisocial and prosocial correlates of classroom social positions: The social network centrality perspective. *Social Development,* 5, 174–188.

Farrington, D. P. (1995). The development of offending and antisocial behaviour from childhood: Key findings from the Cambridge Study in Delinquent Development. *Journal of Child Psychology and Psychiatry,* 360, 929–964.

Fein, G. G., & Fryer, M. G. (1995a). Maternal contributions to early symbolic play competence. *Developmental Review,* 15, 367–381.

Fein, G. G., & Fryer, M. G. (1995b). When theories don't work chuck 'em or change 'em. *Developmental Review,* 15, 401–403.

Fenson, L., Dale, P. S., Reznick, J. S., Bates, E., Thal, D. J., & Pethick, S. J. (1994). Variability in early communicative development. *Monographs of the Society for Research in Child Development,* 59(5, Serial No. 242).

Ferreira, F. (1996). Biography. *American Psychologist,* 51, 315–317.

Festinger, L. (1954). A theory of social comparison processes. *Human Relations,* 7, 117–140.

Fine, G. A. (1981). Friends, impression management, and preadolescent behavior. In S. R. Asher & J. M. Gottman (Eds.), *The development of children's friendships* (pp. 29–52). Cambridge, UK: Cambridge University Press.

Fine, G. A. (1986). The dirty play of little boys. *Society/Transaction,* 24, 63–67.

Fiske, S. T. (1992). Thinking is for doing: Portraits of social cognition from daguerreotype to laserphoto. *Journal of Personality and Social Psychology,* 63, 877–889.

Flaks, D. K., Ficher, I., Masterpasqua, F., & Joseph, G. (1995). Lesbians choosing motherhood: A comparative study of lesbian and heterosexual parents and their children. *Developmental Psychology,* 31, 105–114.

Fletcher, A. C., Darling, N. E., Dornbusch, S. M., & Steinberg, L. (1995). The company they keep: Relations of adolescents' adjustment and behavior to their friends' perceptions of authoritative parenting in the social network. *Developmental Psychology,* 31, 300–310.

Foreman, J. (1997, September 10). Study links parental bond to teenage well-being. *Boston Globe,* p. A1.

Forward, S. (1989). *Toxic parents: Overcoming their hurtful legacy and reclaiming your life.* New York: Bantam Books.

Fox, N. A. (1989). Psychophysiological correlates of emotional reactivity during the first year of life. *Developmental Psychology,* 25, 364–372.

Fox, N. A., Kimmerly, N. L., & Schafer, W. D. (1991). Attachment to Mother/Attachment to Father: A meta-analysis. *Child Development,* 62, 210–225.

Fraiberg, S. H. (1959). *The magic years.* New York: Scribner's.

Freedman, D. G. (1958). Constitutional and environmental interactions in rearing of four dog breeds. *Science,* 127, 585–586.

Freeman, D. (1983). *Margaret Mead and Samoa: The making and unmaking of an anthropological myth.* Cambridge, MA: Harvard University Press.

Freud, A., & Dann, S. (1967). An experiment in group upbringing. In Brackbill, Y., & Thompson, G. G. (Eds.), *Behavior in infancy and early childhood* (pp. 494–514). New York: Free Press. (Originally published in 1951.)

Friedman, R. C., & Downey, J. I. (1994). Homosexuality. *New England Journal of Medicine,* 331, 923–930.

Friend, T. (1995, Aug. 1). A young man goes west to prosper. *New York Times,* pp. B7, B11.

Frisch, R. E. (1988, March). Fatness and fertility. *Scientific American,* 258, 88–95.

Fry, D. P. (1988). Intercommunity differences in aggression among Zapotec children. *Child Development,* 59, 1008–1019.

Galper, A., Wigfield, A., & Seefeldt, C. (1997). Head Start parents' beliefs about their children's abilities, task values, and performances on different activities. *Child Development,* 68, 897–907.

Gardner, H. (1983). *Frames of mind: The theory of multiple intelligences.* New York: Basic Books.

Garvey, C. (1990). *Play* (2nd ed.). Cambridge: Harvard University Press.

Ge, X., Conger, R. D., Cadoret, R. J., Neiderhiser, J. M., Yates, W., Troughton, E., & Stewart, M. A. (1996). The developmental interface between nature and nurture: A mutual influence model of child antisocial behavior and parent behaviors. *Developmental Psychology,* 32, 574–589.

Gesell, A. (1940). *The first five years of life: The preschool years.* New York: Harper & Bros.

Gesell, A., & Ilg, F. (1943). *Infant and child in the culture of today.* New York: Harper & Bros.

Ghodsian-Carpey, J., & Baker, L. A. (1987). Genetic and environmental influences on aggression in 4- to 7-year-old twins. *Aggressive Behavior,* 13, 173–186.

Gibbs, N. (1991, September 30). At the end of their tether. *Time,* p. 34.

Gibran, K. (1978). *The prophet.* New York: Knopf. (Originally published in 1923.)

Gilbert, S. (1997, August 20). Two spanking studies indicate parents should be cautious. *New York Times,* p. C8.

Gilligan, C. (1982). *In a different voice: Sex differences in the expression of moral judgment.* Cambridge: Harvard University Press.

Glick, P. C. (1988). Fifty years of family demography: A record of social change. *Journal of Marriage and the Family,* 50, 861–873.

Glyn, A. (1970). *The British: Portrait of a people.* New York: Putnam's Sons.

Goffman, E. (1961). *Asylums: Essays on the social situation of mental patients and other inmates.* Chicago: Aldine.

Gold, P.-T. (1997, April 21). Bettelheim's legacy (letter to the editor). *The New Yorker,* p. 10.

Golding, W. (1954). *Lord of the flies.* New York: Coward, McCann, & Geoghegan.

Goldsmith, H. H. (1996). Studying temperament via construction of the Toddler Behavior Assessment Questionnaire. *Child Development,* 67, 218–235.

Golombok, S., Cook, R., Bish, A., & Murray, C. (1995). Families created by the new reproductive technologies: Quality of parenting and social and emotional development of the children. *Child Development,* 66, 285–298.

Göncü, A., & Kessel, F. (1988). Preschoolers' collaborative construction in planning and maintaining imaginative play. *International Journal of Behavioral Development,* 11, 327–344.

Goodall, J. (1986). *The chimpanzees of Gombe: Patterns of behavior.* Cambridge: Harvard University Press.

Goodall, J. (1988). *In the shadow of man* (revised ed.). Boston: Houghton Mifflin. (First edition published in 1971.)

Goodenough, F. L. (1945). *Developmental psychology: An introduction to the study of human behavior* (2nd ed.). New York: Appleton-Century-Crofts. (First edition published in 1934).

Goossens, F. A., & van IJzendoorn, M. H. (1990). Quality of infants' attachments to professional caregivers: Relation to infant-parent attachment and day-care characteristics. *Child Development,* 61, 550–567.

Goshen-Gottstein, E. R. (1981). Differential maternal socialization of opposite-sexed twins, triplets, and quadruplets. *Child Development,* 52, 1255–1264.

Gottesman, I. I., Goldsmith, H. H., & Carey, G. (1997). A developmental *and* a genetic perspective on aggression. In N. L. Segal, G. E. Weisfeld, & C. C. Weisfeld (Eds.), *Uniting psychology and biology: Integrating perspectives on human development* (pp. 107–144). Washington, DC: American Psychological Association.

Gottfredson, M. R., & Hirschi, T. (1990). *A general theory of crime.* Stanford, CA: Stanford University Press.

Gottfried, A. W., Gottfried, A. E., Bathurst, K., & Guerin, D. W. (1994). *Gifted IQ: Early developmental aspects: The Fullerton Longitudinal Study.* New York: Plenum Press.

Gottlieb, B. R. (1995, February 23). Abortion—1995. *New England Journal of Medicine,* 332, 532–533.

Gottman, J. M. (1994). Why can't men and women get along? In D. Canary & L. Stafford (Eds.), *Communication and relational maintenance.* San Diego, CA: Academic Press.

Gottman, J. S. (1990). Children of gay and lesbian parents. In F. W. Bozett & M. B. Sussman (Eds.), *Homosexuality and family relations* (pp. 177–196). New York: Harrington Park.

Gould, S. J. (1980). *The panda's thumb.* New York: Norton.

Granleese, J., & Joseph, S. (1994). Self-perception profile of adolescent girls at a single-sex and a mixed-sex school. *Journal of Genetic Psychology,* 154, 525–530.

Gray, P. (1994). *Psychology* (2nd ed.). New York: Worth.

Green, R. (1987). *The "sissy boy syndrome" and the development of homosexuality.* New Haven: Yale University Press.

Greenfield, P. M., & Childs, C. P. (1991). Developmental continuity in biocultural context. In R. Cohen & A. W. Siegel (Eds.), *Context and development* (pp. 135–159). Hillsdale, NJ: Erlbaum.

Grilo, C. M., & Pogue-Geile, M. F. (1991). The nature of environmental influences on weight and obesity: A behavior genetic analysis. *Psychological Bulletin,* 110, 520–537.

Grossman, D. C., Neckerman, H. J., Koepsell, T. D., et al. (1997, May 28). Effectiveness of a violence prevention curriculum among children in elementary school. *Journal of the American Medical Association,* 277, 1605–1611.

Gruenberg, S. M. (Ed.) (1942). *Favorite stories old and new.* Garden City, NY: Doubleday, Doran.

Guisewite, C. (1994, June 19). Cathy. *Asbury Park (N.J.) Press,* Comics section, p. 3.

Gunnoe, M. L., & Mariner, C. L. (1997). Toward a developmental-contextual model of the effects of parental spanking on children's agression. *Archives of Pediatrics and Adolescent Medicine,* 151, 768–775.

Hallinan, M. T. (1992). Determinants of students' friendship choices. In E. J. Lawler, B. Markovsky, C. Ridgeway, & H. A. Walker (Eds.), *Advances in group processes: Vol. 9* (pp. 163–183). Greenwich, CT: JAI Press.

Hallinan, M. T., & Teixeira, R. A. (1987). Students' interracial friendships: Individual characteristics, structural effects, and racial differences. *American Journal of Education,* 95, 563–583.

Halpern, D. F. (1997). Sex differences in intelligence: Implications for education. *American Psychologist,* 52, 1091–1102.

Hareven, T. (1985). Historical changes in the life course: Implications for child development. In A. B. Smuts & J. W. Hagen (Eds.), History and research in child development. *Monographs of the Society for Research in Child Development,* 50(4–5, Serial No. 211).

Harlow, H. F., & Harlow, M. K. (1962, November). Social deprivation in monkeys. *Scientific American,* 207, 136–146.

Harris, J. R. (1995). Where is the child's environment? A group socialization theory of development. *Psychological Review,* 102, 458–489.

Harris, J. R., & Liebert, R. M. (1984). *The child: Development from birth through adolescence.* Englewood Cliffs, NJ: Prentice-Hall.

Harris, J. R., & Liebert, R. M. (1987). *The child: Development from birth through adolescence* (2nd ed.). Englewood Cliffs, NJ: Prentice-Hall.

Harris, J. R., & Liebert, R. M. (1991). *The child: A contemporary view of development* (3rd ed.). Englewood Cliffs, NJ: Prentice Hall.

Harris, J. R., Shaw, M. L., & Bates, M. (1979). Visual search in multicharacter arrays with and without gaps. *Perception & Psychophysics,* 26, 69–84.

Harris, J. R., Shaw, M. L., & Altom, M. J. (1985). Serial position curves for reaction time and accuracy in visual search: Tests of a model of overlapping processing. *Perception & Psychophysics,* 38, 178–187.

Harris, M. (1989). *Our kind: Who we are, where we came from, where we are going.* New York: Harper & Row.

Harter, S. (1983). Developmental perspectives on the self-system. In P. H. Mussen (Series Ed.) & E. M. Hetherington (Vol. Ed.), *Handbook of child psychology: Vol. 4. Socialization, personality, and social development* (4th ed., pp. 275–385). New York: Wiley.

Hartocollis, A. (1998, January 13). 13 Midwood High students take Westinghouse honors. *New York Times,* p. B3.

Hartshorne, H., & May, M. A. (1928). *Studies in the nature of character: Vol. 1. Studies in deceit.* New York: Macmillan.

Hartshorne, H., & May, M. A. (1971). Studies in the organization of character. In H. Munsinger (Ed.), *Readings in child development* (pp. 190–197). New York: Holt, Rinehart and Winston. (Originally published in 1930.)

Hartup, W. W. (1983). Peer relations. In P. H. Mussen (Series Ed.) & E. M. Hetherington (Vol. Ed.), *Handbook of child psychology: Vol. 4. Socialization, personality, and social development* (4th ed., pp. 103–196). New York: Wiley.

Hayakawa, S. I. (1964). *Language in thought and action* (2nd ed.). New York: Harcourt, Brace & World.

Hayden-Thomson, L., Rubin, K. H., & Hymel, S. (1987). Sex preferences in sociometric choices. *Developmental Psychology,* 23, 558–562.

Hayes, D. S., Gershman, E. S., & Halteman, W. (1996). Enmity in males at four developmental levels: Cognitive bases for disliking peers. *Journal of Genetic Psychology,* 157, 153–160.

Heckathorn, D. D. (1992). Collective sanctions and group heterogeneity: Cohesion and polarization in normative systems. In E. J. Lawler, B. Markovsky, C. Ridgeway, & H. A. Walker (Eds.), *Advances in group processes: Vol. 9* (pp. 41–63). Greenwich, CT: JAI Press.

Herbert, B. (1997, December 14). The success taboo. *New York Times,* Section 4, p. 13.

Herrnstein, R. J., & Murray, C. (1994). *The bell curve: Intelligence and class structure in American life.* New York: Free Press.

Hess, E. H. (1970). The ethological approach to socialization. In R. A. Hoppe, G. A. Milton, & E. C. Simmel (Eds.), *Early experiences and the processes of socialization.* New York: Academic Press.

Hetherington, E. M., & Clingempeel, W. G., with Anderson, E. R., Deal, J. E., Hagan, M. S., Hollier, E. A., & Lindner, M. S. (1992). Coping with marital transitions: A family systems perspective. *Monographs of the Society for Research in Child Development,* 57(2–3, Serial No. 227).

Hibbert, C. (1987). *The English: A social history, 1066–1945.* New York: Norton.

Hilgard, E. R., Atkinson, R. L., & Atkinson, R. C. (1979). *Introduction to psychology* (7th ed.). New York: Harcourt Brace Jovanovich.

Hilton, J. L., & von Hippel, W. (1996). Stereotypes. *Annual Review of Psychology,* 47, 237–271.

Hoffman, L. W. (1989). Effects of maternal employment in the two-parent family. *American Psychologist,* 44, 283–292.

Hold, B. (1977). Rank and behavior: An ethological study of preschool children. *Homo,* 28, 158–188.

Holden, C. (1996, November 15). Small refugees suffer the effects of early neglect. *Science,* 274, 1076–1077.

Holden, C. (Ed.) (1995, June 9). Probing nature's hairy secrets. *Science,* 268, 1439.

Horner, M. S. (1969, November). Fail: Bright women. *Psychology Today,* pp. 36–38.

Howell, S. (1988). From child to human: Chewong concepts of self. In G. Jahoda & I. M. Lewis (Eds.), *Acquiring culture: Cross cultural studies in child development* (pp. 147–169). London: Croom Helm.

Howes, C. (1985). Sharing fantasy: Social pretend play in toddlers. *Child Development,* 56, 1253–1258.

Howes, C. (1987). Social competence with peers in young children: Developmental sequences. *Developmental Review,* 7, 252–272.

Howes, C., Matheson, C. C., & Hamilton, C. E. (1994). Maternal, teacher, and child care history correlates of children's relationships with peers. *Child Development,* 65, 264–273.

Howrigan, G. A. (1988). Fertility, infant feeding, and change in Yucatán. In R. A. LeVine, P. M. Miller, & M. M. West (Eds.), *Parental behavior in diverse societies* (pp. 37–50). New Directions for Child Development, No. 40. San Francisco: Jossey-Bass.

Hubbard, J. A., & Coie, J. D. (1994). Emotional correlates of social competence in children's peer relationships. *Merrill-Palmer Quarterly,* 40, 1–20.

Hulbert, A. (1996, May 20). Dr. Spock's baby: Fifty years in the life of a book and the American family. *The New Yorker,* pp. 82–92.

Hunt, M. (1997). *How science takes stock: The story of meta-analysis.* New York: Russell Sage Foundation.

Hur, Y.-M., & Bouchard, T. J., Jr. (1995). Genetic influences on perceptions of childhood family environment: A reared apart twin study. *Child Development,* 66, 330–345.

Ioannidis, J. P. A. (1998, January 28). Effect of the statistical significance of results on the time to completion and publication of randomized efficacy trials. *Journal of the American Medical Association,* 279, 281–286.

Ireys, H. T., Werthamer-Larsson, L. A., Kolodner, K. B., & Gross, S. S. (1994). Mental health of young adults with chronic illness: The mediating effect of perceived impact. *Journal of Pediatric Psychology,* 19, 205–222.

Jacklin, C. N. (1989). Female and male: Issues of gender. *American Psychologist,* 44, 127–133.

Jacobs, F. H., & Davies, M. W. (1991). Rhetoric or reality? Child and family policy in the United States. *Society for Research in Child Development Social Policy Report,* 5, Winter issue (Whole No. 4).

Jacobson, J. L., & Wille, D. E. (1986). The influence of attachment pattern on developmental changes in peer interaction from the toddler to the preschool period. *Child Development,* 57, 338–347.

James, W. (1890). *The principles of psychology: Vol. 1.* New York: Henry Holt.

Jockin, V., McGue, M., & Lykken, D. T. (1996). Personality and divorce: A genetic analysis. *Journal of Personality and Social Psychology,* 71, 288–299.

Jones, M. C. (1957). The later careers of boys who were early or late maturing. *Child Development,* 28, 113–128.

Jones, M. C., & Bayley, N. (1950). Physical maturing among boys as related to behavior. *Journal of Educational Psychology,* 41, 129–148.

Jones, V. E. (1993). Program touts advantages of breast-feeding. *Asbury Park (N.J.) Press,* May 23, pp. AA1, AA5.

Jussim, L. J. (1993). Accuracy in interpersonal expectations: A reflection-construction analysis of current and classic research. *Journal of Personality,* 61, 637–668.

Jussim, L. J., & Fleming, C. (1996). Self-fulfilling prophecies and the maintenance of social stereotypes: The role of dyadic interactions and social forces. In N. C. Macrae, M. Hewstone, & C. Stangor (Eds.), *Stereotypes and stereotyping* (pp. 161–192). New York: Guilford Press.

Jussim, L. J., McCauley, C. R., & Lee, Y.-T. (1995). Why study stereotype accuracy and inaccuracy? In Y.-T. Lee, L. J. Jussim, & C. R. McCauley (Eds.), *Stereotype accuracy: Toward appreciating group differences* (pp. 3–27). Washington, DC: American Psychological Association.

Juvonen, J., & Murdock, T. B. (1993). How to promote social approval: Effects of audience and achievement outcome on publicly communicated attributions. *Journal of Educational Psychology,* 85, 365–376.

Kagan, J. (1978, January). The baby's elastic mind. *Human Nature,* pp. 66–73.

Kagan, J. (1989). Temperamental contributions to social behavior. *American Psychologist,* 44, 668–674.

Kagan, J. (1994). *Galen's prophecy.* New York: Basic Books.

Kaler, S. R., & Freeman, B. J. (1994). Analysis of environmental deprivation: Cognitive and social development in Romanian orphans. *Journal of Child Psychology and Psychiatry,* 35, 769–781.

Karmiloff-Smith, A., Klima, E., Bellugi, U., Grant, J., & Baron-Cohen, S. (1995). Is there a social module? Language, face processing, and theory of mind in individuals with Williams syndrome. *Journal of Cognitive Neuroscience,* 7, 196–208.

Keeley, L. H. (1996). *War before civilization.* New York: Oxford University Press.

Keenan, K., Loeber, R., Zhang, Q., Stouthamer-Loeber, M., & van Kammen, W. B. (1995). The influence of deviant peers on the development of boys' disruptive and delinquent behavior: A temporal analysis. *Development and Psychopathology,* 7, 715–726.

Kegl, J., Senghas, A., & Coppola, M. (1999). Creation through contact: Sign language emergence and sign language change in Nicaragua. In M. DeGraff (Ed.), *Language creation and language change: Creolization, diachrony, and development.* Cambridge: MIT Press.

Kelley, M. L., & Tseng, H.-M. (1992). Cultural differences in child rearing: A comparison of immigrant Chinese and Caucasian American mothers. *Journal of Cross-Cultural Psychology,* 23, 444–455.

Kellogg, W. N., & Kellogg, L. A. (1933). *The ape and the child: A study of environmental influence upon early behavior.* New York: McGraw-Hill.

Kerr, M., Lambert, W. W., Stattin, H., & Klackenberg-Larsson, I. (1994). Stability of inhibition in a Swedish longitudinal sample. *Child Development,* 65, 138–146.

Kindermann, T. A. (1993). Natural peer groups as contexts for individual development: The case of children's motivation in school. *Developmental Psychology,* 29, 970–977.

Kindermann, T. A. (1995). Distinguishing "buddies" from "bystanders": The study of children's development within natural peer contexts. In T. A. Kindermann & J. Valsiner (Eds.), *Development of person–context relations.* Hillsdale, NJ: Erlbaum.

King, C. A., Naylor, M. W., Segal, H. G., Evans, T., & Shain, B. N. (1993). Global self-worth, specific self-perceptions of competence, and depression in adolescents. *Journal of the American Academy of Child & Adolescent Psychiatry,* 32, 745–752.

Klaus, M. H., & Kennell, J. H. (1976). *Maternal-infant bonding: The impact of early separation or loss on family development.* St. Louis: Mosby.

Klinnert, M. D. (1984). The regulation of infant behavior by maternal facial expression. *Infant Behavior and Development,* 7, 447–465.

Klopfer, P. (1971, July/August). Mother love: What turns it on? *American Scientist,* 59, 404–407.

Knight, G. P., Virdin, L. M., & Roosa, M. (1994). Socialization and family correlates of mental health outcomes among Hispanic and Anglo American children: Consideration of cross-ethnic scalar equivalence. *Child Development,* 65, 212–224.

Kochanska, G., Murray, K., & Coy, K. C. (1997). Inhibitory control as a contributor to conscience in childhood: From toddler to early school age. *Child Development,* 68, 263–277.

Kochenderfer, B. J., & Ladd, G. W. (1996). Peer victimization: Cause or consequence of school maladjustment? *Child Development,* 67, 1305–1317.

Kolata, G. (1993, May 25). Brain researcher makes it look easy. *New York Times,* pp. C1, C8.

Kolers, P. A. (1975). Bilingualism and information processing. In R. C. Atkinson (Ed.), *Readings from Scientific American: Psychology in progress* (pp. 188–195). San Francisco: Freeman. (Originally published in 1968.)

Koluchová, J. (1972). Severe deprivation in twins: A case study. *Journal of Child Psychology and Psychiatry,* 13, 107–114.

Koluchová, J. (1976). The further development of twins after severe and prolonged deprivation: A second report. *Journal of Child Psychology and Psychiatry,* 17, 181–188.

Konner, M. J. (1972). Aspects of the developmental ethology of a foraging people. In N. Blurton Jones (Ed.), *Ethological studies of child behavior* (pp. 285–303). London: Cambridge University Press.

Kopp, C. B. (1989). Regulation of distress and negative emotions: A developmental view. *Developmental Psychology,* 25, 343–354.

Kosof, A. (1996). *Living in two worlds: The immigrant children's experience.* New York: Twenty-First Century Books.

Krantz, S. E. (1989). The impact of divorce on children. In A. S. Skolnick & J. H. Skolnick (Eds.), *Families in transition* (6th ed., pp. 341–363). Glenview, IL: Scott, Foresman.

Kristof, N. D. (1995, July 18). Japan's schools: Safe, clean, not much fun. *New York Times,* pp. A1, A6.

Kristof, N. D. (1997, August 17). Where children rule. *New York Times Magazine,* pp. 40–44.

Krueger, J. (1992). On the overestimation of between-group differences. *European Review of Social Psychology,* 3, 31–56.

Krueger, J., & Clement, R. W. (1994). Memory-based judgments about multiple categories: A revision and extension of Tajfel's accentuation theory. *Journal of Personality and Social Psychology,* 67, 35–47.

Kupersmidt, J. B., Griesler, P. C., DeRosier, M. E., Patterson, C. J., & Davis, P. W. (1995). Childhood aggression and peer relations in the context of family and neighborhood factors. *Child Development,* 66, 360–375.

Lab, S. P., & Whitehead, J. T. (1988). An analysis of juvenile correctional treatment. *Crime & Delinquency,* 34, 60–83.

Ladd, G. W. (1983). Social networks of popular, average, and rejected children in school settings. *Merrill-Palmer Quarterly,* 29, 283–307.

Ladd, G. W. (1992). Themes and theories: Perspectives on processes in family–peer relationships. In R. D. Parke & G. W. Ladd (Eds.), *Family–peer relationships: Modes of linkage* (pp. 3–34). Hillsdale, NJ: Erlbaum.

Ladd, G. W., Profilet, S. M., & Hart, C. H. (1992). Parents' management of children's peer relations: Facilitating and supervising children's activities in the peer culture. In R. D. Parke & G. W. Ladd (Eds.), *Family–peer relationships: Modes of linkage* (pp. 215–253). Hillsdale, NJ: Erlbaum.

LaFreniere, P. J., & Sroufe, L. A. (1985). Profiles of peer competence in the preschool: Interrelations between measures, influence of social ecology, and relation to attachment history. *Developmental Psychology,* 21, 56–69.

LaFromboise, T., Coleman, H. L. K., & Gerton, J. (1993). Psychological impact of biculturalism: Evidence and theory. *Psychological Bulletin,* 114, 395–412.

Lahey, B. B., Hartdagen, S. E., Frick, P. J., McBurnett, K., Connor, R., & Hynd, G. W. (1988). Conduct disorder: Parsing the confounded relation to parental divorce and antisocial personality. *Journal of Abnormal Psychology,* 97, 334–337.

Lamb, M. E., & Nash, A. (1989). Infant-mother attachment, sociability, and peer competence. In T. J. Berndt & G. W. Ladd (Eds.), *Peer relationships in child development* (pp. 219–245). New York: Wiley.

Lane, H. (1976). *The wild boy of Aveyron.* Cambridge, MA: Harvard University Press.

Langlois, J. H., Ritter, J. M., Casey, R. J., & Sawin, D. B. (1995). Infant attractiveness predicts maternal behaviors and attitudes. *Developmental Psychology,* 31, 464–472.

Larkin, P. (1989). *Collected poems* (A. Thwaite, Ed.). New York: Farrar, Straus & Giroux.

Laumann, E. O., Gagnon, J. H., Michael, R. T., & Michaels, S. (1994). *The social organization of sexuality.* Chicago: University of Chicago Press.

Leach, G. M. (1972). A comparison of the social behaviour of some normal and problem children. In Blurton Jones, N. (Ed.), *Ethological studies of child behavior* (pp. 249–281). London: Cambridge University Press.

Leach, P. (1995). *Your baby and child: From birth to age five* (2nd ed.). New York: Knopf. (This edition originally published in 1989.)

Leaper, C. (1994a). Editor's notes. In Leaper, C. (Ed.), *Childhood gender segregation: Causes and consequences* (pp. 1–5). New Directions for Child Development, No. 65. San Francisco: Jossey-Bass.

Leaper, C. (1994b). Exploring the consequences of gender segregation on social relationships. In Leaper, C. (Ed.), *Childhood gender segregation: Causes and consequences* (pp. 67–86). New Directions for Child Development, No. 65. San Francisco: Jossey-Bass.

Leary, M. R., Tambor, E. S., Terdal, S. K., & Downs, D. L. (1995). Self-esteem as an interpersonal monitor: The sociometer hypothesis. *Journal of Personality and Social Psychology,* 68, 518–530.

Lee, C.-R. (1995, October 16). Coming home again. *The New Yorker,* pp. 164–168.

Leinbach, M. D., & Fagot, B. I. (1993). Categorical habituation to male and female faces: Gender schematic processing in infancy. *Infant Behavior and Development,* 16, 317–332.

LeLorier, J., Grégoire, G., Benhaddad, A., Lapierre, J., & Derderian, F. (1997, August 21). Discrepancies between meta-analyses and subsequent large randomized, controlled trials. *New England Journal of Medicine,* 337, 536–542.

Lenneberg, E. H. (1972). On explaining language. In M. E. P. Seligman & J. L. Hager (Eds.), *Biological boundaries of learning* (pp. 379–396). New York: Appleton-Century-Crofts.

Lepper, M. R., Greene, D. & Nisbett, R. E. (1973). Undermining children's intrinsic interest with extrinsic reward: A test of the "overjustification" hypothesis. *Journal of Personality and Social Psychology,* 28, 129–137.

Leslie, A. M. (1994). Pretending and believing: Issues in the theory of ToMM. *Cognition,* 50, 211–238.

Levin, H., & Garrett, P. (1990). Sentence structure and formality. *Language in Society,* 19, 511–520.

Levin, H., & Novak, M. (1991). Frequencies of Latinate and Germanic words in English as determinants of formality. *Discourse Processes,* 14, 389–398.

LeVine, R. A., & LeVine, B. B. (1963). Nyansongo: A Gusii Community in Kenya. In B. B. Whiting (Ed.), *Six cultures: Studies of child rearing* (pp. 15–202). New York: Wiley.

LeVine, R. A., & LeVine, S. E. (1988). Parental strategies among the Gusii of Kenya. In R. A. LeVine, P. M. Miller, & M. M. West (Eds.), *Parental behavior in diverse societies* (pp. 27–35). New Directions for Child Development, No. 40. San Francisco: Jossey-Bass.

Levy, G. D., & Haaf, R. A. (1994). Detection of gender-related categories by 10-month-old infants. *Infant Behavior and Development,* 17, 457–459.

Lewicki, P., Hill, T., & Czyzewska, M. (1992). Nonconscious acquisition of information. *American Psychologist,* 47, 796–801.

Lightfoot, C. (1992). Constructing self and peer culture: A narrative perspective on adolescent risk taking. In L. T. Winegar & J. Valsiner (Eds.), *Children's development within social context: Vol. 2. Research and methodology* (pp. 229–245). Hillsdale, NJ: Erlbaum.

Locurto, C. (1990). The malleability of IQ as judged from adoption studies. *Intelligence,* 14, 275–292.

Loehlin, J. C. (1997). A test of J. R. Harris's theory of peer influences on personality. *Journal of Personality and Social Psychology,* 72, 1197–1201.

Loehlin, J. C., & Nichols, R. C. (1976). *Heredity, environment, and personality: A study of 850 sets of twins.* Austin: University of Texas Press.

Lore, R. K., & Schultz, L. A. (1993). Control of human aggression: A comparative perspective. *American Psychologist,* 48, 16–25.

Lung, C.-T., & Daro, D. (1996). *Current trends in child abuse reporting and fatalities: The results of the 1995 annual fifty-state survey.* Chicago: National Center on Child Abuse Prevention Research.

Lykken, D. T. (1995). *The antisocial personalities.* Hillsdale, NJ: Erbaum.

Lykken, D. T. (1997). The American crime factory. *Psychological Inquiry,* 8, 261–270.

Lykken, D. T. (1999). *Happiness.* New York: Golden Books.

Lykken, D. T., McGue, M., Tellegen, A., & Bouchard, T. J., Jr. (1992). Emergenesis: Genetic traits that may not run in families. *American Psychologist,* 47, 1565–1577.

Lykken, D. T., & Tellegen, A. (1996). Happiness is a stochastic phenomenon. *Psychological Science,* 7, 186–189.

Lytton, H., & Romney, D. M. (1991). Parents' differential socialization of boys and girls: A meta-analysis. *Psychological Bulletin,* 109, 267–296.

Maccoby, E. E. (1990). Gender and relationships: A developmental account. *American Psychologist,* 45, 513–520.

Maccoby, E. E. (1992). The role of parents in the socialization of children: An historical overview. *Developmental Psychology,* 28, 1006–1017.

Maccoby, E. E. (1994). Commentary: Gender segregation in childhood. In Leaper, C. (Ed.), *Childhood gender segregation: Causes and consequences* (pp. 87–97). New Directions for Child Development, No. 65. San Francisco: Jossey-Bass.

Maccoby, E. E. (1995). The two sexes and their social systems. In P. Moen, G. H. Elder, Jr., & K. Lüscher (Eds.), *Examining lives in context: Perspectives on the ecology of human development.* Washington, DC: American Psychological Association.

Maccoby, E. E., & Jacklin, C. N. (1974). *The psychology of sex differences.* Stanford, CA: Stanford University Press.

Maccoby, E. E., & Jacklin, C. N. (1987). Gender segregation in childhood. *Advances in Child Development and Behavior,* 20, 239–287.

Maccoby, E. E., & Martin, J. A. (1983). Socialization in the context of the family: Parent–child interaction. In P. H. Mussen (Series Ed.) & E. M. Hetherington (Vol. Ed.), *Handbook of child psychology: Vol. 4. Socialization, personality, and social development* (4th ed., pp. 1–101). New York: Wiley.

Maclean, C. (1977). *The wolf children.* NY: Hill & Wang.

Madon, S., Jussim, L., & Eccles, J. (1997). In search of the powerful self-fulfilling prophecy. *Journal of Personality and Social Psychology,* 72, 791–809.

Main, M., Kaplan, N., & Cassidy, J. (1985). Security in infancy, childhood, and adulthood: A move to the level of representation. In I. Bretherton & E. Waters (Eds.), Growing points of attachment theory and research (pp. 66–104). *Monographs of the Society for Research in Child Development,* 50 (1–2, Serial No. 209).

Main, M., & Weston, D. R. (1981). The quality of the toddler's relationship to mother and to father: Related to conflict behavior and the readiness to establish new relationships. *Child Development,* 52, 932–940.

Malinowsky-Rummell, R., & Hansen, D. J. (1993). Long-term consequences of childhood physical abuse. *Psychological Bulletin,* 114, 68–79.

Mandler, J. M. (1992). How to build a baby: II. Conceptual primitives. *Psychological Review,* 99, 587–604.

Mandler, J. M., & McDonough, L. (1993). Concept formation in infancy. *Cognitive Development,* 8, 291–318.

Mann, C. C. (1994, November 11). Can meta-analysis make policy? *Science,* 266, 960–962.

Mann, T. L. (1997, September). Head Start and the panacea standard. *APS Observer,* pp. 10, 24.

Mar, M. E. (1995, November-December). Blue collar, crimson blazer. *Harvard Magazine,* 98, 47–51.

Marano, H. E. (1995, September/October). Big. Bad. Bully. *Psychology Today,* pp. 50–82.

Maretzki, T. W., & Maretzki, H. (1963). Taira: An Okinawan Village. In B. B. Whiting (Ed.), *Six cultures: Studies of child rearing* (pp. 363–539). New York: Wiley.

Martin, C. L. (1993). Theories of sex typing: Moving toward multiple perspectives. In L. A. Serbin, K. K. Powlishta, & J. Gulko, The development of sex typing in middle childhood (pp. 75–85). *Monographs of the Society for Research in Child Development,* 58(2, Serial No. 232).

Martin, C. L. (1994). Cognitive influences on the development and maintenance of gender segregation. In Leaper, C. (Ed.), *Childhood gender segregation: Causes and consequences* (pp. 35–51). New Directions for Child Development, No. 65. San Francisco: Jossey-Bass.

Martin, J. (1995, August 25). Miss Manners column. United Features (on-line).

Martini, M. (1994). Peer interactions in Polynesia: A view from the Marquesas. In J. L. Roopnarine, J. E. Johnson, & F. H. Hooper (Eds.), *Children's play in diverse cultures* (pp. 73–103). Albany: State University of New York Press.

Masten, A. S. (1986). Humor and competence in school-aged children. *Child Development,* 57, 461–473.

Matas, L., Arend, R. A., & Sroufe, L. A. (1978). Continuity of adaptation in the second year: The relationship between quality of attachment and later competence. *Child Development,* 49, 547–556.

Maternal impressions (1996, November 13). *Journal of the American Medical Association,* 276, 1466. (Originally published in 1896.)

Mathews, J. (1988). *Escalante: The best teacher in America.* New York: Henry Holt.

Maunders, D. (1994). Awakening from the dream: The experience of childhood in Protestant orphan homes in Australia, Canada, and the United States. *Child & Youth Care Forum,* 23, 393–412.

McCall, R. B. (1992). Academic underachievers. *Current Directions in Psychological Science,* 3, 15–19.

McCloskey, L. A. (1996). Gender and the expression of status in children's mixed-age conversations. *Journal of Applied Developmental Psychology,* 17, 117–133.

McCrae, R. R., & Costa, P. T., Jr. (1994). The stability of personality: Observations and evaluations. *Current Directions in Psychological Science,* 3, 173–175.

McDonald, M. A., Sigman, M., Espinosa, M. P., & Neumann, C. G. (1994). Impact of a temporary food shortage on children and their mothers. *Child Development,* 65, 404–415.

McFarland, C., & Buehler, R. (1995). Collective self-esteem as a moderator of the frog-pond effect in reactions to performance feedback. *Journal of Personality and Social Psychology,* 68, 1055–1070.

McGraw, M. B. (1939). Swimming behavior of the human infant. *Journal of Pediatrics,* 15, 485–490.

McGrew, W. G. (1972). *An ethological study of children's behavior.* New York: Academic Press.

McGue, M., & Lykken, D. T. (1992). Genetic influence on risk of divorce. *Psychological Science,* 3, 368-373.

McGuffin, P., & Katz, R. (1993). Genes, adversity, and depression. In R. Plomin & G. E. McClearn (Eds.), *Nature, nurture, & psychology* (pp. 217–230). Washington, DC: American Psychological Association.

McHale, S. M., Crouter, A. C., McGuire, S. A., & Updegraff, K. A. (1995). Congruence between mothers' and fathers' differential treatment of siblings: Links with family relations and children's well-being. *Child Development,* 66, 116–128.

McLanahan, S. (1994, Summer). The consequences of single motherhood. *The American Prospect,* 18, 48–58.

McLanahan, S., & Booth, K. (1989). Mother-only families: Problems, prospects, and politics. *Journal of Marriage and the Family,* 51, 557–580.

McLanahan, S., & Sandefur, G. (1994). *Growing up with a single parent: What hurts, what helps.* Cambridge, MA: Harvard University Press.

Mead, M. (1959). Preface to Ruth Benedict's *Patterns of culture* (2nd ed.). Boston: Houghton Mifflin.

Mead, M. (1963). *Sex and temperament in three primitive societies* (3rd ed.). New York: Dell. (First edition published in 1935.)

Mealey, L. (1995). The sociobiology of sociopathy: An integrated evolutionary model. *Behavioral & Brain Sciences,* 18, 523–599.

Mednick, S. A., Gabrielli, W. F., Jr., & Hutchings, B. (1987). Genetic factors in the etiology of criminal behavior. In S. A. Mednick, T. E. Moffitt, & S. A. Stack (Eds.), *The causes of crime: New biological approaches* (pp. 74–91). Cambridge, UK: Cambridge University Press.

Melson, G. F., Ladd, G. W., & Hsu, H.-C. (1993). Maternal support networks, maternal cognitions, and young children's social and cognitive development. *Child Development,* 64, 1401–1417.

Meredith, W. H., Abbott, D. A., & Ming, Z. F. (1993). Self-concept and sociometric outcomes: A comparison of only children and sibling children from urban and rural areas in the People's Republic of China. *The Journal of Psychology,* 126, 411–419.

Merten, D. E. (1996a). Visibility and vulnerability: Responses to rejection by nonaggressive junior high school boys. *Journal of Early Adolescence,* 16, 5–26.

Merten, D. E. (1996b). Information versus meaning: Toward a further understanding of early adolescent rejection. *Journal of Early Adolescence,* 16, 37–45.

Meyerhoff, B. (1978). *Number our days.* New York: Simon & Schuster.

Minton, R. J. (1971). *Inside: Prison American style.* New York: Random House.

Minturn, L., & Hitchcock, J. T. (1963). The Rajputs of Khalapur, India. In B. B. Whiting (Ed.), *Six cultures: Studies of child rearing* (pp. 202–361). New York: Wiley.

Mitani, J. C., Hasegawa, T., Gros-Louis, J., Marler, P., & Byrne, R. (1992). Dialects in wild chimpanzees? *American Journal of Primatology,* 27, 233–243.

Mitchell, C. M., Libber, S., Johanson, A. J., Plotnick, L., Joyce, S., Migeon, C. J., & Blizzard, R. M. (1986). Psychosocial impact of long-term growth hormone. In B. Stabler & L. E. Underwood (Eds.), *Slow grows the child: Psychosocial aspects of growth delay* (pp. 97–109). Hillsdale, NJ: Erlbaum.

Mitchell, D. E. (1980). The influence of early visual experience on visual perception. In C. S. Harris (Ed.), *Visual coding and adaptability* (pp. 1–50). Hillsdale, NJ: Erlbaum.

Modell, J. (1997, January 31). Family niche and intellectual bent (a review of *Born to Rebel*). *Science,* 275, 624–625.

Moffitt, T. E. (1993). Adolescence-limited and life-course-persistent antisocial behavior: A developmental taxonomy. *Psychological Review,* 100, 674–701.

Money, J., & Ehrhardt, A. A. (1972). *Man & woman, boy & girl.* Baltimore, MD: Johns Hopkins University Press.

Montagu, A. (1976). *The nature of human aggression.* New York: Oxford University Press.

Montour, K. (1977). William James Sidis, the broken twig. *American Psychologist,* 32, 265–279.

Moore, T. R. (1996, September 22). Labor of love. *Asbury Park (N.J.) Press,* pp. D1, D10.

Moran, G. F., & Vinovskis, M. A. (1985). The great care of godly parents: Early childhood in Puritan New England. In A. B. Smuts & J. W. Hagen (Eds.), History and research in child development. *Monographs of the Society for Research in Child Development,* 50(4–5, Serial No. 211).

Morelli, G. A. (1997). Growing up female in a farmer community and a forager community. In M. E. Morbeck, A. Galloway, & A. L. Zihlman (Eds.), *The evolving female: A life-history perspective* (pp. 209–219). Princeton, NJ: Princeton University Press.

Morelli, G. A., Rogoff, B., Oppenheim, D., & Goldsmith, D. (1992). Cultural variation in infants' sleeping arrangements: Questions of independence. *Developmental Psychology,* 28, 604–613.

Morelli, G. A., Winn, S., & Tronick, E. Z. (1987). Perinatal practices: A biosocial perspective. In H. Rauh & H.-C. Steinhausen (Eds.), *Psychobiology and early development.* North Holland: Elsevier.

Morris, J. (1974). *Conundrum.* New York: Harcourt Brace Jovanovich.

Morton, O. (1998, January). Overcoming yuk. *Wired,* pp. 44–48.

Mosteller, F. (1995). The Tennessee study of class size in the early school grades. *The Future of Children,* 5(2), 113–127.

Mounts, N. S., & Steinberg, L. (1995). An ecological analysis of peer influence on adolescent grade point average and drug use. *Developmental Psychology,* 31, 915–922.

Myers, D. G. (1982). Polarizing effects of social interaction. In H. Tajfel (Series Ed.) & H. Brandstätter, J. H. Davis, & G. Stocker-Kreichgauer (Vol. Eds.), *European monographs in social psychology: Vol. 25. Group decision making* (pp. 125–161). New York: Academic Press.

Myers, D. G. (1992). *The pursuit of happiness: Who is happy—and why?* New York: Avon.

Myers, D. G. (1998). *Psychology* (5th ed.). New York: Worth.

Napier, J. R., & Napier, P. H. (1985). *The natural history of the primates.* Cambridge, MA: MIT Press.

Neckerman, H. J. (1996). The stability of social groups in childhood and adolescence: The role of the classroom social environment. *Social Development,* 2, 131–145.

Neifert, M. (1991). *Dr. Mom's parenting guide: Commonsense guidance for the life of your child.* New York: Signet.

Newcomb, A. F., Bukowski, W. M., & Pattee, L. (1993). Childrens' peer relations: A meta-analytic review of popular, rejected, neglected, controversial, and average sociometric status. *Psychological Bulletin,* 113, 99–128.

Newman, L. S., & Ruble, D. N. (1988). Stability and change in self-understanding: The early elementary school years. *Early Child Development and Care,* 40, 77–99.

Norman, G. (1995, May-June). Edward O. Wilson. *Modern Maturity,* 38, 62–71.

Nydegger, W. F., & Nydegger, C. (1963). Tarong, an Ilocos Barrio in the Philippines. In B. B. Whiting (Ed.), *Six cultures: Studies of child rearing* (pp. 692–867). New York: Wiley.

O'Connor, T. G., Hetherington, E. M., Reiss, D., & Plomin, R. (1995). A twin–sibling study of observed parent–adolescent interactions. *Child Development,* 66, 812–829.

Olds, D., Eckenrode, J., Henderson, C. R., et al. (1997, August 27). Long-term effects of home visitation on maternal life course and child abuse and neglect: Fifteen-year follow-up of a randomized trial. *Journal of the American Medical Association,* 278, 637–643.

O'Leary, K. D., & Smith, D. A. (1991). Marital interactions. *Annual Review of Psychology,* 42, 191–212.

Opie, I., & Opie, P. (1969). *Children's games in street and playground.* London: Oxford University Press.

Padden, C., & Humphries, T. (1988). *Deaf in America: Voices from a culture.* Cambridge, MA: Harvard University Press.

Pan, H.-L. W. (1994). Children's play in Taiwan. In J. L. Roopnarine, J. E. Johnson, & F. H. Hooper (Eds.), *Children's play in diverse cultures* (pp. 31–50). Albany: SUNY Press.

Parke, R. D., Cassidy, J., Burkes, V. M., Carson, J. L., & Boyum, L. (1992). Familial contribution to peer competence among young children: The role of interactive and affective processes. In R. D. Parke & G. W. Ladd (Eds.), *Family–peer relationships: Modes of linkage* (pp.107–134). Hillsdale, NJ: Erlbaum.

Parker, I. (1996, September 9). Richard Dawkins's evolution. *The New Yorker,* pp. 41–45.

Parker, J. G., & Asher, S. R. (1987). Peer relations and later personal adjustment: Are low-accepted children at risk? *Psychological Bulletin,* 102, 357–389.

Parker, J. G., & Asher, S. R. (1993). Beyond group acceptance: Friendship and friendship quality as distinct dimensions of children's peer adjustment. In D. Perlman & W. H. Jones (Eds.), *Advances in personal relationships: Vol. 4* (pp. 261–294). London: Jessica Kingsley Publishers.

Parker, J. G., Rubin, K. H., Price, J. M., & DeRosier, M. E. (1995). Peer relations, child development, and adjustment: A developmental psychopathology perspective. In D. Cicchetti & D. Cohen (Eds.), *Developmental psychopathology: Vol. 2. Risk, disorder, and adaptation* (pp. 96–161). New York: Wiley.

Parks, T. (1995). *An Italian education.* New York: Grove Press.

Pastor, D. (1981). The quality of mother–infant attachment and its relationship to toddlers' initial sociability with peers. *Developmental Psychology,* 17, 326–335.

Patterson, C. J. (1992). Children of lesbian and gay parents. *Child Development,* 63, 1025–1042.

Patterson, C. J. (1994). Lesbian and gay families. *Current Directions in Psychological Science,* 3, 62–64.

Patterson, G. R., & Bank, L. (1989). Some amplifying mechanisms for pathologic processes in families. In M. R. Gunnar & E. Thelen (Eds.), *Systems and development: The Minnesota Symposia on Child Psychology: Vol. 22* (pp. 167–209). Hillsdale, NJ: Erlbaum.

Patterson, G. R., & Yoerger, K. (1991, April). A model for general parenting skill is too simple: Mediational models work better. Paper presented at the biennial meeting of the Society for Research in Child Development, Seattle, Washington.

Pedersen, E., Faucher, T. A., & Eaton, W. W. (1978). A new perspective on the effects of first-grade teachers on children's subsequent adult status. *Harvard Educational Review,* 48, 1–31.

Pedersen, N. L., Plomin, R., Nesselroade, J. R., & McClearn, G. E. (1992). A quantitative genetic analysis of cognitive abilities during the second half of the life span. *Psychological Science,* 3, 346–353.

Peeples, F., & Loeber, R. (1994). Do individual factors and neighborhood context explain ethnic differences in juvenile delinquency? *Journal of Quantitative Criminology,* 10, 141–157.

Pelaez-Nogueras, M., Field, T., Cigales, M., Gonzalez, A., & Clasky, S. (1994). Infants of depressed mothers show less "depressed" behavior with their nursery teachers. *Infant Mental Health Journal,* 15, 358–367.

Perez, C. M., & Widom, C. S. (1994). Childhood victimization and longterm intellectual and academic outcomes. *Child Abuse & Neglect,* 18, 617–633.

Perner, J. (1991). *Understanding the representational mind.* Cambridge: MIT Press.

Perry, D. G., & Bussey, K. (1984). *Social development.* Englewood Cliffs, NJ: Prentice-Hall.

Pérusse, D., Neale, M. C., Heath, A. C., & Eaves, L. J. (1994). Human parental behavior: Evidence for genetic influence and potential implications for gene-culture transmission. *Behavior Genetics,* 24, 327–335.

Pfennig, D. W., & Sherman, P. W. (1995, June). Kin recognition. *Scientific American,* 272, 98–103.

Piaget, J. (1952). *The origins of intelligence in children* (M. Cook, Trans.). New York: International Universities Press.

Piaget, J. (1962). *Play, dreams, and imitation in childhood* (C. Gattegno & F. M. Hodgson, Trans.). New York: Norton.

Piaget, J., & Inhelder, B. (1969). *The psychology of the child* (H. Weaver, Trans.). New York: Basic Books.

Pinker, S. (1994). *The language instinct.* New York: HarperCollins.

Pinker, S. (1997). *How the mind works.* New York: Norton.

Pitts, M. B. (1997, March 28–30). The latest on what to feed kids. *USA Weekend,* pp. 22–23.

Pless, I. B., & Nolan, T. (1991). Revision, replication, and neglect: Research on maladjustment in chronic illness. *Journal of Child Psychology & Psychiatry & Allied Disciplines,* 32, 347–365.

Plomin, R. (1990). *Nature and nurture: An introduction to human behavioral genetics.* Pacific Grove, CA: Brooks/Cole.

Plomin, R., Chipuer, H. M., & Neiderhiser, J. M. (1994). Behavioral genetic evidence for the importance of nonshared environment. In E. M. Hetherington, D. Reiss, & R. Plomin (Eds.), *Separate social worlds of siblings: The impact of nonshared environment on development* (pp. 1–31). Hillsdale, NJ: Erlbaum.

Plomin, R., & Daniels, D. (1987). Why are children in the same family so different from one another? *Behavioral and Brain Sciences,* 10, 1–60.

Plomin, R., Fulker, D. W., Corley, R., & DeFries, J. C. (1997). Nature, nurture, and cognitive development from 1 to 16 years: A parent–offspring adoption study. *Psychological Science,* 8, 442–447.

Plomin, R., McClearn, G. E., Pedersen, N. L., Nesselroade, J. R., & Bergeman, C. S. (1988). Genetic influence on childhood family environment perceived retrospectively from the last half of the life span. *Developmental Psychology,* 24, 738–745.

Plomin, R., Owen, M. J., & McGuffin, P. (1994, June 17). The genetic basis of complex human behaviors. *Science,* 264, 1733–1739.

Pogrebin, R. (1996, May 28). For a Bronx teacher, a winning tactic. *New York Times,* pp. B1, B7.

Popenoe, D. (1996). *Life without father: Compelling new evidence that fatherhood and marriage are indispensable for the good of children and society.* New York: Free Press.

Povinelli, D. J., & Eddy, T. J. (1996). What young chimpanzees know about seeing. With commentary by R. P. Hobson & M. Tomasello and a reply by D. J. Povinelli. *Monographs of the Society for Research in Child Development,* 61(3, Serial No. 247).

Powlishta, K. K. (1995a). Gender bias in children's perceptions of personality traits. *Sex Roles,* 32, 17–28.

Powlishta, K. K. (1995b). Intergroup processes in childhood: Social categorization and sex role development. *Developmental Psychology,* 31, 781–788.

Premack, D., & Woodruff, G. (1978). Does the chimpanzee have a theory of mind? *Behavioral and Brain Sciences,* 1, 515–526.

Preston, P. (1994). *Mother father deaf: Living between sound and silence.* Cambridge, MA: Harvard University Press.

Primus IV (1998, March-April). Good Will Sidis. *Harvard Magazine,* p. 80.

Proulx, E. A. (1993). *The shipping news.* New York: Simon & Schuster.

Provine, R. R. (1993). Laughter punctuates speech: Linguistic, social and gender contexts of laughter. *Ethology,* 95, 291–298.

Putnam, F. W. (1989). *Diagnosis and treatment of multiple personality disorder.* New York: Guilford.

Rao, S. C., Rainer, G., & Miller, E. K. (1997, May 2). Integration of what and where in the primate prefrontal cortex. *Science,* 276, 821–824.

Ravitch, D. (1997, September 5). First teach them English. *New York Times,* p. A35.

Readdick, C. A., Grise, K. S., Heitmeyer, J. R., & Furst, M. H. (1996). Children of elementary school age and their clothing: Development of self-perception and of management of appearance. *Perceptual and Motor Skills,* 82, 383–394.

Reader, J. (1988). *Man on earth.* New York: Harper & Row.

Reich, P. A. (1986). *Language development.* Englewood Cliffs, NJ: Prentice-Hall.

Reich, R. (1997, June 13–15). Being a dad: Rewarding labor. *USA Weekend,* pp. 10–11.

Reiss, D. (1997). Mechanisms linking genetic and social influences in adolescent development: Beginning a collaborative search. *Current Directions in Psychological Science,* 6, 100–105.

Resnick, M. D., Bearman, P. S., Blum, R. W., et al. (1997, September 10). Protecting adolescents from harm: Findings from the National Longitudinal Study on Adolescent Health. *Journal of the American Medical Association,* 278, 823–832.

Retherford, R. D., & Sewell, W. H. (1991). Birth order and intelligence: Further tests of the confluence model. *American Sociological Review,* 56, 141–158.

Richman, R. A., Gordon, M., Tegtmeyer, P., Crouthamel, C., & Post, E. M. (1986). Academic and emotional difficulties associated with short stature. In B. Stabler & L. E. Underwood (Eds.), *Slow grows the child: Psychosocial aspects of growth delay* (pp. 13–26). Hillsdale, NJ: Erlbaum.

Rigotti, N. A., DiFranza, J. R., Chang, Y., Tisdale, T., Kemp, B., & Singer, D. E. (1997, October 9). The effect of enforcing tobacco-sales laws on adolescents' access to tobacco and smoking behavior. *New England Journal of Medicine,* 337, 1044–1051.

Riley, D. (1990). Network influences on father involvement in childrearing. In M. Cochran, M. Larner, D. Riley, L. Gunnarsson, & C. R. Henderson, Jr. (Eds.), *Extending families: The social networks of parents and their children* (pp. 131–152). Cambridge, UK: Cambridge University Press.

Roberts, S. V. (1995, August 21). An American tale: Colin Powell is only one chapter in a remarkable immigrant story. *U.S. News & World Report,* pp. 27–30.

Rogers, D. (1977). *The psychology of adolescence.* Englewood Cliffs, NJ: Prentice-Hall.

Roggman, L. A., Langlois, J. H., Hubbs-Tait, L., & Rieser-Danner, L. A. (1994). Infant day-care, attachment, and the "file drawer problem." *Child Development,* 65, 1429–1443.

Rogoff, B., Mistry, J., Göncü, A., & Mosier, C. (1993). Guided participation in cultural activity by toddlers and caregivers. *Monographs of the Society for Research in Child Development,* 58(8, Serial No. 236).

Roitblat, H. L., & von Fersen, L. (1992). Comparative cognition: Representations and processes in learning and memory. *Annual Review of Psychology,* 43, 671–710.

Rosch, E. (1978). Principles of categorization. In E. Rosch & B. B. Lloyd (Eds.), *Cognition and categorization.* Hillsdale, NJ: Erlbaum.

Roth, P. (1967). *Portnoy's complaint.* New York: Bantam Books.

Rothbaum, F., & Weisz, J. R. (1994). Parental caregiving and child externalizing behavior in nonclinical samples: A meta-analysis. *Psychological Bulletin,* 116, 55–74.

Rovee-Collier, C. (1993). The capacity for long-term memory in infancy. *Current Directions in Psychological Science,* 2, 130–135.

Rowe, D. C. (1981). Environmental and genetic influences on dimensions of perceived parenting: A twin study. *Developmental Psychology,* 17, 203–208.

Rowe, D. C. (1994). *The limits of family influence: Genes, experience, and behavior.* New York: Guilford Press.

Rowe, D. C. (1997). Are parents to blame? A look at *The Antisocial Personalities. Psychological Inquiry,* 8, 251–260.

Rowe, D. C., Rodgers, J. L., & Meseck-Bushey, S. (1992). Sibling delinquency and the family environment: Shared and unshared influences. *Child Development,* 63, 59–67.

Rowe, D. C., & Waldman, I. D. (1993). The question "How?" reconsidered. In R. Plomin & G. E. McClearn (Eds.), *Nature, nurture, and psychology* (pp. 355–373). Washington, DC: American Psychological Association.

Rowe, D. C., Woulbroun, E. J., & Gulley, B. L. (1994). Peers and friends as nonshared environmental influences. In E. M. Hetherington, D. Reiss, & R. Plomin (Eds.), *Separate social worlds of siblings: The impact of nonshared environment on development* (pp. 159–173). Hillsdale, NJ: Erlbaum.

Rubin, K. H., Bukowski, W., & Parker, J. (1998). Peer interactions, relationships, and groups. In W. Damon (Series Ed.) and N. Eisenberg (Vol. Ed.), *Handbook of child psychology: Vol. 3. Social, emotional, and personality development* (5th ed., pp. 619–700). New York: Wiley.

Ruble, D. N., & Martin, C. L. (1998). Gender development. In W. Damon (Series Ed.) & N. Eisenberg (Vol. Ed.), *Handbook of child psychology: Vol. 3. Social, emotional, and personality development* (5th ed., pp. 933–1016). New York: Wiley.

Runco, M. A. (1991) Birth order and family size. In M. A. Runco (Ed.), *Divergent thinking* (pp. 13–19). Norwood, NJ: Ablex. (Originally published in 1987.)

Russell, R. J. (1993). *The lemurs' legacy: The evolution of power, sex, and love.* New York: Tarcher/Putnam.

Rutter, M. (1979). Maternal deprivation, 1972–1978: New findings, new concepts, new approaches. *Child Development,* 50, 283–305.

Rutter, M. (1983). School effects on pupil progress: Research findings and policy implications. *Child Development,* 54, 1–29.

Rutter, M. (1997). Nature–nurture integration: The example of antisocial behavior. *American Psychologist,* 52, 390–398.

Rutter, M., & Giller, H. J. (1983). *Juvenile delinquency: Trends and perspectives.* New York: Penguin.

Rybczynski, W. (1986). *Home: A short history of an idea.* New York: Penguin Books.

Rydell, A.-M., Dahl, M., & Sundelin, C. (1995). Characteristics of school children who are choosy eaters. *Journal of Genetic Psychology,* 156, 217–229.

Rymer, R. (1993). *Genie: An abused child's flight from silence.* New York: HarperCollins.

Sacks, O. (1989). *Seeing voices: A journey into the world of the deaf.* Berkeley, CA: University of California Press.

Sadker, M., & Sadker, D. (1994). *Failing at fairness: How America's schools cheat girls.* New York: Scribner's.

St. Pierre, R. G., Layzer, J. I., & Barnes, H. V. (1995, winter). Two generation programs: Design, cost, and short-term effectiveness. *The Future of Children,* 5(3), 76–93.

Salzinger, S. (1990). Social networks in child rearing and child development. *Annals of the New York Academy of Science,* 602, 171–188.

Santrock, J. W., & Tracy, R. L. (1978). Effects of children's family structure status on the development of stereotypes by teachers. *Journal of Educational Psychology,* 70, 754–757.

Sastry, S. V. (1996, May 12). Immigrants face challenge teaching children native languages. *Asbury Park (N.J.) Press,* p. AA5.

Saudino, K. J. (1997). Moving beyond the heritability question: New directions in behavioral genetic studies of personality. *Current Directions in Psychological Science,* 6, 86–90.

Savage, S. L., & Au, T. K. (1996). What word learners do when input contradicts the mutual exclusivity assumption. *Child Development,* 67, 3120–3134.

Savin-Williams, R. C. (1979). An ethological study of dominance formation and maintenance in a group of human adolescents. *Child Development,* 49, 534–536.

Scarr, S. (1992). Developmental theories for the 1990s: Development and individual differences. *Child Development,* 63, 1–19.

Scarr, S. (1993). Biological and cultural diversity: The legacy of Darwin for development. *Child Development,* 64, 1333–1353.

Scarr, S. (1997a, September). Head Start and the panacea standard: A reply to Mann. *APS Observer,* 10, 24–25.

Scarr, S. (1997b). Why child care has little impact on most children's development. *Current Directions in Psychological Science,* 6, 143–148.

Scarr, S., & McCartney, K. (1983). How people make their own environments: A theory of genotype → environment effects. *Child Development,* 54, 424–435.

Schaller, S. (1991). *A man without words.* New York: Summit Books.

Schlegel, A., & Barry, H., III (1991). *Adolescence: An anthropological inquiry.* New York: Free Press.

Schofield, J. W. (1981). Complementary and conflicting identities: Images and interaction in an interracial school. In S. R. Asher & J. M. Gottman (Eds.), *The development of children's friendships* (pp. 53–90). Cambridge, UK: Cambridge University Press.

Schooler, C. (1972). Birth order effects: Not here, not now! *Psychological Bulletin,* 78, 161–175.

Schor, Juliet B. (1992). *The overworked American: The unexpected decline of leisure.* New York: Basic Books.

Schütze, Y. (1987). The good mother: The history of the normative model "mother-love." In P. A. Adler, P. Adler, & N. Mandell (Eds.), *Sociological studies of child development: Vol. 2* (pp. 39–78). Greenwich, CT: JAI Press.

Scott, J. P. (1987). Why does human twin research not produce results consistent with those from nonhuman animals? (Commentary on Plomin & Daniels, 1987.) *Brain & Behavioral Sciences,* 10, 39–40.

Scott, J. P., & Fuller, J. L. (1965). *Genetics and the social behavior of the dog.* Chicago: University of Chicago Press.

Segal, N. L. (1993). Twin, sibling, and adoption methods: Tests of evolutionary hypotheses. *American Psychologist,* 48, 943–956.

Seligman, D. (1992). *A question of intelligence: The IQ debate in America.* New York: Birch Lane Press.

Seligman, M. E. P., with Reivich, K., Jaycox, L., & Gillham, J. (1995). *The optimistic child.* Boston: Houghton Mifflin.

Seligman, M. E. P. (1994). *What you can change and what you can't.* New York: Knopf.

Senghas, A. (1995). Conventionalization in the first generation: A community acquires a language. *Journal of Contemporary Legal Issues,* 6, 501–519.

Senghas, R. J., & Kegl, J. (1994). Social considerations in the emergence of Idioma de Signos Nicaragüense (Nicaraguan Sign Language). *SignPost,* 7(1), 40–46.

Serbin, L. A., Moller, L. C., Gulko, J., Powlishta, K. K., & Colburne, K. A. (1994). The emergence of gender segregation in toddler playgroups. In Leaper, C. (Ed.), *Childhood gender segregation: Causes and consequences* (pp. 7–17). New Directions for Child Development, No. 65. San Francisco: Jossey-Bass.

Serbin, L. A., Powlishta, K. K., & Gulko, J. (1993). The development of sex typing in middle childhood. *Monographs of the Society for Research in Child Development,* 58(2, Serial No. 232).

Serbin, L. A., Sprafkin, C., Elman, M., & Doyle, A. (1984). The early development of sex differentiated patterns of social influence. *Canadian Journal of Social Science,* 14, 350–363.

Sherif, M., Harvey, O. J., White, B. J., Hood, W. R., & Sherif, C. W. (1961). *Intergroup cooperation and competition: The Robbers Cave experiment.* Norman, OK: University Book Exchange.

Shibutani, T. (1955). Reference groups as perspectives. *American Journal of Sociology,* 60, 562–569.

Sidransky, R. (1990). *In silence: Growing up hearing in a deaf world.* New York: St. Martin's.

Skinner, B. F. (1938). *The behavior of organisms.* New York: Appleton-Century-Crofts.

Smart, M. S., & Smart, R. C. (1978). *School-age children: Development and relationships* (2nd ed.). New York: Macmillan.

Smetana, J. G. (1994). Parenting styles and beliefs about parental authority. In J. G. Smetana (Ed.), *Beliefs about parenting and developmental implications* (pp. 21–36). San Francisco: Jossey-Bass.

Smetana, J. G. (1995). Parenting styles and conceptions of parental authority during adolescence. *Child Development,* 66, 299–316.

Smith, G. E., Gerrard, M., & Gibbons, F. X. (1997). Self-esteem and the relation between risk behavior and perceptions of vulnerability to unplanned pregnancy in college women. *Health Psychology,* 16, 137–146.

Smith, M. S. (1987). Research in developmental sociobiology: Parenting and family behavior. In K. B. MacDonald (Ed.), *Sociobiological perspectives on human development* (pp. 271–292). New York: Springer-Verlag.

Smith, M. S. (1988). Modern childhood: An evolutionary perspective. In K. Ekberg & P. E. Mjaavatn (Eds.), *Growing into a modern world* (pp. 1057–1069). Dragvoll, Norway: The Norwegian Centre for Child Research.

Smith, S. E., Snow, C. W., Ironsmith, M., & Poteat, G. M. (1993). Romantic dyads, friendships, and the social skill ratings of preschool children. *Early Education and Development,* 4, 59–67.

Smolowe, J. (1996, May 20). Parenting on trial. *Time,* p. 50.

Snow, C. (1991). A new environmentalism for child language acquisition. *Harvard Graduate School of Education Bulletin,* 36(1), 15–16.

Somit, A., Arwine, A., & Peterson, S. A. (1996). *Birth order and political behavior.* Lanham, MD: University Press of America.

Sommerfeld, D. P. (1989). The origins of mother blaming: Historical perspectives on childhood and motherhood. *Infant Mental Health Journal,* 10, 14–24.

Sommers, C. H. (1994). *Who stole feminism? How women have betrayed women.* New York: Simon & Schuster.

Sorce, J. F., Emde, R. N., Campos, J. J., & Klinnert, M. D. (1985). Maternal emotional signaling: Its effect on the visual cliff behavior of one-year-olds. *Developmental Psychology,* 21, 195–200.

Spock, B. (1968). *Baby and child care.* New York: Pocket Books. (First edition published in 1946.)

Sroufe, L. A. (1985). Attachment classification from the perspective of infant–caregiver relationships and infant temperament. *Child Development,* 56, 1–14.

Sroufe, L. A., Bennett, C., Englund, M., & Urban, J. (1993). The significance of gender boundaries in preadolescence: Contemporary correlates and antecedents of boundary violation and maintenance. *Child Development,* 64, 455–466.

Stabler, B., Clopper, R. R., Siegel, P. T., Stoppani, C., Compton, P. G., & Underwood,

L. E. (1994). Academic achievement and psychological adjustment in short children. *Developmental and Behavioral Pediatrics,* 15, 1–6.

Stanton, W. R., & Silva, P. A. (1992). A longitudinal study of the influence of parents and friends on children's initiation of smoking. *Journal of Applied Developmental Psychology,* 13, 423–434.

Stavish, S. (1994, May/June). On the biology of temperament development. *APS Observer,* pp. 7, 35.

Steele, C. M. (1997). A threat in the air: How stereotypes shape intellectual identity and performance. *American Psychologist,* 52, 613–629.

Steen, L. A. (1987, July 17). Mathematics education: A predictor of scientific competitiveness. *Science,* 237, 251–252, 302.

Steinberg, L., Dornbusch, S. M., & Brown, B. B. (1992). Ethnic differences in adolescent achievement: An ecological perspective. *American Psychologist,* 47, 723–729.

Stern, M., & Karraker, K. H. (1989). Sex stereotyping of infants: A review of gender labeling studies. *Sex Roles,* 20, 501–520.

Stevenson, H. W., & Stevenson, N. G. (1960). Social interaction in an interracial nursery school. *Genetic Psychology Monographs,* 61, 41–75.

Stevenson, H. W., Chen, C., & Uttal, D. H. (1990). Beliefs and achievement: A study of black, white, and Hispanic children. *Child Development,* 61, 508–523.

Stevenson, M. R., & Black, K. N. (1988). Paternal absence and sex-role development: A meta-analysis. *Child Development,* 59, 793–814.

Stipek, D. (1992). The child at school. In M. H. Bornstein & M. E. Lamb (Eds.), *Developmental psychology: An advanced textbook* (pp. 579–625). Hillsdale, NJ: Erlbaum.

Stocker, C., & Dunn, J. (1990). Sibling relationships in childhood: Links with friendships and peer relationships. *British Journal of Developmental Psychology,* 8, 227–244.

Stocker, C., Dunn, J., & Plomin, R. (1989). Sibling relationships: Links with child temperament, maternal behavior, and family structure. *Child Development,* 60, 715–727.

Stone, L. J., & Church, J. (1957). *Childhood and adolescence: A psychology of the growing person.* New York: Random House.

Straus, M. A., Sugarman, D. B., & Giles-Sims, J. (1997). Spanking by parents and subsequent antisocial behavior of children. *Archives of Pediatrics and Adolescent Medicine,* 151, 761–767.

Strayer, F. F., & Santos, A. J. (1996). Affiliative structures in preschool play groups. *Social Development,* 5, 117–130.

Study cites teen smoking risks (1995, April 24). Associated Press (on-line).

Sulloway, F. J. (1996). *Born to rebel: Birth order, family dynamics, and creative lives.* New York: Pantheon.

Suomi, S. J. (1997). Early determinants of behaviour: Evidence from primate studies. *British Medical Bulletin,* 53, 170–184.

Suomi, S. J., & Harlow, H. F. (1975). The role and reason of peer relationships in rhesus monkeys. In M. Lewis & L. A. Rosenblum (Eds.), *Friendship and peer relations* (pp. 153–186). New York: Wiley.

Swim, J. K. (1994). Perceived versus meta-analytic effect sizes: An assessment of the accuracy of gender stereotypes. *Journal of Personality and Social Psychology,* 66, 21–36.

Tajfel, H. (1970, November). Experiments in intergroup discrimination. *Scientific American,* 223, 96–102.

Tannen, D. (1990). *You just don't understand: Women and men in conversation.* New York: Ballantine Books.

Tanner, J. M. (1978). *Foetus into man: Physical growth from conception to maturity.* Cambridge: Harvard University Press.

Tate, D. C., Reppucci, N. D., & Mulvey, E. P. (1995). Violent juvenile delinquents: Treatment effectiveness and implications for future action. *American Psychologist,* 50, 777–781.

Taubes, G. (1995, July 14). Epidemiology faces its limits. *Science,* 269, 164–169.

Tellegen, A., Lykken, D. T., Bouchard, T. J., Jr., Wilcox, K. J., Segal, N. L., & Rich, S. (1988). Personality similarity in twins reared together and apart. *Journal of Personality and Social Psychology,* 54, 1031–1039.

Terrace, H. S. (1985). *Nim.* New York: Knopf.

Tesser, A. (1988). Toward a self-evaluation maintenance model of social behavior. In L. Berkowitz (Ed.), *Advances in experimental social psychology: Vol. 21* (pp. 81–227). San Diego, CA: Academic Press.

Thigpen, A. E., Davis, D. L., Gautier, T., Imperato-McGinley, J., & Russell, D. W. (1992, October 22). Brief report: the molecular basis of steroid 5α-reductase deficiency in a large Dominican kindred. *New England Journal of Medicine,* 327, 1216–1219.

Thigpen, C. H., & Cleckley, H. (1954). *The three faces of Eve.* Kingsport, TN: Kingsport Press.

Thomas, J. R., & French, K. E. (1985). Gender differences across age in motor performance: A meta-analysis. *Psychological Bulletin,* 98, 260–282.

Thorne, B. (1993). *Gender play: Girls and boys in school.* New Brunswick, NJ: Rutgers University Press.

Thornton, Y. S., as told to Coudert, J. (1995). *The ditchdigger's daughters: A black family's astonishing success story.* New York: Birch Lane Press.

Tiger, L. (1969). *Men in groups.* New York: Vintage.

Toman, W. (1971). The duplication theory of social relationships as tested in the general population. *Psychological Review,* 78, 380–390.

Tomasello, M. (1995). Commentary. *Human Development,* 38, 46–52.

Townsend, F. (1997). Rebelling against *Born to Rebel. Journal of Social and Evolutionary Systems,* 20, 191–204.

Trevathan, W. R. (1993, February). Evolutionary obstetrics. Paper presented at the annual meeting of the American Association for the Advancement of Science, Boston.

Triandis, H. C. (1994). *Culture and social behavior.* New York: McGraw-Hill.

Trivers, R. (1985). *Social evolution.* Menlo Park, CA: Benjamin/Cummings.

Turner, J. C., with Hogg, M. A., Oakes, P. J., Reicher, S. D., & Wetherell, M. S. (1987). *Rediscovering the social group: A self-categorization theory.* Oxford, UK: Basil Blackwell.

Umbel, V. M., Pearson, B. Z., Fernández, M. C., & Oller, D. K. (1992). Measuring bilingual children's receptive vocabulary. *Child Development,* 63, 1012–1020.

Ungar, S. J. (1995). *Fresh blood: The new American immigrants.* New York: Simon & Schuster.

Valero, H., as told to E. Biocca (1970). *Yanoáma: The narrative of a white girl kidnapped by Amazonian Indians* (D. Rhodes, Trans.). New York: Dutton.

van den Oord, E. J. C. G., Boomsma, D. I., & Verhulst, F. C. (1994). A study of problem behaviors in 10- to 15-year-old biologically related and unrelated international adoptees. *Behavior Genetics,* 24, 193–205.

Vasta, R. (1982). Physical child abuse: A dual-component analysis. *Developmental Review,* 2, 125–149.

Veenhoven, R., & Verkuyten, M. (1989). The well-being of only children. *Adolescence,* 24, 155–166.

Vernberg, E. M. (1990). Experiences with peers following relocation during early adolescence. *American Journal of Orthopsychiatry,* 60, 466–472.

Vogel, G. (1996, November 22). Asia and Europe top in world, but reasons are hard to find. *Science,* 274, 1296.

Volling, B. L., Youngblade, L. M., & Belsky, J. (1997). Young children's social relationships with siblings and friends. *American Journal of Orthopsychiatry,* 67, 102–111.

Wagner, B. M. (1997). Family risk factors for child and adolescent suicidal behavior. *Psychological Bulletin,* 121, 246–298.

Waller, N. G., & Shaver, P. R. (1994). The importance of nongenetic influences on romantic love styles: A twin-family study. *Psychological Science,* 5, 268–274.

Wallerstein, J. S., & Blakeslee, S. (1989). *Second chances: Men, women, and children a decade after divorce.* New York: Ticknor & Fields.

Wallerstein, J. S., & Kelly, J. B. (1980). *Surviving the breakup: How children and parents cope with divorce.* New York: Basic Books.

Wallis, C. (1996, March 25). The most intimate bond. *Time,* pp. 60–64.

Waring, N.-P. (1996, July 3). Social pediatrics. *Journal of the American Medical Association,* 276, 76.

Wasserman, E. A. (1993). Comparative cognition: Toward a general understanding of cognition in behavior. *Psychological Science,* 4, 156–161.

Watson, J. B. (1924). *Behaviorism.* New York: Norton.

Watson, J. B. (1928). *Psychological care of infant and child.* New York: Norton.

Weinstein, C. S. (1991). The classroom as a social context for learning. *Annual Review of Psychology,* 42, 493–525.

Weisfeld, C. C., Weisfeld, G. E., & Callaghan, J. W. (1982). Female inhibition in mixed-sex competition among young adolescents. *Ethology and Sociobiology,* 3, 29–42.

Weisfeld, G. E., & Billings, R. L. (1988). Observations on adolescence. In K. B. MacDonald (Ed.), *Sociobiological perspectives on human development* (pp. 207–233). New York: Springer-Verlag.

Weisner, T. S. (1986). Implementing new relationship styles in American families. In W. W. Hartup & Z. Rubin (Eds.), *Relationships and development* (pp. 185–205). Hillsdale, NJ: Erlbaum.

Weiss, L. H., & Schwarz, J. C. (1996). The relationship between parenting types and older adolescents' personality, academic achievement, adjustment, and substance use. *Child Development,* 67, 2101–2114.

Weissman, M. M., & Olfson, M. (1995, August 11). Depression in women: Implications for health care research. *Science,* 269, 799–801.

Weisstein, N. (1971). Psychology constructs the female. *Journal of Social Education,* 35, 362–373.

Weisstein, N. (1977). "How can a little girl like you teach a great big class of men?" the chairman said, and other adventures of a woman in science. In S. Ruddick & P. Daniels (Eds.), *Working it out* (pp. 241–250). New York: Pantheon.

Wellman, H. M. (1990). *The child's theory of mind.* Cambridge: MIT Press.

White, K. R., Taylor, M. J., & Moss, V. D. (1992). Does research support claims about the benefits of involving parents in early intervention programs? *Review of Educational Research,* 62, 91–125.

Whiting, B. B., & Edwards, C. P. (1988). *Children of different worlds: The formation of social behavior.* Cambridge, MA: Harvard University Press.

Wilder, D. A. (1986). Cognitive factors affecting the success of intergroup contact. In S. Worchel & W. G. Austin (Eds.), *Intergroup relations* (pp. 49–66). Chicago: Nelson-Hall.

Wilder, L. I. (1971). *Little house on the prairie.* New York: Harper & Row. (Originally published in 1935.)

Williams, J. E., & Best, D. L. (1986). Sex stereotypes and intergroup relations. In S. Worchel & W. G. Austin (Eds.), *Intergroup relations* (pp. 244–259). Chicago: Nelson-Hall.

Winitz, H., Gillespie, B., & Starcev, J. (1995). The development of English speech patterns of a 7-year-old Polish-speaking child. *Journal of Psycholinguistic Research, 24,* 117–143.

Winner, E. (1996). *Gifted children: Myths and realities.* New York: Basic Books.

Winner, E. (1997). Exceptionally high intelligence and schooling. *American Psychologist,* 52, 1070–1081.

Wolfe, D. A. (1985). Child-abusive parents: An empirical review. *Psychological Bulletin,* 97, 462–482.

Wolff, P. H., Tesfai, B., Egasso, H., & Aradom, T. (1995). The orphans of Eritrea: A comparison study. *Journal of Child Psychology and Psychiatry,* 36, 633–644.

Wood, D., Halfon, N., Scarlata, D., Newacheck, P., & Nessim, S. (1993, September 15). Impact of family relocation on children's growth, development, school function, and behavior. *Journal of the American Medical Association,* 270, 1334–1338.

Wrangham, R., & Peterson, D. (1996). *Demonic males: Apes and the origins of human violence.* Boston: Houghton Mifflin.

Wright, L. (1995, August 7). Double mystery. *The New Yorker,* pp. 45–62.

Wright, R. (1994). *The moral animal.* New York: Pantheon.

WuDunn, S. (1996, September 8). For Japan's children, a Japanese torment. *New York Times,* p. 3.

Yamamoto, K., Soliman, A., Parsons, J., & Davies, O. L., Jr. (1987). Voices in unison: Stressful events in the lives of children in six countries. *Journal of Child Psychology & Psychiatry,* 28, 855–864.

Yang, B., Ollendick, T. H., Dong, Q., Xia, Y., & Lin, L. (1995). Only children and children with siblings in the People's Republic of China: Levels of fear, anxiety, and depression. *Child Development,* 66, 1301–1311.

Youngblade, L. M., Park, K. A., & Belsky, J. (1993). Measurement of young children's close friendship: A comparison of two independent assessment systems and their associations with attachment security. *International Journal of Behavioral Development,* 16, 563–587.

Young-Hyman, D. (1986). Effects of short stature on social competence. In B. Stabler & L. E. Underwood (Eds.), *Slow grows the child: Psychosocial aspects of growth delay* (pp. 27–45). Hillsdale, NJ: Erlbaum.

Youniss, J. (1992). Parent and peer relations in the emergence of cultural competence. In H. McGurk (Ed.), *Childhood social development: Contemporary perspectives* (pp. 131–147). Hove, UK: Erlbaum.

Zajonc, R. B. (1983). Validating the confluence model. *Psychological Bulletin,* 93, 457–480.

Zervas, L. J., & Sherman, M. F. (1994). The relationship between perceived parental favoritism and self-esteem. *Journal of Genetic Psychology,* 155, 25–33.

Zimbardo, P. G. (1993). Pathology of imprisonment. In B. Byers (ed.), *Readings in Social Psychology* (pp. 15–19). Boston: Allyn & Bacon. (Originally published in 1972.)

Zimmerman, L., & McDonald, L. (1995). Emotional availability in infants' relationships with multiple caregivers. *American Journal of Orthopsychiatry,* 65, 147–152.

Zimmerman, M. A., Salem, D. A., & Maton, K. I. (1995). Family structure and psychosocial correlates among urban African-American adolescent males. *Child Development,* 66, 1598–1613.

Zuckerman, M. (1984). Sensation seeking: A comparative approach to a human trait. *Behavioral and Brain Sciences,* 7, 413–471.

Zuger, B. (1988). Is early effeminate behavior in boys early homosexuality? *Comprehensive Psychiatry,* 29, 509–519.

Zukow, P. G. (1989). Siblings as effective socializing agents: Evidence from Central Mexico. In P. G. Zukow (Ed.), *Sibling interaction across cultures: Theoretical and methodological issues* (pp. 79–104). New York: Springer-Verlag.

INDEX